13th International X-ray Microscopy Conference (XRM 2016)

Journal of Physics: Conference Series Volume 849

Oxford, United Kingdom
15 – 19 August 2016

ISBN: 978-1-5108-4266-3
ISSN: 1742-6588

Printed from e-media with permission by:

Curran Associates, Inc.
57 Morehouse Lane
Red Hook, NY 12571

Some format issues inherent in the e-media version may also appear in this print version.

Printed by Curran Associates, Inc. (2017)

Institute of Physics
Dirac House, Temple Back
Bristol BS1 6BE UK

Phone: 44 1 17 929 7481
Fax: 44 1 17 920 0979

techtracking@iop.org

Additional copies of this publication are available from:

Curran Associates, Inc.
57 Morehouse Lane
Red Hook, NY 12571 USA
Phone: 845-758-0400
Fax: 845-758-2633
Email: curran@proceedings.com
Web: www.proceedings.com

Table of contents

Volume 849

X-Ray Microscopy Conference 2016 (XRM 2016)

15–19 August 2016, Oxford University, United Kingdom

Accepted papers received: 16 May 2017
Published online: 14 June 2017

Preface

Bio-imaging, Multi-modal imaging, Environmental and geo sciences

Elemental contrast and chemistry, Energy and Materials

Diffraction imaging

New sources and facilities, Novel techniques & applications, Optics, detectors and instrumentation, Data processing

IOP Conf. Series: Journal of Physics: Conf. Series **849** (2017) 011001 doi:10.1088/1742-6596/849/1/011001

Preface

On behalf of the Program Committee I would like to thank all the participants of the 13th International X-ray Microscopy Conference, XRM2016, for their contributions. The conference was hosted by the Diamond Light Source and took place in the nearby historic city of Oxford, United Kingdom from the 15th to the19th August, 2016. The goal of this biennial conference is to address the most recent advances in X-ray microscopy by bringing together experts in the development and the application of X-ray microscopes. The conference also explored the position of X-ray microscopy alongside related techniques and disciplines.

The present proceedings contain over 60 contributions, providing a representative selection of the conference content. Overall there were more than 380 participants in this conference, with a total of 72 oral presentations and 250 posters contributed. In addition we had three sessions of early career flash talks, which were well received. The manuscripts submitted for these proceedings were reviewed by a large team of referees. I thank them for their rapid and thorough work on the manuscripts as well as the authors for their contributions.

The conference contained ten different topics. They are categorized into four groups here for a better overview:
- Bio-imaging, Multi-modal imaging, Environmental and geosciences
- Elemental contrast and chemistry, Energy and materials
- Diffraction imaging
- New sources and facilities, Novel techniques & applications, Optics, detectors and instrumentation, Data processing.

Following the tradition of the XRM conference series, the Werner Meyer-Ilse Memorial Award (WIMA) rewards young scientists for exceptional contributions to the advancement of X-ray microscopy. The WIMA committee awarded the prize to Junjing Deng (Northwestern University, USA) and Matias Kagias (ETH Zurich / PSI, Switzerland). The winners for the poster prizes, who presented their work during the early career flash talk sessions, were Burcu Kepsutlu (Helmholtz Zentrum Berlin, Germany) Simone Sala (UCL London, UK) and Ottó Márkus (KIT Karlsruhe, Germany).

One highlight of the conference was the honoring of Janos Kirz and Günther Schmahl for their outstanding contributions and leadership over the years and especially for the conference series. We all remember the song composed by Chris Jacobsen that we sang together during the Wednesday special event presentation.

Multiple sponsors had kindly contributed to the conference. I would like to acknowledge the following sponsors: Zeiss (conference dinner and conference AV); Silson (delegate giveaway); Bruker micro CT (literature package); Dectris Ltd (lanyards); and in alphabetic order: Applied Nanotools; Axilion; Axo Dresdon/Hubor Diffraction; Excillum AB; Hamamatsu Photonics; Lyncean Technologies Inc.; Norcada Inc; NTT Advanced Technology Corporation; PI miCos; Quantum Detectors; Queensgate; Rigaku; SGX Sensortech; Suna-Precision GmbH; Toyama Co. Ltd; XIA. Many of these sponsors provided exhibits during the conference offering the latest X-ray microscopy related products.

Finally I would like to express my gratitude to all those who have contributed to XRM2016: the International Advisory Committee and the local Scientific Programme Committee for their support and advice, all the speakers for preparing and presenting insriping talks and, last but by no means least, the Local Organising Committee for its enthusiastic efforts that have turned this conference into reality.

I acknowledge my co-chairs, Liz Duke and Burkhard Kaulich, and our former Physical Sciences Director, Trevor Rayment, for the numerous and helpful discussions regarding the content and details of the conference. My special thanks go to Emma Clarke for her relentless work and effort. The support of the Diamond helpers and the communications team are also acknowledged.

It has been a great honour to host the XRM2016 conference and we hope that all the participants have found the conference to be enjoyable and rewarding. We look forward to seeing you all at XRM2018 in Saskatoon, Canada.

Christoph Rau
Editor XRM2016 proceedings
Chair XRM2016 conference

Peer review statement

All papers published in this volume of *Journal of Physics: Conference Series* have been peer reviewed through processes administered by the proceedings Editors. Reviews were conducted by expert referees to the professional and scientific standards expected of a proceedings journal published by IOP Publishing.

IOP Conf. Series: Journal of Physics: Conf. Series **849** (2017) 012001 doi:10.1088/1742-6596/849/1/012001

Phase-contrast tomography of sciatic nerves: image quality and experimental parameters

M Töpperwien[1,2,a)], M Krenkel[1], T Ruhwedel[3], W Möbius[2,3], A Pacureanu[4], P Cloetens[4] and T Salditt[1,2,b)]

[1] Institut für Röntgenphysik, Friedrich-Hund-Platz 1, 37077 Göttingen, Germany
[2] Center for Nanoscopy and Molecular Physiology of the Brain, Göttingen, Germany
[3] Max-Planck-Institut für Experimentelle Medizin, Hermann-Rein-Str. 3, 37075 Göttingen, Germany
[4] European Synchrotron Radiation Facility, Grenoble, France

E-mail: [a)]`mtoeppe@gwdg.de`, [b)]`tsaldit@gwdg.de`

Abstract. We present propagation-based phase-contrast tomography of mouse sciatic nerves stained with osmium, leading to an enhanced contrast in the myelin sheath around the axons, in order to visualize the threedimensional (3D) structure of the nerve. We compare different experimental parameters and show that contrast and resolution are high enough to identify single axons in the nerve, including characteristic functional structures such as Schmidt-Lanterman incisures.

Assessing the three-dimensional structure of myelinated nerves is required in order to fully understand the function of the nerve based on its anatomical features. A standard tool for imaging of nerve tissue is electron microscopy which leads to high resolution reconstructions of the inner structure of the nerve but is limited to small volumes. (Hard) x-ray computed tomography is a promising approach for 3D imaging of relatively large volumes, especially by using the phase shift instead of the absorption of the beam, as this leads to superior contrast in soft tissue. High resolution reconstructions can be achieved via propagation-based phase contrast exploiting the near-field intensity modulations resulting from free space propagation between the object and the detector [1, 2]. By applying suitable phase retrieval algorithms the quantitative phase distribution of the object can be obtained [3]. In a previous study of our group, propagation-based imaging of high-pressure frozen nerves was performed at the ID22NI beamline at the ESRF in Grenoble, yielding high contrast reconstructions of the axons within the nerve and making anatomical features like Schmidt-Lanterman incisures visible in 3D [4]. However, since then the new ESRF upgrade beamline ID16A [5], as well as an upgrade of the GINIX endstation at the P10 beamline at DESY in Hamburg [6] have been commissioned. The scope of this paper is to assess the image quality of the reconstructions obtained at these two setups and to evaluate whether a comparable resolution and contrast can be achieved.

Reconstruction starts with phase retrieval for the individual projections recorded at each angle of a tomographic scan. The phase retrieval schemes applied in this work are based on the contrast transfer function (CTF) which is valid for weak or non-absorbing objects with a slowly varying phase. The theory underlying these reconstruction schemes can be found, e.g., in [7, 8, 9]. Crucial parameters for the reconstruction are the number of propagation distances between the

Table 1. Experimental parameters.

	GINIX overview	GINIX zoom	ID16A
energy [keV]	13.8	13.8	17.05
z_1/z_2 [mm]	140/5187	50/5187	53/337
detector	sCMOS (Hamamatsu)	sCMOS (Hamamatsu)	FReLoN (ESRF)
pixel size/eff. pixel size [nm]	6500/182.8	6500/62.7	845/131.8
# distances/$\frac{\text{\# projections}}{\text{distance}}$	4/1000	4/1000	4/1995
expos. time [s]	1	1	0.2
CTF $\frac{\delta}{\beta}/\alpha_1/\alpha_2$	6/-/1e-04	10/-/1e-06	-/1e-05/0.01

sample and the detector as well as the choice of the regularization parameters α_1 or $\frac{\delta}{\beta}$ and α_2 which are defined as in [4]. As a final step in data analysis, tomographic reconstruction is carried out using the Matlab implemented *iradon* function with a standard Ram-Lak filter.
The sciatic nerves of two adult mice were prepared freshly and fixed by immersion into 4% formaldehyde, 2.5% glutaraldehyde in 0.1 M phosphate buffer pH 7.3 and 0.5% NaCl. After postfixation in 2% OsO_4 in 0.1 M phosphate buffer for 4h at 4 °C and dehydration with ethanol and propylenoxide, the nerves were embedded in EPON, transferred into a Kapton tube and polymerized for 24h. Experiments have been carried out at the ID16A beamline at the ESRF in Grenoble, as well as the GINIX setup at the P10 beamline at DESY in Hamburg. In the case of the ID16A beamline the photons are focussed by a set of Kirkpatrick-Baez (KB) mirrors to a focus size of 26×39 nm^2 (FWHM, h×v) whereas in the case of the GINIX setup the beam is prefocussed by a KB system to a size of about 300×300 nm^2 and further shaped by placing a waveguide into the focus, minimizing the focal spot size to approximately 18×18 nm^2 (FWHM) while at the same time increasing the coherence [10]. Due to the cone-beam geometry in both setups, the effective pixel size can be changed by varying the geometrical magnification $M = \frac{z_2}{z_1}$, with z_2 the focus-detector-distance and z_1 the focus-sample-distance. The experimental parameters for each measurement are summarized in Tab. 1.

In Fig. 1 the basic steps of data analysis are illustrated for the example of the high resolution measurement at the P10 beamline. Data analysis starts with the empty-beam corrected

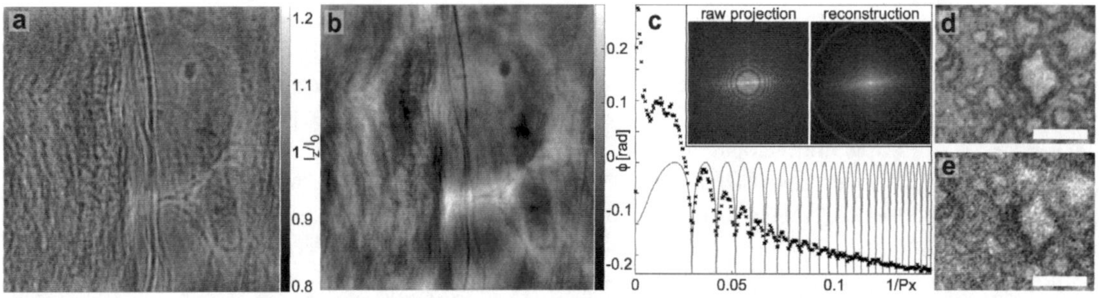

Figure 1. Schematics of the phase-retrieval procedure. (a) Empty-beam corrected projection of the sample. (b) Reconstructed phase of the same projection. (c) Inset: Central part of the power spectral density (PSD) of the projections from (a) and (b). Graph: Logarithmic PSD of the experimental data (black) compared to the theoretical shape according to the CTF (red, for visibility shifted and scaled). (d) Part of an exemplary slice through the volume using 4 distances for phase retrieval. (e) The same slice using only a single distance. Scalebars: 10 μm

projection shown in (a). For each of the individual projections phase retrieval with one of the two CTF approaches has to be performed, leading to the reconstructed phase distribution in (b). Due to the different magnifications at different propagation distances, images have to be rescaled and aligned to the projection with the smallest effective pixel size prior to phase retrieval. When looking at the power spectral density (PSD) of the empty-beam corrected raw projection in (c, left) the typical oscillations predicted by the CTF are visible. However, a comparison with the theory shows that the visibility of the modulations is decreasing, especially for higher spatial frequencies, which can be explained by the violation of the weak object assumption [11]. This leads to the introduction of high-frequency artifacts in the phase reconstruction as can be seen in the corresponding PSF (c, right). As the CTF approach nevertheless provides reasonable results, it is in the following used for reconstruction of the 3D density distribution. One exemplary virtual slice through the volume with several propagation distances taken into account is shown in (d) whereas for the slice in (e) only one propagation distance was used. As can be seen in the insets, a reasonable reconstruction of the general shape of the object is still possible. However, when aiming for highest resolution and contrast the use of several distances is inevitable.

At the GINIX setup two datasets with different magnifications were recorded, one with a pixel size of 182.8 nm, in the following denoted as the 'overview' scan, and one with a pixel size of 62.7 nm, in the following denoted as the 'zoom' dataset. Phase retrieval was performed using the parameters listed in Tab. 1. The results of both scans are visualized in Fig. 2. One virtual transverse slice through the reconstructed volume of the overview scan is shown in (a), revealing the typical structure of the mylin sheath around the individual axons in the nerve. Note that the contrast is decreasing towards the middle of the nerve which can be explained by an unequal penetration of the osmium throughout the whole nerve. The inset shows part of the slice in higher detail as well as the result of the zoom dataset as a comparison, showing the advantage in resolution. In order to get a higher signal-to-noise ratio (SNR) both datasets were filtered with a Gaussian with a 0.7 pixel (overview) and a 2 pixel kernel (zoom), respectively. The resulting SNRs are approximately 7.6 and 4.7 for the outer and inner region in the overview scan and 6.1

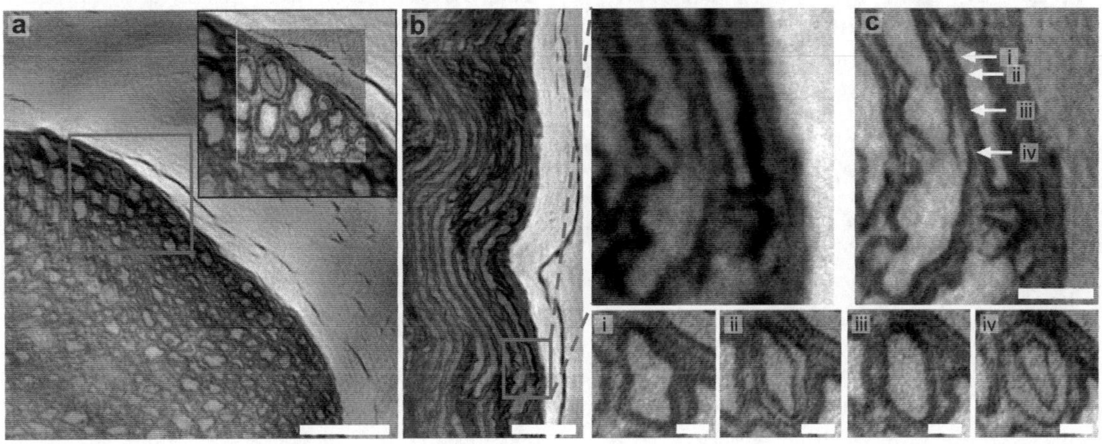

Figure 2. Measurement at the GINIX setup at the P10 beamline. (a) Virtual transverse slice through the reconstructed 3D density distribution of the large overview scan. The inset shows a comparison between the image quality of the overview and the high resolution scan. (b) Virtual longitudinal slice. (c) Same slice for the zoom dataset showing the inset in (b) in higher detail. (i-iv) Parts of transverse slices at the positions indicated by the arrows. Scalebars: (a,b) 50 μm, (c) 10 μm, (i-iv) 5 μm

for the zoom dataset. To estimate the resulting resolution, an error function is fitted to the edge of an object within the reconstructed volume. This gives 465.8 nm and 291.6 nm (FWHM), respectively. The longitudinal slice in (b) reveals the orientation of the axons within the sciatic nerve, showing the typical shevron pattern observed previously [4]. Additionally, anatomical features like the Schmidt-Lanterman incisures, clefts in the myelin sheath around the axons, can be identified in the reconstructed slices, see (c, i-iv).
In Fig. 3 the results of the dataset recorded at the ID16A setup with a pixel size of 131.8 nm are shown. The transverse slice in (a) again reveals the typical shape of the myelin sheath around the axons, though more uniformly stained compared to the nerve in Fig. 2 and at a higher SNR of approximately 9.7 (no filter was applied). The resolution can be estimated as 303.1 nm (FWHM), again by fitting an error function to the edge of an object. The longitudinal slice in (b) shows the course of the axons within the nerve whereas in the inset in (c) as well as the transverse slices at the positions indicated by the arrows the typical Schmidt-Lanterman incisures can again be clearly recognized.
The reconstructions from both experimental setups provide a data quality high enough for identification of the individual axons in the nerve as well as typical anatomical features like the Schmidt-Lanterman incisures. In contrast to the previous study samples were not high pressure frozen but prepared according to standard protocols for electron microscopy, still yielding at least comparable contrast within the tissue.
The next step would still be an automatic rendering of the data including the quantification of the number of axons within the nerve as well as the recognition of anatomical features like Schmidt-Lanterman incisures or nodes of Ranvier and their frequency of occurrence and spatial distribution along the nerve. This will pave the way for biomedical studies aiming for the quantification of structural differences between different phenotypes in relatively large volumes of interest.

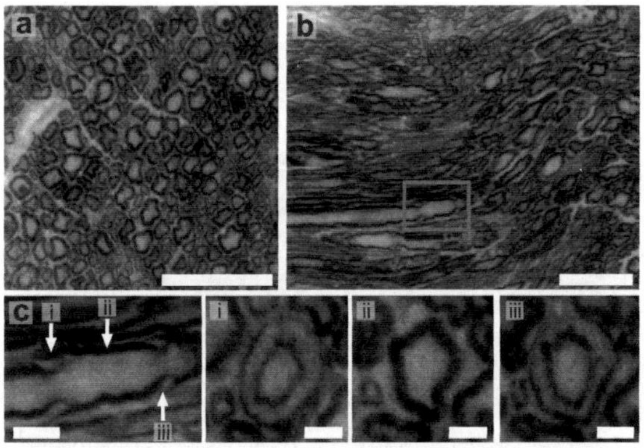

Figure 3. Measurement at ID16A beamline. (a) Virtual transverse slice through the reconstructed 3D density distribution. (b) Virtual longitudinal slice. (c) Inset of the longitudinal slice in (b) showing part of the myelin sheath around the one axon in higher detail. (i-vi) Parts of transverse slices at the positions indicated by the arrows. Scalebars: (a,b) 50 μm, (c) 10 μm, (i-iii) 5 μm

Acknowledgments
We thank Michael Sprung for excellent support during our beamtimes at P10. We particularly thank Matthias Bartels for discussions, and the establishment of methods in a preceeding collaboration on nerve tomography. This work was supported by the Collaborative Research Center 755 *Nanoscale Photonic Imaging* of the German science foundation (DFG).

References

[1] Zanette I, Lang S, Rack A, Dominietto M, Langer M, Pfeiffer F, Weitkamp T and Müller B 2013 *Appl. Phys. Lett.* **103** 244105
[2] Bartels M, Krenkel M, Haber J, Wilke R N and Salditt T 2015 *Phys. Rev. Lett.* **114**(4) 048103

[3] Paganin D M 2006 *Coherent X-Ray Optics* (New York: Oxford University Press)
[4] Bartels M, Krenkel M, Cloetens P, Möbius W and Salditt T 2015 *Journal of structural biology* **192** 561–568
[5] Morawe C, Barrett R, Cloetens P, Lantelme B, Peffen J C and Vivo A 2015 *SPIE Optical Engineering+ Applications* (International Society for Optics and Photonics) pp 958803–958803
[6] Salditt T, Osterhoff M, Krenkel M, Wilke R N, Priebe M, Bartels M, Kalbfleisch S and Sprung M 2015 *J. Synchrotron Rad.* **22** 867–878
[7] Cloetens P, Ludwig W, Baruchel J, Van Dyck D, Van Landuyt J, Guigay J P and Schlenker M a 1999 *Appl. Phys. Lett.* **75** 2912–2914
[8] Zabler S, Cloetens P, Guigay J P, Baruchel J and Schlenker M 2005 *Rev. Sci. Instrum.* **76** 073705
[9] Turner L, Dhal B, Hayes J, Mancuso A, Nugent K, Paterson D, Scholten R, Tran C and Peele A 2004 *Opt. Express* **12** 2960–2965
[10] Krüger S P, Neubauer H, Bartels M, Kalbfleisch S, Giewekemeyer K, Wilbrandt P J, Sprung M and Salditt T 2012 *J. Synchrotron Rad.* **19** 227–236
[11] Moosmann J, Hofmann R and Baumbach T 2011 *Optics express* **19** 12066–12073

IOP Conf. Series: Journal of Physics: Conf. Series **849** (2017) 012002 doi:10.1088/1742-6596/849/1/012002

X-ray micro computed tomography for the visualization of an atherosclerotic human coronary artery

Sofiya Matviykiv[1], Marzia Buscema[1], Hans Deyhle[1], Thomas Pfohl[1], Andreas Zumbuehl[2], Till Saxer[3], and Bert Müller[1]

[1]Biomaterials Science Center, Department of Biomedical Engineering, University of Basel, Allschwil, Switzerland
[2]Department of Chemistry, University of Fribourg, Fribourg, Switzerland
[3]Cardiology Division, University Hospitals of Geneva, Geneva, Switzerland

Email: sofiya.matviykiv@unibas.ch

Abstract. Atherosclerosis refers to narrowing or blocking of blood vessels that can lead to a heart attack, chest pain or stroke. Constricted segments of diseased arteries exhibit considerably increased wall shear stress, compared to the healthy ones. One of the possibilities to improve patient's treatment is the application of nano-therapeutic approaches, based on shear stress sensitive nano-containers. In order to tailor the chemical composition and subsequent physical properties of such liposomes, one has to know precisely the morphology of critically stenosed arteries at micrometre resolution. It is often obtained by means of histology, which has the drawback of offering only two-dimensional information. Additionally, it requires the artery to be decalcified before sectioning, which might lead to deformations within the tissue. Micro computed tomography (μCT) enables the three-dimensional (3D) visualization of soft and hard tissues at micrometre level. μCT allows lumen segmentation that is crucial for subsequent flow simulation analysis. In this communication, tomographic images of a human coronary artery before and after decalcification are qualitatively and quantitatively compared. We analyse the cross section of the diseased human coronary artery before and after decalcification, and calculate the lumen area of both samples.

1. Introduction

The development of atherosclerosis is a complex process. It causes thickening of the vessel wall and a constriction of the vessel lumen that can block the blood flow for example known as heart attack or stroke. To counter the constriction in case of an infarct, vasodilator drugs are often employed. The first aid ambulance medication given to the patient includes intravenous injection of nitroglycerine to treat clotted blood vessels and to restore the blood flow in critically constricted parts of the arteries. Systemic administration of such vasodilator, however, leads to undesired hypotension and decreased blood perfusion of the heart. Up to 15 or even 20 % of people struck by heart attack die before arriving to hospital. Critically stenosed segments of arteries (e.g. 80 % stenosis) reveal an at least 10 times higher wall shear stress, compared to the healthy parts [1]. One can take advantage of such blood flow-specific changes and employ them as purely mechanical trigger for the delivery of vasodilators. The therapeutic approach focused on the application of shear stress sensitive nano-containers, which become mechanically activated by shear forces, has been proposed [2, 3]. Such nanometre-sized lipid-based liposomes can offer local, rapid and non-invasive treatment, before the patient reaches the hospital.

To optimize the mechano-sensitive liposomes, the morphology of atherosclerotic vessels including its lumen is required. These data can be used to simulate the blood flow [4]. The standard procedure of characterization is histology, which provides sub-micrometre resolution in the lateral dimensions.

The tissue morphology, however, can be altered owing to preparation artefacts, which can originate from decalcification, cutting and embedding, leading to tissue shrinkage or deformation. In particular, the histological evaluation provides only limited information in third dimension. This drawback significantly limits the application in blood flow simulations. X-ray tomography can master this challenge, providing non-destructive, quantitative 3D visualization of both, soft and hard tissues containing specimens [4-9]. A variety of operation modes of μCT can be applied for specimen characterization, depending on the defined task and type of information required. One of them is absorption contrast, which is mostly used to discriminate anatomical structures with reasonable differences in X-ray absorption, including the determination of plaque morphology within artery and to determine the extent of stenosis resulting in lumen segmentation. X-ray absorption is described by the imaginary part of the refractive index β, which is related to the attenuation coefficient [6]. Since the difference in absorption contrast within soft tissues is weak, differentiation among tissue types is challenging. Therefore, phase contrast (PC) modalities are preferred for such purpose. Here, the decrement of the real part of the refractive index is measured, which is two to three orders of magnitude larger than β at the X-ray energies considered. Thus, PC modalities provide significantly improved contrast for soft tissue differentiation [5, 7, 8] compared to the absorption contrast mode.

In this study, we present the tomography data of a diseased human coronary artery, acquired using conventional laboratory μCT in absorption contrast mode. A human artery containing a calcified atherosclerotic plaque was measured before and after the decalcification process to evaluate the cross section of the lumen, to calculate the non-constricted area of the artery, and to determine the cross-section changes due to decalcification procedure.

2. Materials and methods

2.1. *Human coronary artery preparation*

The artery was obtained from a donated body (Institute of Anatomy, University of Basel). The responsible Ethical Committee approved the request and written consent of the donor has been obtained. A 2.2 cm-long section was cut from the atherosclerotic human coronary artery. The artery was cleaned from surrounding tissues using a scalpel and fixed in 4 % paraformaldehyde (PFA). Afterwards, the sample was dehydrated by soaking, firstly, in ethanol, then in xylene overnight at reduced pressure (Tissue-Tek VIP E300 Histokinetic automated dehydrator, Sakura). Subsequently, the artery was immersed in molten paraffin at a temperature of 60 °C. After the paraffin solidification, the shape and size of the block were adjusted with a scalpel to the size of artery-containing region.

To decalcify the sample, the paraffin embedded artery was heated to a temperature of 60 °C and immersed in decalcifier solution (distilled water: formic acid: PFA, 87: 8: 5 v/v) at a temperature of 37 °C for a period of two days. The dehydration steps were repeated as described above to re-embed.

2.2. *Advanced laboratory X-ray microtomography system*

The laboratory μCT system nanotom® m (Phoenix|x-ray, GE Sensing & Inspection Technologies GmbH, Wunstorf, Germany) operating in absorption contrast mode was used to visualize the calcified and decalcified parts of the human artery. The instrument is equipped with a 180 kV/15 W nanofocus X-ray source with a W target. An accelerating voltage of 60 kV, which results in a mean photon energy of 40 to 50 keV, and a beam current of 310 μA were used. The sample was mounted on a high-precision manipulator and rotated through the incoming X-ray beam. During the rotation over an angular range of 360° 1,000 equiangular projections were recorded. For each projection three images were acquired with a total exposure time of 3.0 s (calcified sample) and 3.5 s (decalcified sample), respectively. A pixel size of 17.78 μm (calcified sample) and 13.73 μm (decalcified sample) as well as a field of view of 1,300 × 2,400 pixels (calcified sample) and 2,000 × 2,400 pixels (decalcified sample) were used, respectively. The acquired projections were reconstructed using the cone-beam, filtered back-projection algorithm of the phoenix datos|x 2.0.1-RTM software. Subsequently, the lumen of the artery was visualized. For the segmentation purpose of the artery lumen, we have selected dedicated slices to start the region-growing algorithm tool in VG Studio Max 2.1 (Volume Graphics, Heidelberg, Germany).

IOP Conf. Series: Journal of Physics: Conf. Series **849** (2017) 012002 doi:10.1088/1742-6596/849/1/012002

3. Results and discussions

3.1. Visualization of the plaque containing coronary artery

The preparation steps during fixation and embedding as well as decalcification can significantly modify the vessel morphology. Therefore, each step has to be carefully investigated. In this work, we visualized the human coronary artery using μCT. It is beneficial, because one can decrease X-ray energy or accelerating voltage, which provides increased contrast between soft tissue components. It allows us to visualize the precise geometry of the vessel wall and observe changes in lumen morphology in comparison to the originally calcified artery.

Figure 1 compares slices of plaque-containing artery (A – top view, C – lateral view) and the same artery after decalcification process (B – top view, D – lateral view). In Figure 1A, the artery walls and plaque are clearly visible, however, the contrast in the surrounding fatty tissue is low. On the contrary, in Figure 1B, better contrast in the soft tissue is present, and the artery wall is clearly visible. However, the hard tissue residuals from the plaque decalcification are still present in the surrounding tissue (shown by yellow arrows). One can also notice the presence of preparation artefacts including air bubble in the left lumen of calcified artery (Figure 1A), and small paraffin cracks inside the lumens of the decalcified artery (Figure 1B).

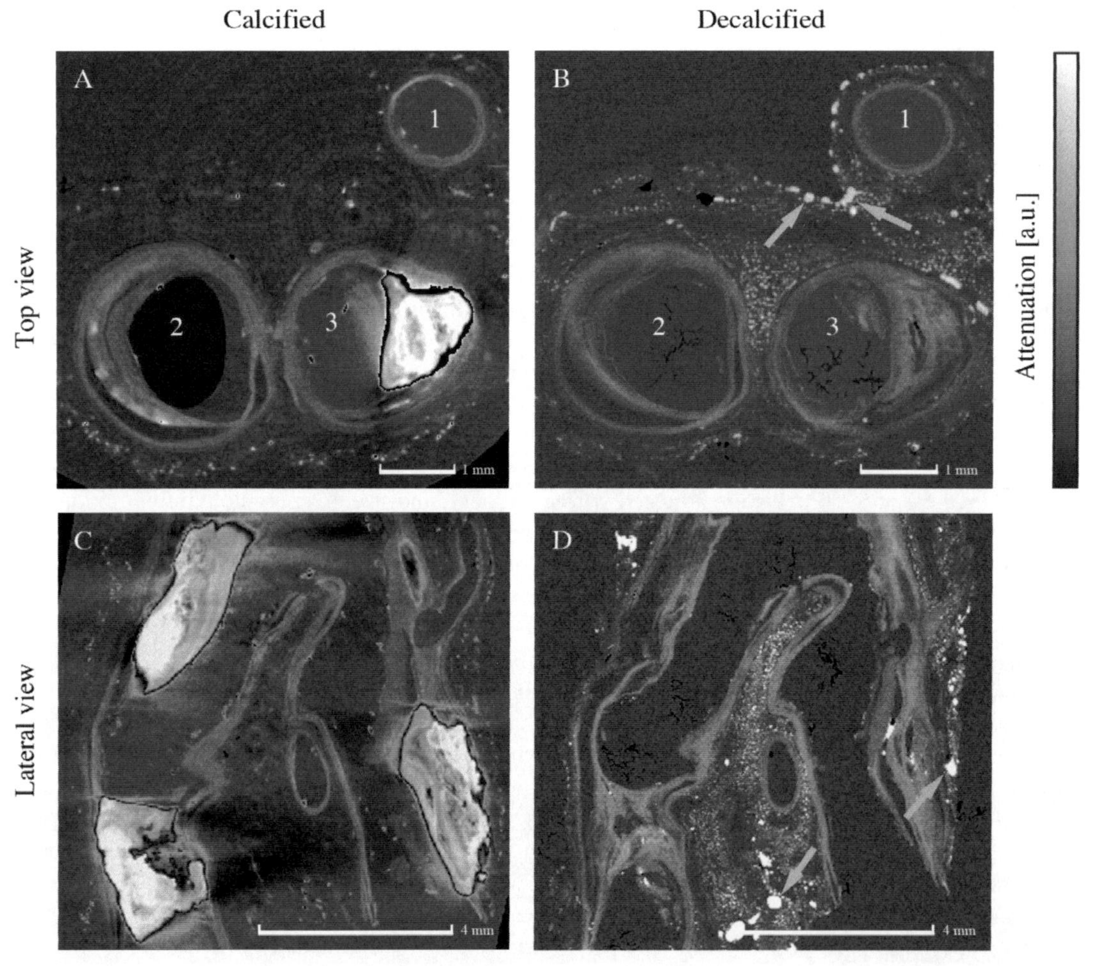

Figure 1. Axial slices of a calcified (A – top view, C – lateral view) and decalcified (B – top view, D – lateral view) human coronary artery embedded in paraffin, measured in absorption contrast μCT. Region 1 of images A and B corresponds to the cross-sectional area of lumen coloured in green, Region 2 coloured in blue and region 3 coloured in red, respectively. Region 3 of image A reveals presence of calcified plaque, depicted in bright white colour.

Streak artefacts (whitish colour on red background) around the highly X-ray absorbing calcified plaque were observed in the calcified sample (Figure 1A). These artefacts overlap with the morphology of the lumen and make segmentation more challenging.

The cross-sectional area of both calcified and decalcified samples was extracted using the region-growing segmentation tool of VG Studio Max 2.1, which is a fast and convenient method to obtain the morphology of the lumen. The lumen area is used for flow simulations. The cross-sectional area was calculated and the results are presented in Table 1.

Table 1. The cross-sectional area of the artery lumen.

	Calcified artery area, mm^2	Decalcified artery area, mm^2
Region 1 (green)	0.84 ± 0.02	0.80 ± 0.02
Region 2 (blue)	2.42 ± 0.02	2.40 ± 0.02
Region 3 (red)	1.86 ± 0.02	1.82 ± 0.02

Numbers and related colour indicate ach region of the artery. The cross-sectional area of the segmented slice was calculated and found to be within the error bars between the two samples before and after decalcification process: from 0.84 to 0.80 mm^2 (Region 1), from 2.42 to 2.40 mm^2 (Region 2), and from 1.86 to 1.82 mm^2 (Region 3). The shrinkage of the artery, hence decrease in all lumen areas, could be caused by the chemical treatment within the decalcification process. The area of Region 3, which contained an atherosclerotic plaque, was not notably decreased at the location of the removed plaque, because the soft tissue was still present. The area of the calcified plaque was calculated to be 1.07 mm^2, whereas the area of the soft tissue residuals after decalcification was equal to 1.02 mm^2. Comparing these two slices from calcified and decalcified datasets, the decalcification process does not significantly change the cross section of the lumen.

The authors have realized the surprisingly high contrast for unstained soft tissue imaged. One can reasonably assume that the dehydration is the key factor, as it leads to increased tissue density. This increase gives a rise to the observed attenuation differences. Recently, it has been shown that paraffin-embedded human brain tissue yields significant contrast even in laboratory-based μCT [8].

4. Conclusion

Atherosclerosis is a complex inflammatory disease, related to the damage of endothelium and its inflammation as well as to the accumulation of the lipids and white blood cells, proliferation of intimal-smooth-muscle cells, followed by tissue calcification that lead to a fibro-fatty plaque formation. Here, we show the difference in lumen morphology between calcified and decalcified human coronary artery. We present the quantified cross-sectional area of the lumen region for both samples. In future, it will be beneficial to compare artery lumen areas between samples fixed in water, formalin, paraffin, or after a decalcification process, and histology in order to better understand how each processing affects the artery wall. Micro computed tomography is a valuable tool to study the morphology of diseased atherosclerotic artery, containing soft and hard tissues, as previously demonstrated for formalin fixation of the human brain [10].

References

[1] Yin W *et al* 2009 *Conf. Proc. IEEE Eng. Med. Biol. Soc.* 4003
[2] Holme M N *et al* 2012 *Nat. Nanotech.* **7** 536
[3] Saxer T *et al* 2013 *Cardiovasc. Res.* **99** 32
[4] Holme M N *et al* 2014 *Nat. Protoc.* **9** 1401
[5] Langheinrich A C *et al* 2004 *Radiol* **231** 675
[6] Holme M N *et al* 2012 *Proc. of SPIE* **8506** 850609
[7] Buscema M *et al* 2014 *Proc. of SPIE* **9212** 921203
[8] Khimchenko A *et al* 2016 *Proc. SPIE* **9797** 97970B
[9] Müller B *et al* 2016 *AIP Conf. Proc.* **1696** 0200101
[10] Schulz G *et al* 2011 *J. Neurosci. Meth.* **202** 17-27.

IOP Conf. Series: Journal of Physics: Conf. Series **849** (2017) 012003 doi:10.1088/1742-6596/849/1/012003

Quantitative study of mammalian cells by scanning transmission soft X-ray microscopy

K Shinohara[1], T Ohigashi[2], S Toné[3,5], M Kado[4] and A Ito[1]

[1]School of Engineering,Tokai Univ, Hiratsuka, Kanagawa 259-1292, Japan
[2]UVSOR Synchrotron, Inst Molecular Science, Okazaki, Aichi 444-8585, Japan
[3]Dept Biochem, Kawasaki Medical School, Kurashiki, Okayama 701-0192, Japan
[4]Kansai Photon Science Institute, QST, Kizugawa, Kyoto 619-0215, Japan
[5]Present address: Tokyo Denki Univ, Hatoyama, Saitama 350-0394, Japan

E-mail: kshino-tky@umin.net

Abstract. Molecular distribution in mammalian cells was studied by soft X-ray scanning transmission microscopy with respect to the quantitative aspect of analysis. NEXAFS profiles at the C, N and O K-absorption edges were combined and used for the analysis. For the estimation of quantity for nucleic acids and proteins, NEXAFS profiles of DNA and bovine serum albumin (BSA) at the N K-absorption edge were applied assuming that those were their representatives. The method has a potential to explore the other molecular components than nucleic acids and proteins.

1. Introduction

Scanning transmission X-ray microscope (STXM) has been established and used in a wide area of science. In a biological field, distribution of DNA in a bean chromosome [1] and a bull sperm [2] has been imaged by the spectromicroscopy. In addition, quantitative approach has been developed for the analysis of molecules and elements in biological specimens such as microbial biofilms [3], quinoa [4] and bacterial endospores [5] using STXM. For the quantitative analysis of molecular distribution of DNA and proteins, near edge X-ray absorption fine structure (NEXAFS) at carbon K-absorption edge has been applied. We have proposed that NEXAFS at the N K-absorption edge has a merit of clear separation between DNA and proteins in addition to the exclusion of biomolecules containing no N [6].

2. Materials and methods

Cultured CHO cells at interphase were collected, fixed with ethanol-acetic acid (3:1) solution or glutaraldehyde (2.5% in 0.075M KCl) and dropped on the commercially available formvar membrane (#10-1008, Okenshoji. Japan). The cells and molecules such as DNA and bovine serum albumin (BSA; protein) were observed with scanning transmission X-ray microscope (STXM) installed at BL4U, UVSOR-III, Okazaki, Japan. The results were analyzed using the aXis2000 program. For quantitative analysis, we use NEXAFS spectra of DNA and BSA at the N K-absorption edge, since nucleic acids and proteins are major components of cells containing nitrogen and were clearly separated at 1s-π* transition.

Calculation has been made as follows: At the first step, each spectrum of image, DNA or BSA at the C, N and O K-absorption edges was combined to form one absorption spectrum covering the energy range from the C K-absorption edge to O K-absorption edge. Combined image spectrum was named as spect1. DNA spectrum was adjusted to spect1 so that it fitted the optical density (OD) difference between ODs at 399.6 eV and 398 eV (the peak and the valley in DNA NEXAFS) of spect1 and named as DNAspect. DNAspect was subtracted from spect1 resulting in a spectrum named as spect2. BSA spectrum was adjusted to spect2 so that it fited the OD difference between ODs at 401.4 eV and 398 eV (the peak and the valley in BSA NEXAFS) of spect2 and named as Prspect. Prspect was subtracted from spect2 resulting in a spectrum named as spect3. ODs in DNAspect and Prspect were converted to mass absorption coefficient. For comparison of the profile of each spectrum, it was rescaled so that the difference of maximum level and minimum level was adjusted to be the same OD difference and set to be the same at the left end of the energy.

3. Results and Discussion

3.1. NEXAFS of DNA and BSA

Figure 1 shows the NEXAFS profiles of DNA and BSA at the C, N and O K-absorption edges. Peaks were referred to the papers for DNA [7] and proteins [8]. NEXAFS of DNA at the N K-absorption edge shows a peak of 1s to π^* transition of N=C bond at 399.6 eV, which is clearly separated from a peak of 1s to π^* transition of amide at 401.4 eV in BSA. It should be noted that DNA and RNA include the same type of N=C bond, hence the present analysis of DNA corresponds to nucleic acids. In addition, amide is common in proteins, so that the analysis of BSA is for proteins.

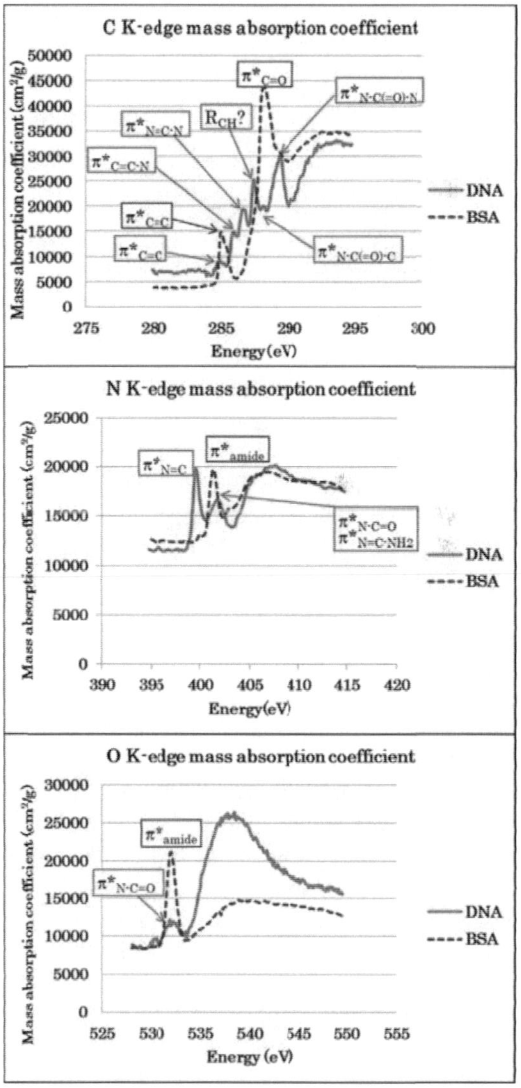

Figure 1. Mass absorption coefficient of DNA (solid line) and BSA (dotted line) at the C, N and O K-absorption edges.

3.2. Images of CHO cells

Figure 2 shows the optical density images at 285 eV and RGB maps for interphase cells fixed with either ethanol-acetic acid or glutaraldehyde obtained with STXM. The spect3 was estimated from the average absorption spectrum (spect1) of a whole imaging area by the method shown in section 2. RGB maps

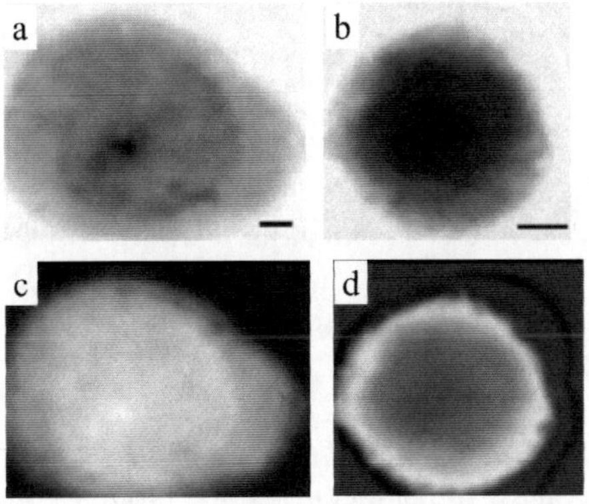

Figure 2. Optical density images (a,b) and their RGB maps (c,d) of CHO cells fixed with ethanol-acetic acid (a,c) and glutaraldehyde (b,d). In RGB maps (c,d), red, green and blue are for nucleic acid, protein, and spect3, respectively. Bar, 2 μm.

were obtained by the aXis2000 program with the mass absorption coefficient spectra of DNA and BSA, and OD of spect3.

The results in figure 2 show different RGB maps due to the difference in the fixatives of the cells, ethanol-acetic acid and glutaraldehyde. In the center of the cells (nucleus) fixed with glutaraldehyde (figure 2d), color is blue meaning that the main component was spect3 (residual component other than nucleic acids and proteins). The result is not reasonable to our knowledge that nucleus mainly consists of DNA and proteins. In contrast, the result for the cells fixed with ethanol-acetic acid shows the reasonable feature. The reason for this discrepancy is attributable to the glutaraldehyde. The observed data for the cells fixed with glutaraldehyde may not be sensitive enough to separate the spectra of DNA and proteins at the present condition. The results suggested that fixatives of the cells should be selected for the purpose of the analysis, though tomographic sliced-image will clear the problem.

3.3. NEXAFS of local areas in CHO cells

The NEXAFS profiles for the local areas in CHO cells were studied as the next step. The areas were chosen for nucleus (1, green), nucleus containing nucleolus (2, blue) and cytoplasm (3, pink) as shown in figure 3. It should be noted that the areas for region 1 and 2 in the cells fixed with glutaraldehyde was not adopted because of the reason mentioned above. Figure 4 shows the spect1 of each local area for comparison. In figure 4, each part in figure 3 corresponds to the same colour and number, i.e., part1 for green (solid line), part2 for blue (broken line), part3 for pink (broken at long distance) of the cells fixed with ethanol-acetic acid (figure 3a), and g-part3 for purple (dotted line) of the cells fixed with glutaraldehyde (figure 3b).

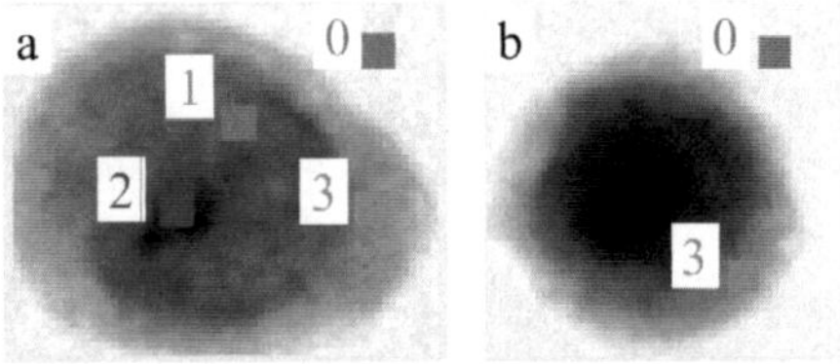

Figure 3. Local areas in CHO cells fixed with ethanol-acetic acid (a), and glutaraldehyde (b) for the analysis. Parts of 0 (red) for outside the cells (background), 1 (green) for nucleus, 2 (blue) for nucleus including mainly nucleolus, and 3 (pink) for cytoplasm. The sizes of the areas were a, 2.25 μm^2 each and b, 0.73μm^2.

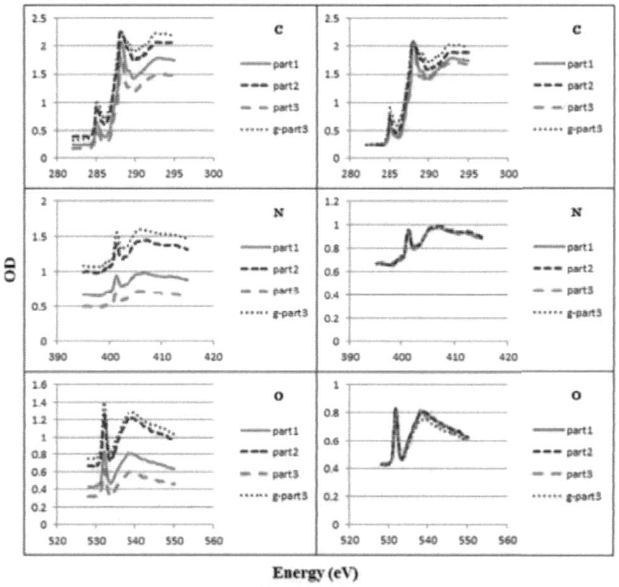

Figure 4. Spect1 of three local areas (figure 3a parts1-3) in a CHO cell fixed with ethanol-acetic acid and one area (figure 3b part3; g-part3) in a CHO cell fixed with glutaraldehyde for the spectra at the C (top), N (middle) and O (bottom) K-absorption edges. Left half shows direct comparison of observed spect1 and right half is for the rescaled spectra.

It is interesting to note that the profiles of each area are quite similar at the N and O K-absorption edges (right half in figure 4). However, the profiles of the spectra from different position are not the same at the C K-absorption edge which may be applicable to identify the molecules other than nucleic acids and proteins, or even among the nucleic acids and proteins. Left half in figure 4 reflects the comparison of the density difference of the molecules. The density is low at cytoplasm and high at nucleus for the cells fixed with ethanol-acetic acid. In contrast, the density in cytoplasm (figure3b, 3) for the cells fixed with glutaraldehyde was much higher compared to the cells fixed with ethanol-acetic acid.

3.4. NEXAFS of spect3 for local areas in CHO cells

Figure 5 shows the spect3 of each local area for comparison. The spect3 indicated the spectrum for the molecules other than nucleic acids and proteins. Therefore, it should be useful to find the molecular components in the cells. Figure 5 suggests that there is not much difference at all the local areas in cells for the distribution of molecules containing nitrogen and oxygen, but carbon, at which significant difference in the NEXAFS was observed. A sharp decrease in the spectra may be caused by either a small shift of the peak for proteins or the presence of component(s) which have a peak close to the C=O amide absorption peak.

With the present results for DNAspect and Prspect, the amount of nucleic acids and proteins at the local areas were estimated (Table1). The ratio of nucleic acids/proteins were almost the same regardless of the local areas in the cells fixed with ethanol-acetic acid in addition to the cytoplasm of the cells fixed with glutaraldehyde, though molecular species should be different, i.e., mainly DNA for nucleus and RNA for cytoplasm.

Table1. Ratio of nucleic acids per proteins estimated from DNAspect and Prspect for local areas in CHO cells shown in figure 3.

part	nucleic acids(N)	Proteins(P)	N/P
	μg/cm^2		
1	6.71	33.1	0.20
2	10.3	48.4	0.21
3	5.5	23.5	0.23
g-part3	11.1	57.7	0.19

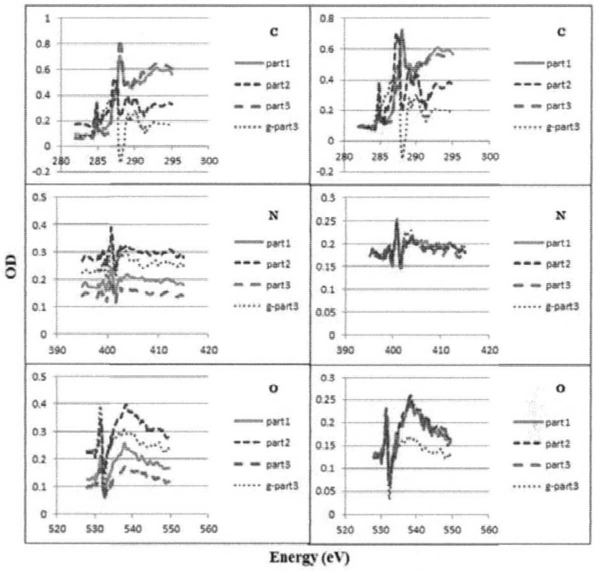

Figure 5. Spect3 of three local areas (figure 3a parts1-3) in a CHO cell fixed with ethanol-acetic acid and one area (figure 3b part3; g-part3) in a CHO cell fixed with glutaraldehyde for the spectra at the C (top), N (middle) and O (bottom) K-absorption edges. Left half shows direct comparison of the spectra and right half is for the rescaled spectra.

In conclusion, quantitative analysis using combined NEXAFS profiles at C, N, and O K-absorption edges showed that the ratios of nucleic acids per proteins were the same level in all the cell regions.

References

[1] Ade H, Zhang X, Cameron S, Costello C, Kirz J and Willoams S 1992 *Science* **258** 972
[2] Zhang X, Balhorn R, Mazrimas J and Kirz J 1996 *J. Struct. Biol.* **116** 335
[3] Lawrence J R, Swerhone G D W, Leppard G G, Araki T, Zhang X, West M M and Hitchcock A P 2003 *Applied Environmental Microbiology* **69** 5543
[4] Yangquanwei Z, Neethirajan S and Karunakaran C 2013 *Nanoscale Res. Lett.* **8** 463
[5] Jamroskovic J, Shao P P, Suvorova E, Barak I and Bernier-Latmani R 2014 *FEMS Microbiol. Lett.* **355** 188
[6] Shinohara K, Ito A and Kobayashi K 1998 *X-Ray Microscopy and Spectromicroscopy* (Berlin/Heidelberg: Springer-Verlag) p III-157
[7] Kummer K, Vyalikh D V, Gavrila G, Preobrajenski A B, Kick A, Bönsch M, Mertig M and Molodtsov S L 2010 *J. Phys. Chem.* B **114** 9645
[8] Stewart-Ornstein J, Hitchcock A P, Cruz D H, Henklein P, Overhage J, Hilpert K, Hale J D and Hancock R E W 2007 *J. Phys. Chem.* B **111** 7691; Zubavichus Y, Shaporenko A, Grunze M and Zharnikov M 2008 *J. Phys. Chem.* B **112** 4478

Imaging samples larger than the field of view: the SLS experience

Ioannis Vogiatzis Oikonomidis[1,2], Goran Lovric[2,3], Tiziana P Cremona[1], Filippo Arcadu[2,4], Alessandra Patera[2,4], Johannes C Schittny[1] and Marco Stampanoni[2,4]

[1] Institute of Anatomy, University of Bern, 3012 Bern, Switzerland
[2] Swiss Light Source, Paul Scherrer Institute, 5234 Villigen, Switzerland
[3] Center for Biomedical Imaging (CIBM), EPFL, 1015 Lausanne, Switzerland
[4] Institute for Biomedical Engineering, University and ETH Zurich, 8092 Zurich, Switzerland

E-mail: ioannis.vogiatzis@ana.unibe.ch

Abstract. Volumetric datasets with micrometer spatial and sub-second temporal resolutions are nowadays routinely acquired using synchrotron X-ray tomographic microscopy (SRXTM). Although SRXTM technology allows the examination of multiple samples with short scan times, many specimens are larger than the field-of-view (FOV) provided by the detector. The extension of the FOV in the direction perpendicular to the rotation axis remains non-trivial. We present a method that can efficiently increase the FOV merging volumetric datasets obtained by region-of-interest tomographies in different 3D positions of the sample with a minimal amount of artefacts and with the ability to handle large amounts of data. The method has been successfully applied for the three-dimensional imaging of a small number of mouse lung acini of intact animals, where pixel sizes down to the micrometer range and short exposure times are required.

1. Introduction

Synchrotron X-ray tomographic microscopy (SRXTM) is an effective technique that can provide the three-dimensional (3D) structural information of samples in a non-destructive manner and with high throughput. When high spatial resolutions are employed (i.e. very small pixel sizes), the detector's field of view (FOV) is often smaller than the object's size. This is the case in respiratory studies, where the feature sizes of interest, i.e. the aleveolar septa inside the lung, are in the order of micrometers while the sample size is typically 3 orders of magnitude bigger.

A number of approaches allowing to overcome this restriction exist. Obviously, the simplest method is to combine single X-ray projection images (radiographs) that cover the whole sample and use the larger composite images to conduct the 3D reconstruction [1]. However, this leads to long acquisition times and increases computational complexity for the CT reconstruction. In addition, such an approach is very prone to misalignment artefacts in the radiography space, since they are propagated along the entire volume when reconstructing the 3D volume with the standard filtered back projection method [2]. Another approach is the so-called region of interest tomography or local tomography, where parts of the object are only partially illuminated during the entire scan [2]. The partial sinograms are then reconstructed separately and the resulting volumes are merged together. Not having the entire object within the FOV however, leads to a change in the mean value of the image and low-frequency artefacts [3].

Here we present a method to efficiently merge together different 3D volumes that have been acquired through multiple local tomographies. The overlap of the individual tomographic slices is automatically registered and then combined together through graph cuts, thus minimizing stitching artefacts.

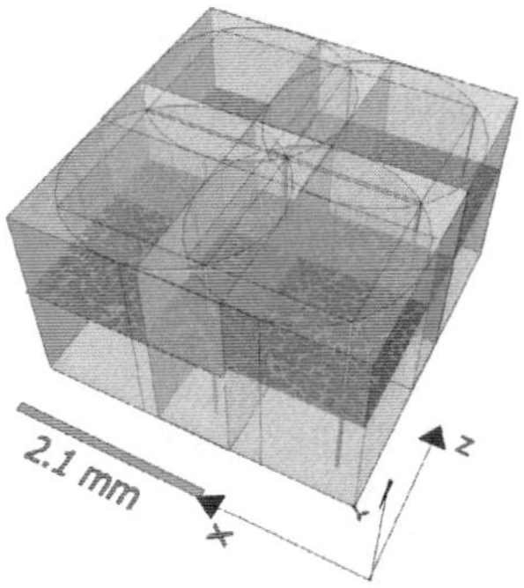

Figure 1: 2 × 2 spatial grid of different parts of the lung parenchyma. The areas outside of the resolution circle suffer from poor signal and contrast to noise ratio. An overlap of roughly 800 pixels was selected in order to be able to align the volumes automatically.

2. Materials and Methods

2.1. Animal Preparation, Image Acquisition and CT-reconstructions

Balb-C mice were anesthetized, placed in an upright position into a custom-made sample holder and ventilated by a small-animal ventilator (SCIREQ FlexiVent) via an endotracheal cannula. During imaging the inter-pulmonary pressure was maintained at a certain level while acquiring the tomographic scan. Animals were sacrificed by an overdose of anesthesia and images of the intact animals were taken no longer than 20 minutes afterwards. The experiments were approved and supervised by the Veterinary Service of the Cantons Aargau and Bern. Data were acquired at the X02DA TOMCAT beamline of the Swiss Light Source (SLS) at the Paul Scherrer Institute (Villigen, Switzerland). The beam was monochromatized at 21 keV by a multilayer crystal providing a good trade-off between contrast to noise ratio and resolution [4]. The in-house developed high-speed CMOS GigaFRoST detector coupled to visible-light optics of 10× magnification and a 20 μm thick LuAG:Ce scintillator leading to an effective pixel size of 1.1 μm was used for the lung imaging. The tomographic scans consist of 1500 views for each volume and each projection is acquired with an exposure time of 5 ms. A single-shot propagation-based phase retrieval algorithms [5] was applied to every projection and subsequently CT-reconstructed with the *gridrec* algorithm [6] with no ring artefact removal or lens distortions corrections, since the quality of the data did not require it.

2.2. Stitching

Image stitching is performed in most cases by using invariant feature detection and matching on image pairs [7, 8]. These methods extract keypoints on the images and compute a descriptor vector for each of them. The extracted keypoints need to be detectable under changes in image scale, orientation, noise and illumination. As a result, they are usually detected on high contrast regions and near edges. The descriptor vector is then computed for every extracted feature point and normalized to unit length in order to prevent any bias due to the varying illumination.

Consequently, matching is performed between the two sets of extracted vectors, usually by means of Euclidean distance metrics. The matching is used to estimate the registration transform of the images. Outliers are detected and discarded from the set of matches in order to compute a model for the shift between the two images using least-squares fitting. The model is updated iteratively every time a new set of outliers is detected. When the probability of a correct interpolation of the images is greater than a given number, the affine transform to register the two images has been successfully estimated. Weighted averages are performed on the basis of distance transforms, that are computed with respect to the overlapping region, and the final merged image is obtained.

In order to overcome the above problems, the following steps are here proposed to stitch the reconstructed volumes in an efficient way, while minimizing the artefacts:

(i) The tomographic scans $I_1^{m \times n \times l}$, $I_2^{m \times n \times l}$ with m, n, l the number of voxels along the 3 directions, are placed on a 2D spatial grid according to their position in real space so that the adjacent volumes can be defined. This position has not been used for the exact overlap estimation.

(ii) The two slides are registered by computing the phase correlation (PC) $R = \frac{\mathscr{F} I_1^{m \times n} . \mathscr{F} I_2^{m \times n *}}{|\mathscr{F} I_1^{m \times n} . \mathscr{F} I_2^{m \times n *}|}$ [9]. The shift is given by the position of the minimum value of the inverse Fourier transform of PC, $(\Delta x, \Delta y) = \arg\max_{(x,y)} \mathscr{F}^{-1} R$, and the same is applied on all the slices along the tomography axis, resulting in the shifted images $I_3^{o \times p \times l}$, $I_4^{o \times p \times l}$

(iii) The absolute difference image $G^{m \times n} = \left| I_3^{o \times p} - I_4^{o \times p} \right|$ of the adjacent shifted slices is computed, and the result is treated as an undirected graph with the difference values determining the cost of each node.

(iv) Two seed points v amd v' on the two opposite edges of the graph G are defined and the minimum cost path is determined as the sequence of vertices $P = (g_1, g_2, g_3, \ldots, v_{o \times p}) \in G \times G \times G \times \ldots \times G$ such that g_i is adjacent to $g_{(i+1)}$ for $1 \leq i \leq (o \times p)$ [10].

(v) The computed path is turned into a binary mask and the parts of the images I_3 and I_4 that are on each side of the path respectively are merged together, resulting in the least amount of stitching errors. The results of the masking can be seen in figure 2a - figure 2d.

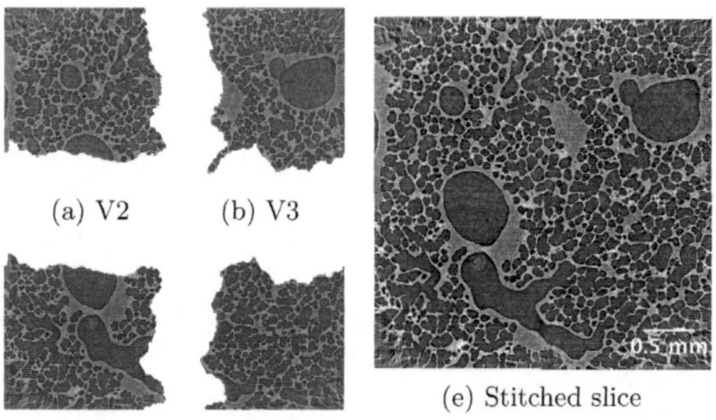

(a) V2 (b) V3 (c) V1 (d) V4 (e) Stitched slice

Figure 2: Figures (a) to (d) represent the graph cuts of the seperate volumes of the lung parenchyma, and (e) the final result of the stitching.

3. Results and Discussion

We discuss the case of a mouse lung that resulted in approximately 4500 × 4500 pixels shown in figure 2e. The undersampled numbered of projections per single scan (1500), is a result of the limited time window before lung collapse after death. In addition, the thickness of our scintilator and the numerical aperture of the visible light optics limit our resolution [11], thus the Nyquist tomographic sampling criterio can be relaxed. The approach is based on automatic registration of the tomographic slices of the overlap regions inside the resolution circle as depicted in figure 1. The individual slices are then combined with the use of graph cuts seen in figure 2a - figure 2d. Our approach is though general and can work on larger datasets, being limited only by the workstation memory. The method offers minimal amount of stitching artefacts, a prerequisite when the datasets are characterized by a low contrast to noise ratio and further quantifications need to be performed. Since the validity of the final result is mostly dependent on the estimated overlap, a robust and accurate registration technique is required. Biological samples have a complex structure with no clear shape descriptor and no background because of the local tomography acquisition, making feature extraction based stitching inapplicable. In the majority of the cases the feature matching fails, since there are no clear edges between the different structures and the contrast of the image is not high enough, resulting in wrongly estimated affine registration transforms. Due to the large data size, more complex registration techniques that maximize mutual information or are based on a similar optimization approach, are also inapplicable. The fourier based registration approach provides robust and accurate results and is characetrized by computational efficiency. Furthermore, the low contrast that characterizes our datasets, makes them even more prone to the so-called *seam* artefact [12], which is produced by blending the registered overlapping images. This artefact is caused by the different illumination of the images and results in a visible border separating the original input images. The only way to mitigate it, within the context of a blending approach, is with more sophisticated weighting approaches that, compromising however the resolution. By employing a graph cut approach, the images are stitched together like different pieces of a puzzle. Thus, we completely avoid any kind of blending and along with it the *seam* artefact.

4. Conclusion

In this work we have presented a methodology for increasing the FOV of SXRTM. Neighbouring reconstructed volumes are stitched together to increase the FOV by means of an effective graph-cuts based algorithm. The *seam* artefact, present in weighted blending merged datasets, is suppressed without compromising the local resolution. Complex weighting averages approaches that are used to suppress seam artefacts are computationally expensive and make the stitching procedure slower as the slice size increases. With the proposed method single slices are combined together like a puzzle without having to recompute weighted blending every time an adjacent slice is to be merged with the rest and thus altering the contrast of the initial datasets.

Acknowledgments

We would like to acknowledge Dr. Christian M Schlepütz for his support at the TOMCAT beamline and for carefully reading the manuscript. We are thankful for the grants 310030-153468 and CR23I2-135550 of the Swiss National Science Foundation.

References

[1] Haberthür D, Hintermüller C, Marone F, Schittny J C and Stampanoni M 2010 *Journal of synchrotron radiation* **17** 590–599
[2] Kyrieleis A, Titarenko V, Ibison M, Connolley T and Withers P 2011 *Journal of microscopy* **241** 69–82
[3] Rashed E A and Kudo H 2013 *Journal of synchrotron radiation* **20** 116–24
[4] Lovric G, Barré S F, Schittny J C, Roth-Kleiner M, Stampanoni M and Mokso R 2013 *Journal of applied crystallography* **46** 856–860
[5] Moosmann J, Hofmann R and Baumbach T 2011 *Optics express* **19** 12066–73
[6] Marone F and Stampanoni M 2012 *Journal of synchrotron radiation* **19** 1029–37
[7] Lowe D G 1999 *Computer vision, 1999. The proceedings of the seventh IEEE international conference on* vol 2 (Ieee) pp 1150–57
[8] Brown M and Lowe D G 2007 *International journal of computer vision* **74** 59–73
[9] Guizar-Sicairos M, Thurman S T and Fienup J R 2008 *Optics letters* **33** 156–58
[10] Dijkstra E W 1959 *Numerische mathematik* **1** 269–71
[11] Stampanoni M, Borchert G, Wyss P, Abela R, Patterson B, Hunt S, Vermeulen D and Rüegsegger P 2002 *Nuclear Instruments and Methods in Physics Research Section A: Accelerators, Spectrometers, Detectors and Associated Equipment* **491** 291–301
[12] Levin A, Zomet A, Peleg S and Weiss Y 2004 *European Conference on Computer Vision* (Springer) pp 377–89

IOP Conf. Series: Journal of Physics: Conf. Series **849** (2017) 012005 doi:10.1088/1742-6596/849/1/012005

Determination of carbon-to-nitrogen ratio in the filamentous and heterocystous cyanobacterium *Anabaena* sp. PCC 7120 with single-cell soft X-ray imaging

T Teramoto[1], M Yoshimura[2], C Azai[3], K Terauchi[3] and T Ohta[2]

[1] *College of Science & Engineering, Ritsumeikan University, Shiga, 525-8577, JAPAN*
[2] *SR center, Ritsumeikan University, Shiga, 525-8577, JAPAN*
[3] *College of Life Sciences, Ritsumeikan University, Shiga, 525-8577, JAPAN*

E-mail: tteramo@fc.ritsumei.ac.jp

Abstract. Vegetative cells and heterocysts in the filamentous cyanobacterium *Anabaena* sp. PCC 7120 were observed by soft X-ray microscopy. Carbon-to-nitrogen (C/N) ratio of each cell was estimated by the difference of the absorbance of the images below and above the nitrogen K-edge absorption. It was revealed that the C/N ratios in vegetative cells and heterocysts are 4.54 and 2.46, respectively.

1. Introduction

Natural or anthropogenic conversion of unreactive N_2 to other reactive nitrogen compounds such as ammonia, is called as nitrogen fixation and this is one of the main issues for ecosystem and food production on the earth [1]. Although the chemical Haber–Bosch process produces ammonia and fills the industrial demand, a biological nitrogen fixation is expected to displace the industrial production of ammonia in terms of a sustainable and environment-friendly, low-carbon society.

Filamentous cyanobacterium *Anabaena* sp. PCC 7120 is one of its promising candidates [2]. It forms a filament of "vegetative cells", which is responsible for oxygenic photosynthesis. When starved for reactive nitrogen compounds, one out of about ten vegetative cells differentiates to a specialized cell for nitrogen fixation, called "heterocyst". The heterocyst expresses the nitrogenase enzyme to catalyze the nitrogen fixation reaction, while it loses photosystem II, which is the water-oxidizing enzyme of photosynthesis, to avoid degradation of the active metal center of nitrogenase [3]. The vegetative cells cannot grow without reactive nitrogen compounds and the heterocyst is not capable of photosynthetic life maintenance; the vegetative cell and heterocyst are thus mutually dependent on their metabolites. Vegetative cells provide the photosynthesis products, such as sugars or organic acids, to the neighbor heterocyst. Instead, heterocysts give amino acids to vegetative cells. The metabolites are transported through a cellular junction of the vegetable cell and heterocyst.

The C/N ratio is a nominal parameter of biological samples. This is estimated conventionally by the combustion analysis, which quantitates amounts of CO_2 and N_2 by burning dried cells. It has been revealed that the heterocyst differentiation is triggered by a high carbon-to-nitrogen (C/N) ratio of a parental vegetative cell [3, 4]. However, the C/N ratio in the single cell level *in vivo* had never been observed so far. In this study, we observed nitrogen distribution in the heterocysts and vegetative cells of *Anabaena* sp. PCC 7120 by soft X-ray microscopy at the wavelengths shorter and longer than the nitrogen K-edge energy and estimated the C/N ratio of each cell *in vivo*.

2. Experimental

The cells of *Anabaena* sp. PCC 7120 were grown on a nitrogen-depleted agar plate. The filaments including heterocysts were suspended in pure water and tightly packed between two silicon nitride membranes with the thickness of 100 nm. The absolute position of each cell in the sandwich was determined by optical and fluorescence microscopy. The soft X-ray microscopic observation was performed at the wavelengths (λ) of 2.98 and 3.11 nm (above and below N1s absorption threshold, respectively) using the full-field transmission soft X-ray microscopy beamline (BL12) in the SR centre of Ritsumeikan University [5]. The soft X-ray microscope is composed of a SiO_2 plane mirror for cutting higher-order X-rays, a condenser Fresnel zone plate, an order-sorting aperture, a sample, an objective Fresnel zone plate, and a back-illuminated CCD camera. The sample was located in air, separated from the vacuum chamber by two silicon nitride membranes. The spatial resolution was 110 nm at the magnification with 600 that was estimated by knife-edge method. The effective pixel size in the image was 40×40 nm^2. The exposure time was fixed to 3 min.

3. Results and Discussion

Figures 1 (a) and (b) are the images of heterocysts and vegetative cells of *Anabaena* sp. PCC 7120 taken by optical and fluorescence microscopy, respectively. Because heterocysts do not include photosystem II, whose chlorophylls are fluorescent even at room temperature, they do not emit the chlorophyll fluorescence (for example, the white circle in figure 1(b)). The heterocysts and vegetative cells were observed as a monolayer both in the optical and fluorescence images. Thus, their absolute positions could be determined in the silicon nitride membranes. Each cell has a barrel shape with the major and minor axes of approximately 4.5 and 3.3 μm, respectively.

The soft X-ray microscopic images corresponding to the centre part of Fig. 1 are shown in Figs. 2 (a) and (b) with below and above threshold of N K-edge absorption, respectively. From these images, we could observe clear absorbance differences between heterocysts and vegetative cells.

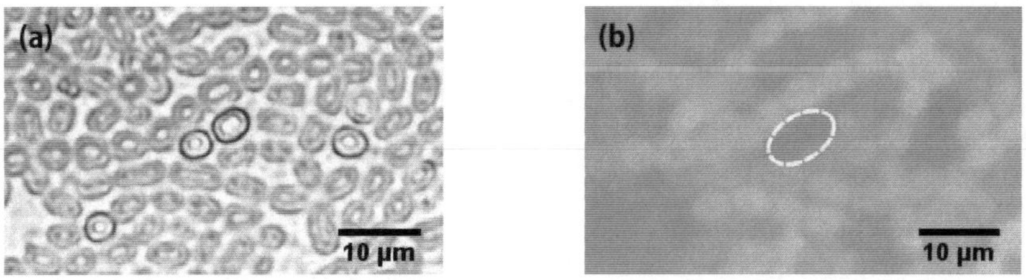

Figure 1. (a) Optical and (b) fluorescence images of heterocysts and vegetative cells

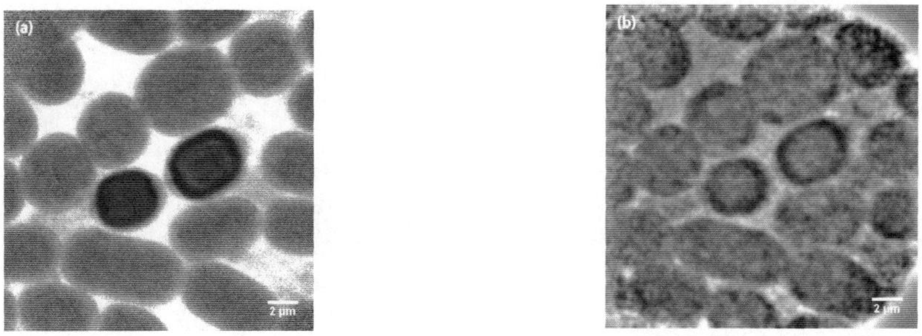

Figure 2. Soft X-ray images of heterocysts and vegetative cells at (a) λ = 3.11 nm and (b) λ = 2.98 nm

According to the Lambert-Beer law ($A = -\log(I/I_0)$), we could get the absorbance (A) at each cell from the soft X-ray image. Here, I and I_0 are transmitted intensities with/without the sample, respectively. The intensity of a cell image in Fig. 2 corresponds to I. On the other hand, I_0 could be assumed as the intensity of an area without cells, since the membrane is large enough (250×250 μm^2) to show several areas without cells, as shown in Fig. 2.

To statistically analyse the difference in the soft X-ray absorbance between heterocysts and vegetative cells, we collected several images of the cells. The resultant absorbances are summarized as the histograms, which is shown in Fig. 3. The absorbance used here is the area average of the cell images. The standard deviation to determine the absorbance in this way is ~15%. Totally, 131 vegetative cells and 54 heterocysts were sampled. The mean values and variances of the histogram are different between heterocysts and vegetative cells at both 3.11 and 2.98 nm of the illumination wavelengths. The mean values of absorbance of heterocyst and vegetative cell were 0.73 and 0.53 at λ = 3.11 nm, respectively, while those were 0.51 and 0.44 at λ = 2.98 nm.

At each wavelength, absorbance of heterocyst is higher than vegetative cell, indicating that the contents of cellular compounds are different.

The total absorbance with different several components at specific wavelength is defined as follows:

$$A(\lambda) = \sum_i A_i(\lambda) = \sum_i \varepsilon_i c_i \ell \quad (1)$$

Here $A(\lambda)$ is absorbance at wavelength λ. $A_i(\lambda)$, ε_i, c_i are absorbance, absorption cross section, concentration of the *i*-th element, respectively. ℓ is the optical path length and is the thickness of the cell in the present study. The main elements contributing to the images below and above of N1s absorption-edge are C and N in the cells, because the O1s absorption threshold is far higher than N1s absorption threshold. We thus assumed the absorption of O-element as a negligible constant at the observed wavelengths and ignored it in the following analysis. From these considerations, the formula (1) can be simply rewritten as follows:

$$A(\lambda) = (\varepsilon_c(\lambda)c_c + \varepsilon_N(\lambda)c_N)\ell \quad (2)$$

The relationship between absorbance and concentration ratios of two elements, C and N, can be derived from absorbance at two different wavelengths (λ_1, λ_2), as follows:

$$\frac{A(\lambda_1)}{A(\lambda_2)} = \frac{\varepsilon_c(\lambda_1)R + \varepsilon_N(\lambda_1)}{\varepsilon_c(\lambda_2)R + \varepsilon_N(\lambda_2)} \quad (3)$$

$$R = \frac{c_c}{c_N} \quad (4)$$

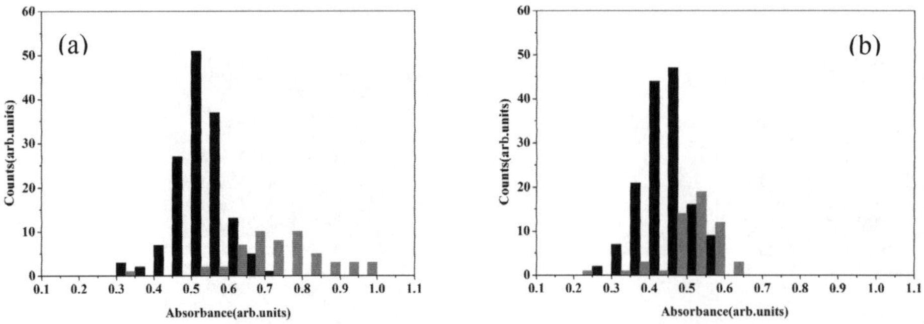

Figure 3. Histograms of the absorbance of heterocysts (red) and vegetative cells (black) from soft X-ray imaging at (a) λ=3.11nm and (b) λ=2.98nm

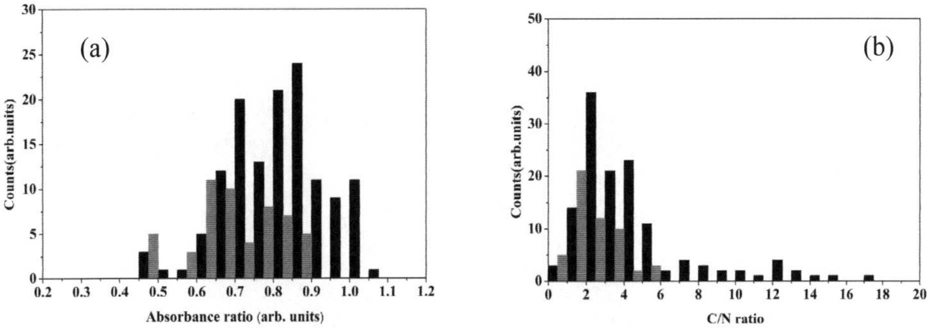

Figure 4. (a) Histogram of the absorbance ratio $\frac{A(\lambda=3.11nm)}{A(\lambda=2.98nm)}$ and (b) C/N ratio (R) of heterocysts (red) and vegetative cells (black) from soft X-ray images

Here R is the same as the well-known C/N ratio. From the observed soft X-ray images, we calculated the absorbance ratio of the individual cells between 3.11 and 2.98 nm-wavelengths, and summarized as a histogram in Fig. 4(a), which exhibits a distinct difference between vegetative cells and heterocysts. We could obtain the C/N ratio by using formula (3) and reference values of absorption cross-sections of C and N in the database [6]. Figure 4(b) is the histogram of C/N ratios in each cell. The mean values of the C/N ratio in vegetative cells and heterocysists are 4.54 and 2.46, respectively. Since there is no report about the C/N ratio of *Anabaena* sp. PCC 7120, we compared with the same kind of photosynthetic bacteria from the reference [7]. The C/N ratio of *Anabaena cylindrica*, which is a closely related species of *Anabaena* sp. PCC 7120, is around 4.5 to 8 and this value is close to our results, indicating that our technique with soft X-ray imaging is useful to determine the cellular C/N ratio *in vivo*.

4. Conclusions

The filamentous and heterocystous cyanobacterium *Anabaena* sp. PCC 7120 was observed by soft X-ray microscopy. We derived the simple formula to estimate the carbon-to-nitrogen (C/N) ratios of the individual cells *in vivo* in this study and applied it to the observed images. The estimated C/N ratios of the vegetative cells and heterocysts were 4.54 and 2.46, respectively.

5. Acknowledgement

This study was supported in part by Grants-in-Aid for Scientific Research from the Japan Society for the Promotion of Science (nos. 15H00886 (to C.A.), 26102543(to K.T.) and 16KT0167 (to T.T) and the ALCA Program of the Japan Science and Technology Agency (JST) (to K.T.)

References

[1] J.N. Galloway *et al.*, 2004 *Biogeochem.*, **70**, 153
[2] A.M. Muro-Pastor and W. Hess, 2012 *Trends Microbiol.*, **20**, 548
[3] H.BÖhme, 1998 *Trends in plant science*, **3**, 346
[4] J. Muñoz-García and S.Ares, 2016 *Proc.Natl.Acad.Sci.*, **113**, 6218
[5] K. Takemoto *et al.*, 2013 *J.Pys.Conf.Ser.*, **463**, 012009
[6] The Centre for X-Ray Optics http://www.cxro.lbl.gov/
[7] Kulasooriya *et al.*, *Proc. Roy. Soc. Lon. B:* **181,** 199(1972).

Effective segmentation of fresh *post-mortem* murine lung parenchyma in phase contrast X-ray tomographic microscopy images

Ioannis Vogiatzis Oikonomidis[1,2], Tiziana P Cremona[1], Goran Lovric[2,3], Filippo Arcadu[2,4], Marco Stampanoni[2,4] and Johannes C Schittny[1]

[1] Institute of Anatomy, University of Bern, 3012 Bern, Switzerland
[2] Swiss Light Source, Paul Scherrer Institute, 5234 Villigen, Switzerland
[3] Center for Biomedical Imaging (CIBM), EPFL, 1015 Lausanne, Switzerland
[4] Institute for Biomedical Engineering, University and ETH Zurich, 8092 Zurich, Switzerland

E-mail: ioannis.vogiatzis@ana.unibe.ch

Abstract. The acinus represents the functional unit of the mammalian lung. It is defined as the small tree of gas-exchanging airways, which is fed by the most distal purely conducting airway. Different hypotheses exist on how the fine structure of the acinus changes during ventilation and development. Since in classical 2-dimensional (2D) sections of the lung the borders of the acini are not detectable, every study of acini requires 3-dimensional (3D) datasets. As a basis for further studies of pulmonary acini we imaged rodent lungs as close to life as possible using phase contrast synchrotron radiation-based X-ray tomographic microscopy (SRXTM), and developed a protocol for the segmentation of the alveolar septa. The method is based on a combined multilevel filtering approach. Seeds are automatically defined for separate regions of tissue and airspace during each 2D filtering level and then given as input to a 3D random walk segmentation. Thus, the different types of artifacts present in the images are treated separately, taking into account the sample's structural complexity. The proposed procedure yields high-quality 3D segmentations of acinar microstructure that can be used for a reliable morphological analysis.

1. Introduction

The mammalian lung is a vital organ facilitating the gas-exchange between the ambient air and the blood. The gas-exchange area consists of functional units called acini which represent small trees of gas-exchanging airways. There is limited knowledge about the 3D microscopic dynamics and development of the acini, because their borders may not be recognized in classical 2D-sections. Therefore, the link between microscopic observations and specific lung function is still incomplete. Furthermore, the connection between lung diseases and the changes in the acinar microstructure is not fully understood.

3D imaging and morphological quantifications have become essential to further understand the function and development of the lung and in addressing the questions outlined above. The imaging setup needs to feature sufficient temporal and spatial resolution to correctly segment and recognize the acinus.

Here, we introduce an automatic segmentation method that can correctly differentiate the thin septa between the borders of small alveoli from air. Our technique is based on a combined multilevel filtering approach that takes into account the complex 3D structure of the lung. The method offers an unsupervised platform for the automatic extraction of the skeleton of the alveolar septa, which is essential for analyzing the deformation of inflated lungs as well as to track their development.

2. Materials and Methods

2.1. Image Acquisition, Animal Preparation and CT-reconstructions

Data were acquired at the X02DA TOMCAT beamline of the Swiss Light Source (SLS) at the Paul Scherrer Institute (Villigen, Switzerland). The experimental setup is illustrated in detail in [1]. A sample-to-source distance of $25\,m$ at $21\,keV$ was chosen as a good trade-off between contrast-to-noise ratio (CNR) and resolution of the acquired tomographic projections [1]. We used the in-house developed high-speed CMOS GigaFRoST detector coupled to visible-light optics of $10\times$ magnification and a $20\,\mu m$ thick LuAG:Ce scintillator leading to an effective pixel size of $1.1\,\mu m$. Tomographic scans were acquired with 1501 projections of $5\ ms$ exposure time. Including dark and flat field images, the total scan time per volume was approximately 40 seconds. Balb-C mice were first anesthetized with a mixture of Medetomidin, Midazolam and Fentanyl and an endotracheal cannula was installed after tracheotomy. The animal was placed in an upright position into a custom-made sample holder. A small-animal ventilator (SCIREQ FlexiVent) was employed for ventilation and to maintain the internal lung pressure, while acquiring the tomographic scan. Animals were sacrificied by an overdose of anaesthesia and images were taken no longer than 20 minutes afterwards to preserve *quasi-in-vivo* conditions. All animal experiments were approved and supervised by the Veterinary Service of the Cantons Aargau and Bern. Propagation-based phase-contrast was employed in order to increase the CNR of our individual projections. A single-distance phase retrieval algorithm by Moosmann *et al.* [2] was applied to every X-ray projection image and subsequently CT-reconstructed with the *gridrec* algorithm [3].

2.2. Segmentation

Since the data are acquired in interior/local tomography of a whole animal, the differentiation between air and tissue in the reconstructions varies from region to region of the volume and a gradient artifact also appears slice-wise, with tissues appearing darker on the one side of the image than the other. In addition, single-distance phase retrieval with no boundary conditions results in a super-imposed low frequency noise on the reconstructions [4]. The $5-10\,\mu m$ thick interairspace walls (alveolar septa) can be barely distinguished from the surrounding air on the basis of the grey level. An approach combining the random walk segmentation [5], a vesselness filter [6], local thresholding, and morphological closing based on the 2D average size in pixels of single alveoli has been implemented both in Python and MATLAB, and has been applied to the 8 bit depth reconstructed volume under study. The segmentation procedure can be summarized in the following steps:

Step 1) Otsu [7] thresholding results in figure (b), which is used as a basis to determine the segmentation seeds.

Step 2) 2D Connected component analysis [8] and separation between regions above and below 400 pixels (size of small alveoli with low contrast) yields images (c) and (d) respectively.

Step 3) Independent local thresholding is applied to each small local connected component (c) separately. The rest of the connected components are eroded and then combined with the locally thresholded ones, resulting in different seeds for different airspace regions, as seen in figure (d).

Step 4) A low-pass filtered images is subtracted from the raw data to enhance the edges. A 2D vesselness filter is then applied twice to obtain a single seed for all the septa resulting in the binary image of figure (e).

Step 5) Combine images (d) and (e) into a single image (f) to create the input image for the random walk.

Step 6) Random walker segmentation using the previously computed seeds as input to classify the remaining undetermined pixels. The final result can be seen in (g).

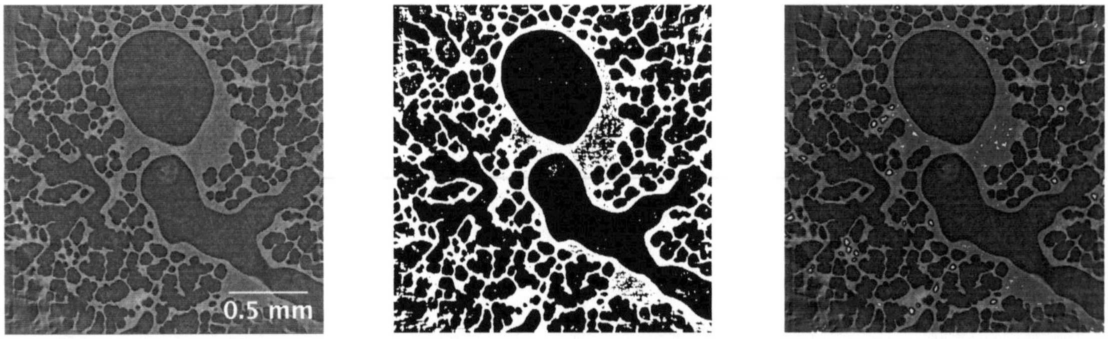

(a) Reconstructed slice (b) Otsu thresholded slice (c) Locally thresholded areas

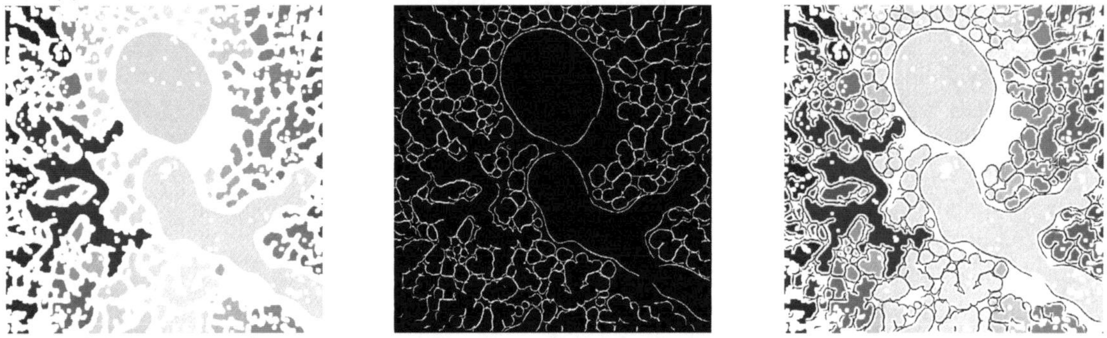

(d) Seeds for airspace (e) Seed for the thin septa (f) Random walk input

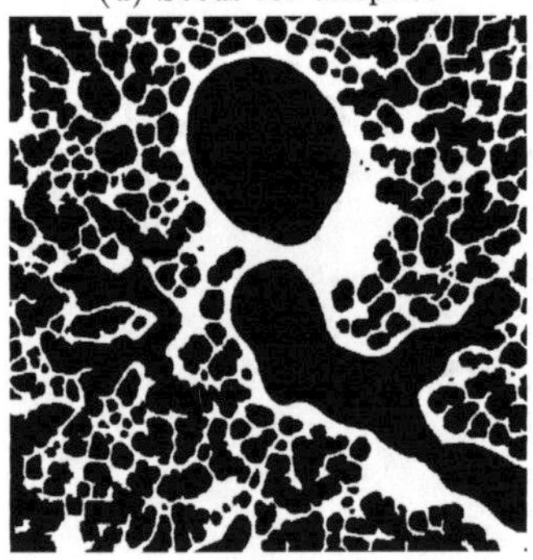

(g) Final result

Figure 1: Multilevel segmentation process. The bright pixels in (a) represent tissue and the dark ones airspace. Different colors represent different seeds with different ID (numbering). The seeds for the thin septa in (e) are all given the same ID along the whole volume. By running the final step in 3D, in a predefined number of adjacent slices, we are essentially mimicking the morphological decision done by eye whether unassigned pixels should be assigned to air or tissue i.e. by scrolling through adjacent slices.

3. Results and Discussion

In this work, we propose a novel methodology to segment 3D microstructures of fresh *post-mortem* murine lung tissue imaged inside the closed thorax. Our goal was to develop a methodology that does not require manual input or supervision. The proposed multi-level filtering approach allows us to accurately recognise lung parenchyma, despite its structural complexity, the low CNR and the presence of local tomography and phase retrieval artifacts. Using standard thresholding methods, the thin septa, as can be seen in the top middle and bottom left of figure (b) are wrongly segmented because of low CNR, and their size being close to the resolution limit of the setup. In the final result in figure (g), they are now correctly classified as tissue by taking into account the 3D local structure of the datasets. The large septa regions that suffer from local tomography artifacts and can be seen in the middle of figure (b) are also correctly segmented. The proposed segmentation was tested on full volumes and different samples. Our method does not require any preprocessing with filters for noise removal, which is necessary for many other segmentation methods. The structural complexity of the lung parenchyma is very high and any such approach would result in altering the shape of the thin alveolar septa. Watershed based segmentation always leads to over-segmentation, due to local tomography artifacts (speckles), structural complexity and phase retrieval artifacts. Approaches, that are based on histogram models also lead to inaccurate results due to their inability of taking into account the complex structural shape of the lung parenchyma. With the proposed method, both the small and large features of the lung parenchyma are correctly segmented as seen by human eye comparison. Validation and comparison to classical histology is not straightforward due to the different thickness of the sliced sections to the tomographic slices, and also because histological sample preparation alters the microstructural thickness in a non uniform way. An automatic approach for quantifying the general improvement of the algorithm would be to simulate the whole imaging pipeline. This, however, requires apriori knowledge of the full noise model of the setup. Finally, an artificial phantom that resembles the lung parenchyma is very difficult to create due to the structural complexity of the lung and it has been a subject of research. To the best of our knowledge, no such model exists at the moment that can include multiple acini and airways into the resulting computer generated phantom.

4. Conclusion

The proposed framework provides reliable 3D segmentations of the lung microstructure, that can be used as input for any subsequent morphological analysis/quantification of the lungs at micrometer scale. In particular the described segmentation of the alveolar septa may be applied as basis for the segmentation of individual acini. An easy and reliable segmentation of the acini bears the potential for many new insights into lung function and development which are thus far hindered by the inability of classical histology to recognize individual acini. Finally, our method is also applicable in samples where feature sizes vary by more than one order of magnitude within the tomographic slices.

Acknowledgments

We would like to acknowledge Dr. Christian M. Schlepütz for his support at the TOMCAT beamline and for carefully reading the manuscript. We are thankful for the grants 310030-153468 and CR23I2-135550 of the Swiss National Science Foundation.

References

[1] Lovric G, Barré S F, Schittny J C, Roth-Kleiner M, Stampanoni M and Mokso R 2013 *Journal of applied crystallography* **46** 856–860
[2] Moosmann J, Hofmann R and Baumbach T 2011 *Optics express* **19** 12066–73
[3] Marone F and Stampanoni M 2012 *Journal of synchrotron radiation* **19** 1029–37
[4] Huang L, Zuo C, Idir M, Qu W and Asundi A 2015 *Optics letters* **40** 1976–79
[5] Grady L 2006 *IEEE transactions on pattern analysis and machine intelligence* **28** 1768–83
[6] Frangi A F, Niessen W J, Vincken K L and Viergever M A 1998 *International Conference on Medical Image Computing and Computer-Assisted Intervention* (Springer) pp 130–37
[7] Otsu N 1975 *Automatica* **11** 23–27
[8] Samet H and Tamminen M 1988 *IEEE Transactions on Pattern Analysis and Machine Intelligence* **10** 579–586

IOP Conf. Series: Journal of Physics: Conf. Series **849** (2017) 012007 doi:10.1088/1742-6596/849/1/012007

Scanning Transmission X-ray Microscopy with X-ray Fluorescence Detection at the XUV Beamline P04, PETRA III, DESY

K Andrianov[1], L Lühl[2], T Nisius[1], A Haidl[1], R Gnewkow[2], L Lötgering[1], H Dierks[2], B Kanngießer[2] and T Wilhein[1]

[1] Institute for X-Optics, University of Applied Sciences Koblenz Rhein-Ahr-Campus, Joseph-Rovan-Allee 2, 53424 Remagen, Germany

[2] Institute of Optics and Atomic Physics, Technical University of Berlin, Hardenbergstr. 36, 10623 Berlin, Germany

E-mail: andrianov@hs-koblenz.de

Abstract. The presented scanning transmission x-ray microscope (STXM), build on top of our existing modular platform (FlexIX) for high resolution imaging experiments, allows versatile investigations of different samples. The FlexIX endstation allows to switch between a Full Field and a STXM mode. For the STXM mode we use a spatialy resolved detector together with an energy dispersive detector, this allows to investigate the morphology and the chemical or elemental distribution of the sample simultaneous. The combination of the nanoscopy endstation and the XUV beamline P04 results in a powerful tool for investigations of life science samples.

1. Motivation

Classical optical microscopy techniques are fundamentally limited by the diffraction limit, which implies that spatial resolutions below 200 nm can only be achieved by the use of short-wavelength radiation. While electron microscopy enables two- and three-dimensional imaging of materials in the sub-nanometer range [1], there is a gap in spatial resolution in the 10th nm range, which can be covered with X-ray Microscopy (XRM). For state of the art investigations in biomedicine this spatial resolution range is of a great interest [2]. In addition, the penetration depth of XRM is much higher as compared to electron microscopy, which facilitates sample preparation.

The $P04$ beamline of Petra III is a variable polarisation XUV beamline with an energy range of 250 eV up to 3000 eV [3]. This energy range is suitable to identify most of the interesting elements typically found in biological samples [2]. The beamline provides exceptionally high photon flux of up to 10^{12} photons per second in combination with a very high resolving power ($E/\Delta E > 10.000$) [3]. In contrast to most other beamlines, $P04$ has no fixed endstation and the experimental setups are provided by the user. In order to perform XRM at the $P04$ beamline the variable FlexIX endstation was developed and tested in the laboratory.

The FlexIX endstaion consists out of up to three separate vacuum chambers and custom designed high resolution piezo stages. These piezo stages are able to move precisely in x-, y- and z-direction, which makes it possible to adjust components for different imaging modes.

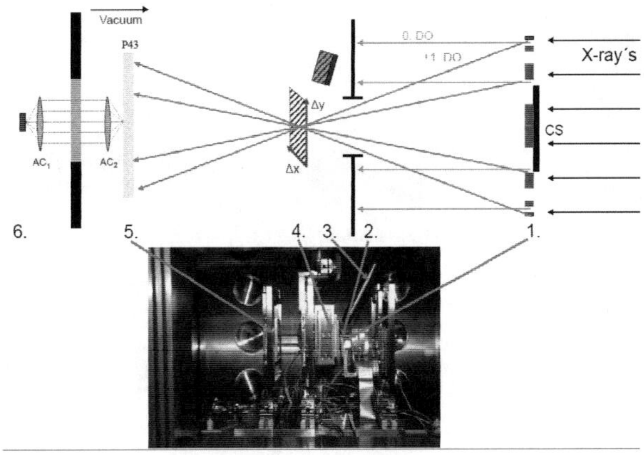

Figure 1. Experimental setup of the scanning microscope. Top: Sketch of the setup, bottom: real setup insight the vacuum chamber. 1. Fresnel zone plate; 2. Order sorting aperture (OSA); 3. Fluorescence detector; 4. Sample on scanner; 5. Phosphor screen; 6. CCD

The endstation is designed for quick changes between full-field (FF) mode and STXM/ fluorescence mode.
The main advantage of FF mode is that a large area of the sample can be investigated with very high spatial resolution ≈ 30 nm with a single exposure. During recent experiments the endstation was operated as a transmission x-ray microscope (TXM) in FF mode [4].

First experiments with the STXM mode were done in October 2015 and June 2016. In STXM mode the spot size on the sample defines the spatial resolution of the system. Therefore a zone plate (ZP) is used to focus the radiation on a small spot. Then the sample is moved through the focus point in a raster grid or in a continuous on-the-fly scan to retrieve an image.
In STXM similar resolutions with respect to FF can be achieved, but since scanning is needed, investigations of large areas are quite time consuming. However, in contrast to FF different imaging options such as phase-contrast or dark-field are easier to implement [5] and elemental contrast by the detection of characteristic fluorescence radiation is accessible.

2. Experimental Setup

A sketch of the STXM setup is shown in Figure 1. X-Ray radiation from the beamline is focussed by a zone plate with 333 μm in diameter and an outer most zone width of 40 nm onto the sample. The zone plate consists out of 150 nm tungsten deposited on a 100 nm thick SiN membrane and possesses a central stop of gold with 160 μm in diameter. Between ZP and sample is an order sorting aperture (OSA) made of 100 μm thick stainless steel with a 150 μm hole. It is mounted on a small aluminium tube in order to prevent the fluorescence radiation from the OSA to reach the detector.
The sample is scanned by a $P733$ scanning stage from Physik Instruments (PI). It has a travel range of 100 μm x 100 μm with a resolution of 0.1 nm.
The sample holder is designed to have space for up to eight samples prepared on common transmission electron microscopy (TEM) SiN membranes or grids.
The detection of the transmitted intensity distribution is done indirect with a $P43$ phosphor screen which is used to convert the X-rays into visible light. An optical path is used to transfer the visible light out of the vacuum chamber and an objective is used to image the signal from the phosphor screen on the Andor iXon3 EMCCD detector with 128 x 128 pixel and a pixel size of 24μm x 24 μm.

IOP Conf. Series: Journal of Physics: Conf. Series **849** (2017) 012007 doi:10.1088/1742-6596/849/1/012007

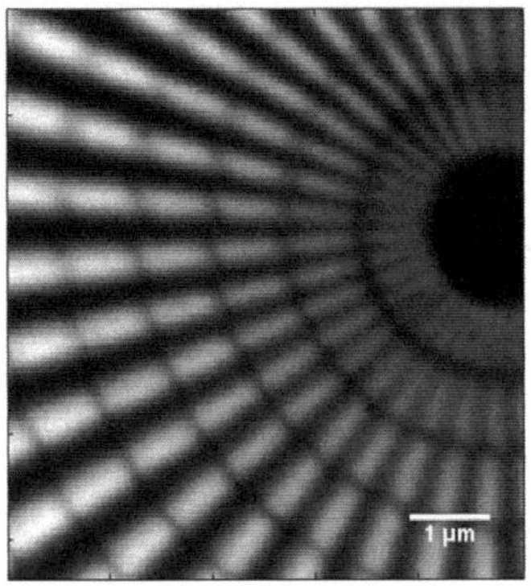

Figure 2. Transmission image of a test pattern, taken at 720 eV photon energy, smallest structures 50 nm. On-the-fly scan, step size: 20 nm, field of view: 5.4 μm x 5.4 μm, scan time: 72 min. (without electron multiplier)

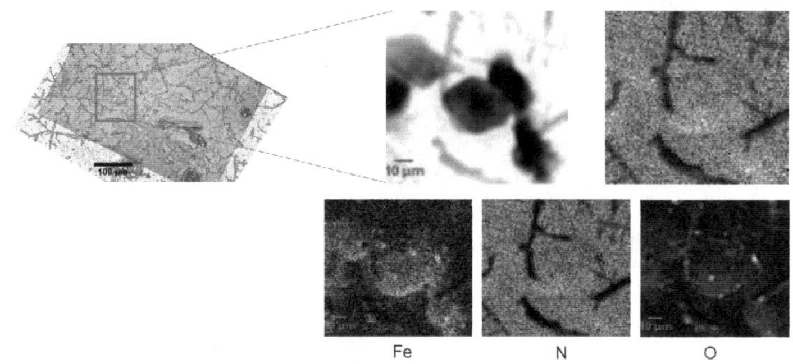

Figure 3. Mouse liver cells grown on a 200 nm thick SiN membrane, after air drying and fixation the sample was investigated by 720 eV photon energy, top left: Light microscope image of the sample; top middle: Transmission image, top right: Fluorescence image, (red) iron, (blue) oxygen, (green) nitrogen; bottom: Fluorescence signal of (Fe) iron, (N) nitrogen, (O) oxygen; Step size: 1 μm, field of view: 100 μm x 100 μm, scan time: 34 min. (without electron multiplier)

Using a spatially resolved detector for the transmitted intensity allows to gather several imaging modes such absorption contrast and phase contrast simultaneously [5]. This is done without any change in the optical path only by different treatment of the experimental data from the transmission detector. For each of the several imaging modes specific evaluation of the data is required to generate the differnet images.

The STXM is also suitable to investigate the elemental distribution. Characteristic X-ray fluorescence radiation from the sample is detected by a windowless 30 mm^2 SDD detector (Brucker Xflash 5030) with an enrgy resolution of < 127 eV FWHM @ $MnK\alpha$.

For each beamtime the endstation has to be assembled and aligned, this takes typically one day. After the alignment procedure is completed, usually a resolution test is performed. An exemplary STXM result is shown in Figure 2. The used test object was a Siemens star with 50 nm smallest structures consisting of 200 nm thick layer of tungsten on a 100 nm thick SiN membrane.

During the last phase of the commissioning of the experimental setup, it was possible to perform on-the-fly scans. This was a great and important step, since it reduces scanning time significantly.

3. Biological imaging

The STXM setup was used for biological samples with interest to the transport mechanism of lipids through the blood circuit into adipose tissue.

In first tests mouse liver cells marked with iron nano particless (SPIO) grown on 200 nm thick SiN windows were investigated. After air drying and fixation with 2.5 % formaldehyde, the samples were investigated with the STXM endstation with a photon energy of 720 eV (Figure 3). During the first beamtime the resolution was limited, so it was not possible to resolve the agglomerations of iron particles which are expected to be in the 100 nm range. Nevertheless, the combination of fluorescence and transmission microscopy is working, as can be seen in figure 3.

In the second beamtime tissue slices from brown adipose tissues of a mouse were investigated.

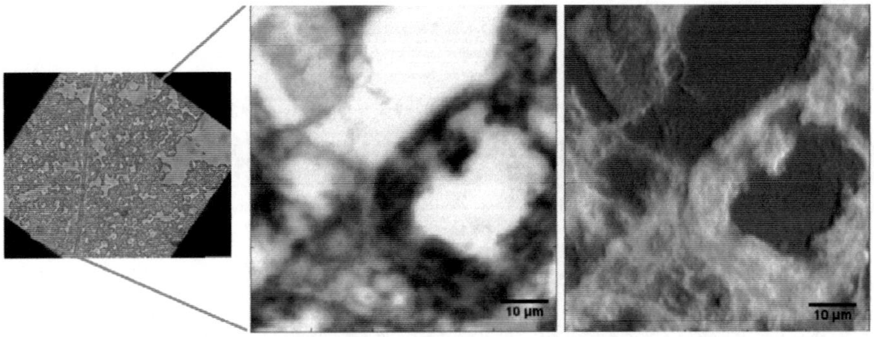

Figure 4. Tissue slices from brown adipose tissues of a mouse. The tissue is marked with iron nano particles (SPIOs). The slices are ≈ 3 μm thick, placed on a 200 nm thik SiN membrane and fixed with 2.5 % formaldehyde. After air drying the samples were investigated with a photon energy of 750 eV. Left: Light microscope image of the sample; Midle: Transmission image; Right: Differential phase contrast (horizontal); On-the-fly scan; 270 x 270 Pixel; Field of view: 54 μm x 54 μm; Scan time: 72 min. (without electron multiplier)

The ≈ 3 μm thick slices were deposited on 200 nm thick SiN windows (TEM windows) and were fixated with 2.5 % formaldehyde. After air drying, the samples were investigated using a photon energy of 750 eV. Some results are shown in figure 4.

4. Outloock

In order to reduce radiation damages for biological samples, a cryo environment including a transfer system will be implemented. The bottleneck of the system with respect to faster scanning times is the solid angle of the fluorescence detector. A new detector with approximately ten times larger solid angle of detection will be implemented. For this detector a semi-quantitative approach for the analysis of the fluorescence data will be developed.

Acknowledgments

This project is financially supported by the German Federal Ministry for Education and Research (BMBF) at the DLR (05K13UL1).

References

[1] Chen, Chien-Chun, et al. "Three-dimensional imaging of dislocations in a nanoparticle at atomic resolution." Nature 496.7443 (2013): 74-77.

[2] Kirz, J., Jacobsen, C., & Howells, M. (1995). Soft X-ray microscopes and their biological applications. Quarterly reviews of biophysics, 28(01), 33-130

[3] Viefhaus, Jens, et al. "The Variable Polarization XUV Beamline P04 at PETRA III: Optics, mechanics and their performance." Nuclear Instruments and Methods in Physics Research Section A: Accelerators, Spectrometers, Detectors and Associated Equipment 710 (2013): 151-154.

[4] Wessels, Philipp, et al. "Time-resolved imaging of domain pattern destruction and recovery via nonequilibrium magnetization states." Physical Review B 90.18 (2014): 184417.

[5] Feser, M., et al. "'Scanning transmission X-ray microscopy with a segmented detector."' Journal de Physique IV (Proceedings). Vol. 104. EDP sciences, 2003.

Combined use of X-ray fluorescence microscopy, phase contrast imaging for high resolution quantitative iron mapping in inflamed cells

C Gramaccioni[1,4], A Procopio[2], G Farruggia[2], E Malucelli[2], S Iotti[2], A Notargiacomo[3], M Fratini[4,5], Y Yang[6], A Pacureanu[6], P Cloetens[6], S Bohic[6], L Massimi[4], A Cutone[7], P Valenti[7], L Rosa[7], F Berlutti[7] and S Lagomarsino[4]

[1] Department of Physics, University of Cosenza, Arcavata di Rende, (Cosenza), Italy

[2] Department of Pharmacy and biotechnology, University of Bologna, Bologna, Italy

[3] Institute for Photonics and Nanotechnologies - CNR, Roma Italy

[4] CNR-Nanotec, c/o Department of Physics University of Sapienza, Roma, Italy

[5] Fondazione Santa Lucia, Roma, Italy

[6] ESRF, Grenoble, France

[7] Department of Public Health and Infectious Diseases, University of Sapienza, Roma, Italy

Abstract. X-ray fluorescence microscopy (XRFM) is a powerful technique to detect and localize elements in cells. To derive information useful for biology and medicine, it is essential not only to localize, but also to map quantitatively the element concentration. Here we applied quantitative XRFM to iron in phagocytic cells. Iron, a primary component of living cells, can become toxic when present in excess. In human fluids, free iron is maintained at 10^{-18} M concentration thanks to iron binding proteins as lactoferrin (Lf). The iron homeostasis, involving the physiological ratio of iron between tissues/secretions and blood, is strictly regulated by ferroportin, the sole protein able to export iron from cells to blood. Inflammatory processes induced by lipopolysaccharide (LPS) or bacterial pathoge inhibit ferroportin synthesis in epithelial and phagocytic cells thus hindering iron export, increasing intracellular iron and bacterial multiplication. In this respect, Lf is emerging as an important regulator of both iron and inflammatory homeostasis. Here we studied phagocytic cells inflamed by bacterial LPS and untreated or treated with milk derived bovine Lf. Quantitative mapping of iron concentration and mass fraction at high spatial resolution is obtained combining X-ray fluorescence microscopy, atomic force microscopy and synchrotron phase contrast imaging.

1. Introduction

Iron is an essential element for cell growth and proliferation being involved in fundamental processes, such as DNA replication and energy production. However, iron can also be toxic when present in excess because of its capacity to donate electrons to oxygen. Therefore, iron homeostasis is strongly regulated both at tissue/secretion and cell levels [1]. In human fluids free iron is maintained at low concentration (10^{-18}M) thanks to iron binding proteins as lactoferrin (Lf) in secretions and transferrin in blood [1]. In fact, Lf is the most important iron binding glycoprotein of natural immunity [1, 2]. Iron homeostasis is also tightly related to inflammation and infection [1, 3]. In particular, macrophages are specialized cells involved in these processes regulating iron influx in human fluids [4]. In inflammation and infection induced by lipopolysaccharide (LPS) or bacterial pathogens, the synthesis of ferroportin, the sole protein able to export iron from macrophages to blood, is inhibited thus hindering iron export, increasing intracellular iron, bacterial multiplication and inflammatory processes. Recently, Lf has been found to be an important regulator of iron homeostasis and a potent anti-inflammatory factor [1,5]. Conversely, very little (if anything) is known about the effect of Lf on the concentration and distribution of iron in inflamed and non-inflamed macrophages. The present work has two main aims: i) to demonstrate the possibility to quantitatively map the intracellular iron mass fraction and concentration at nanoscale spatial resolution in macrophages using combination of several techniques, i.e. X-ray Fluorescence Microscopy (XRFM), Atomic Force Microscopy (AFM) and Phase Contrast Imaging (PCI); ii) to compare the iron maps of the cells inflamed by bacterial LPS and untreated or treated with milk derived bovine Lf (bLf). In previous works [6, 7], we utilized X-ray absorption microscopy instead of PCI. However, in this case the relatively high energy photons (17 keV) would give no contrast in absorption mode, and phase contrast mode, which is sensitive to tiny variations of refraction index, has been used to obtain visible contrast.

2. Experimental methods

2.1 Sample preparation

J774.2 macrophages (Sigma Aldrich) were cultured in RPMI 1640 medium supplemented with 20% heat-inactivated foetal calf serum (FCS), 100 unit/ml of penicillin and 100 µg/ml of streptomycin on 1×1 mm^2, 200-nm-thick silicon nitride (Si_3N_4) membrane windows, mounted on a 5×5 mm^2 Si frame (purchased from www.Silson.co.uk) previously sterilized in ethanol. Cells were inflamed using *E. coli* LPS (1µg/ml) (InvivoGen). The inflamed cells were untreated or treated with bLf (100µg/ml) (Morinaga Milk Industries, Japan). After 48h, cells were rinsed with 100 mM ammonium acetate, cryofixed by plunge freezing in liquid ethane bath cooled with liquid nitrogen and then dehydrated in vacuum at low temperature overnight.

2.2 XRFM, PCI and AFM measurements

Dehydrated cells mounted on the Si_3N_4 membrane windows were carefully examined with optical microscope and selected with the following criteria: integrity, dimensions and distance from other cells. AFM measurements were performed on selected cells before XRFM and PCI measurements. AFM topography maps were collected off-line by using a Digital Instruments D3100 microscope equipped with a Nanoscope IIIa controller. The AFM was operated in ambient air using the tapping mode at a resonance frequency of about 260 kHz. Commercial monolithic silicon tips were employed, with an apex curvature radius in the 5-10 nm range and a nominal force constant of 40 Nm^{-1}. AFM maps with size of the order of 50 µm were collected with lateral spatial resolution of ~100 nm and accuracy in the cell thickness measurement of about 30 nm. Images were collected without any real-time filtering or flattening, and were post-processed to eliminate all the possible artefacts [6]. The Fig. 1 shows the AFM topographic data of a cell treated with LPS in the absence of bovine Lf (bLf).

The X-ray measurements were carried out at the beamline ID16A-NI at the European synchrotron radiation facility (ESRF, Grenoble, France) with spatial resolution of 100 nm (dwell time 50 ms) for XRFM and 50 nm for PCI. In this work, the energy was set to 17 keV for all experiments. The samples were positioned in the focal plane of the KB system, translated by piezo-stages in the directions perpendicular to the beam. The X-ray fluorescence emission lines of several elements were collected by the silicon drift energy dispersive detectors positioned in the horizontal plane at 95° of the incoming X-ray beam. After acquisition the fluorescence spectra were analysed using the PyMCA free program [8], and the total counts of the Fe K lines, after subtraction of the backgrounds, were mapped as a function of the sample position with respect to the incoming beam. The resulting fluorescence map for the same cell of Figure 1 is displayed in Fig. 2.

The PC projections were acquired at four different distances (29 mm, 30 mm, 34 mm, 44 mm) downstream of the X-ray focus resulting in magnified Fresnel patterns collected by a highly sensitive charge-coupled device (CCD) camera. For each 4 distances, we have acquired 17 images, 10 references, with 0.25 s, and 20 dark-fields. A complex procedure to retrieve the phase has been carried out [9, 10]. The result for the same cell is shown in Fig. 3.

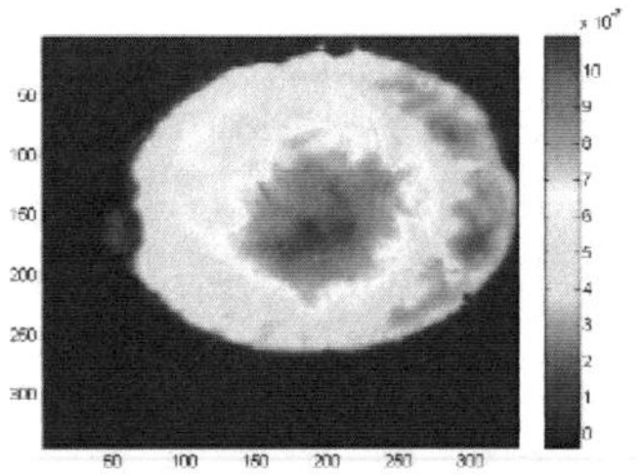

Figure 1. AFM topographic map of LPS-treated cell [μm]

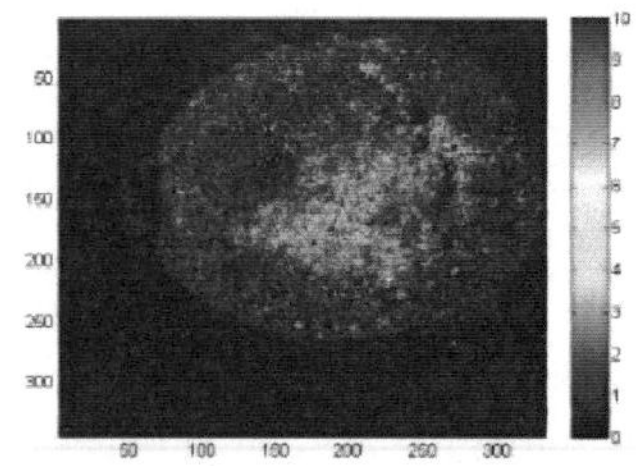

Figure 2. Fluorescence intensity iron map of LPS-treated cell

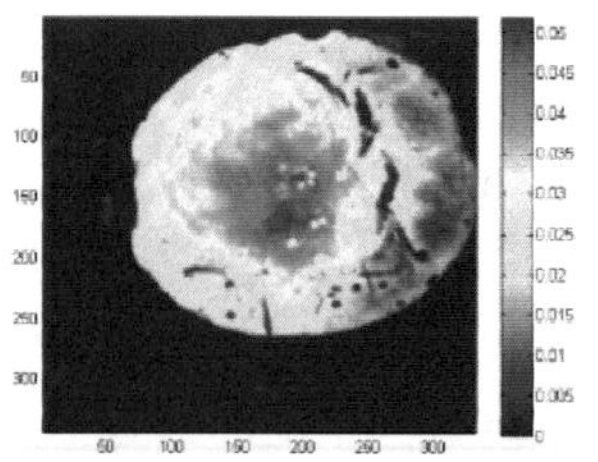

Figure 3. 2D Phase Reconstruction map of LPS-treated cell

3. Results and conclusion

The experimental results presented in Figs 1, 2 and 3, allowed to calculate both the iron mass fraction and the concentration. The mass fraction (Fig. 4) is obtained normalizing the fluorescence map with the projected electron density derived by the analysis of the PCI. The concentration (Fig. 5) is instead obtained by normalizing the fluorescence map with the volume of each pixel, as derived by the AFM. Both Figs 4 and 5 refer to a cell treated with LPS in the absence of bLf.

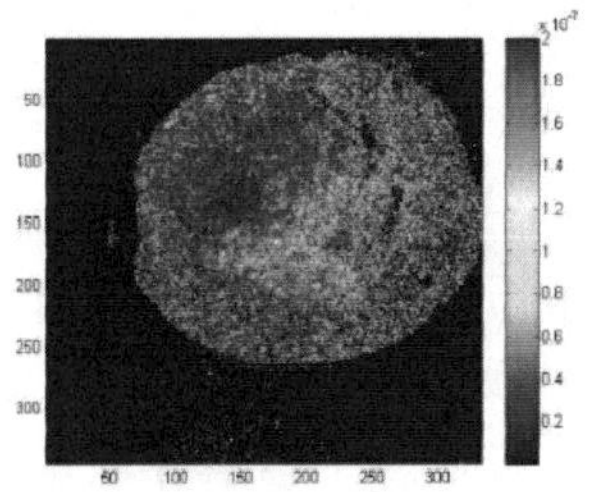

Figure 4. Iron mass fraction map of LPS-treated cell in the absence of bLf

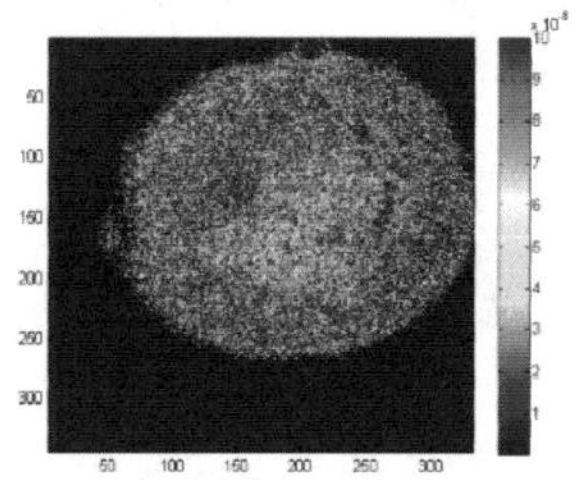

Figure 5. Iron concentration map of LPS-treated cell in the absence of bLf

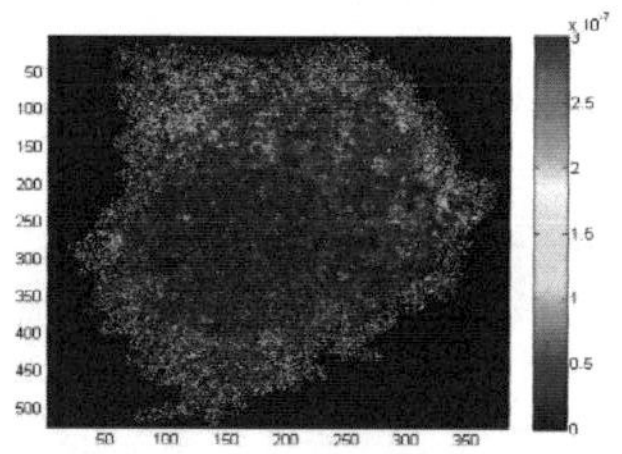

Figure 6. Iron concentration map of LPS-treated cell in the presence of bLf

It is noteworthy that the distribution maps of the iron mass fraction and of the iron concentration in LPS-treated cell without Lf are very similar (Figures 4 and 5), confirming the validity of the experimental approach. In Fig. 6 the iron concentration map of LPS-treated cell in the presence of Lf is shown. Preliminary quantitative evaluation indicated that intracellular iron concentrations in LPS-inflamed cells untreated or treated with bLf were quite different (35 and 7 nM, respectively) in agreement with previously reported data on ferroportin synthesis [1, 4, 5]. Interestingly, the cellular iron distribution in LPS-treated cell (Fig.5) appeared different respect to that of LPS-treated cell in the presence of bLf (Fig.6) being located around the cytoplasm membrane cell, suggesting the reduction of its export probably associated to the inhibition of ferroportin [4, 5].

In conclusion, for the first time we have shown that with proper combination of different techniques, it is possible to obtain quantitative information at nanoscale spatial resolution about iron concentration and distribution in inflamed cells, opening the way to more complex studies on the role and therapeutic efficacy of Lf in restoring iron and inflammatory homeostasis disorders. This study has been conducted on freeze dried cells, but we have also analyzed frozen hydrated cells, the closest ones to the living cells, using nanotomography instead of AFM to measure the volume of the cells. The analysis here reported has great potentiality to improve the comprehension of the mechanisms of iron homeostasis in conditions close to the living one.

References

[1] Paesano R, Natalizi T, Berlutti Fand Valenti P 2012 Body iron delocalization: the serious drawback in iron disorders in both developing and developed countries Pathog. Glob. Health 106 200-16
[2] Valenti P and Antonini G 2005 Lactoferrin: an important host defence against microbial and viral attack *Cell Mol Life Sci.* **62**(22) 2576-87.
[3] Valenti P and al. 2015 Lactoferrin and cystic fibrosis airway infection *Diet and Exercise in Cystic Fibrosis* **30** 259
[4] Cairo G, Recalcati S, Mantovani A andLocati M 2011 Iron trafficking and metabolism in macrophages: contribution to the polarized phenotype *Trends Immunol* **32** 241–47
[5] Cutone A, Frioni A, Berlutti F, Valenti P, Musci G and Bonaccorsi di Patti MC 2014 Lactoferrin prevents LPS-induced decrease of the iron exporter ferroportin in human monocytes/macrophages. *Biometals* **27** 807-13
[6] Lagomarsino S, Iotti S, Farruggia G, Cedola A, Trapani V, Fratini M and al. 2011 Intracellular concentration map of magnesium in whole cells by combined use of X-ray fluorescence microscopy and atomic force microscopy *Spectrochim Acta Part B At Spectrosc.* **66** (11–12) 834–40
[7] Malucelli E, Iotti S, Gianoncelli A, Fratini M, Merolle L, Notargiacomo A and al. 2014 Quantitative Chemical Imaging of the Intracellular Spatial Distribution of Fundamental Elements and Light Metals in Single Cells *Anal. Chem.* **86** (10) 5108–15
[8] Sole V, Papillon E, Cotte M, Walter P and Susini J 2007 A multiplatform code for the analysis of energy dispersive X-ray fluorescence spectra *Spectrochim. ActaPart B At. Spectrosc.* **62** 63–8
[9] Cloetens P, Ludwig W, Baruchel J, Dyck, J V, Landuyt J V and al. 1999 Holotomography: quantitative phase tomography with micrometer resolution using hard synchrotron radiation X-rays *Appl. Phys. Lett.* **75** 2912–14
[10] Langer M, Cloetens P, Guigay JP and Peyrin F 2008 Quantitative comparison of direct phase retrieval algorithms in in-line phase tomography Med. *Phys.* **35** (10) 4556–66
[11] E. Kosior , Bohic S, Suhonen H, Ortega R, Devès G, Carmona A and al. 2012 Combined use of hard X-ray phase contrast imaging and X-ray fluorescence microscopy for sub-cellular metal quantification *J Struct Biol.* **177** (2) 239–47
[12] Malucelli E, Fratini M, Notargiacomo A, Gianoncelli A, Merolle L and al. 2016 Where is it and how much? Mapping and quantifying elements in single cells *Analyst.* **141** 5221-35

The research was granted by Sapienza University of Rome to PV and FB.
Authors thank the ID16-A beamline at the ESRF for the provision of beamtime and the beamline staff for the experimental support. The COST Action X-ray MP1207 "Tomographic Reconstruction: Experiment, Modeling, and Algorithms" and the COST Action MP1203 "Advanced X-ray spatial and temporal metrology" are acknowledged for networking support.

IOP Conf. Series: Journal of Physics: Conf. Series **849** (2017) 012009 doi:10.1088/1742-6596/849/1/012009

Organelle Distribution in a Hydrated Bio-cell by Correlation between Soft X-ray and Fluorescence Images

T. Ejima[1], M. Kado[2], M. Aoyama[3], K. Yasuda[3], and S. Tamotsu[3]

[1] Institute of Multidisciplinary Research for Advanced Materials, Tohoku University, Sendai 980-8577, JPN

[2] Kansai Photon Science Institution, National Inst. for Quantum and Radiological Sci. Tech., Kyoto 619-0215, JPN

[3] Faculty Division of Natural Sciences, Nara Women's University, Nara 630-8506, JPN

Corresponding author's e-mail address: ejima@tagen.tohoku.ac.jp

Abstract. Fine structures of bio-specimens obtained by soft X-ray (SX) imaging are clearer compared with those obtained by visible imaging owing to the difference in extinction coefficients between the two wavelength regions. However, it is difficult to identify the fine structure imaged in the 2.3 – 4.5 nm wavelength region by appearance alone. Here, we obtain and compare SX and fluorescence images of Leydig cells of a mouse testis loaded with fluorophores. Identification of the fine structures in the SX image is carried out by comparison with the fluorescence images with the use of principal component analysis (PCA). The result shows that the common structures between the SX- and the fluorescence- images.

1. Introduction

The soft X-rays between carbon and oxygen K-edges (λ=2.3 – 4.5 nm) are less absorbed by water and are more absorbed by carbon, nitrogen and other elements of biological specimens [1]. When bio-cells are imaged with an SX microscope in this wavelength region, the appearance of the small organelles of the bio-cells in the images depends on both the spatial resolution of the SX microscope and absorption coefficients of the organelles. Mouse 3T3 cells have been observed by a SX microscope using a zone plate as the objective lens. Many organelles, such as nuclei and nucleoli in the nucleus, can be observed in an SX image [2]. As the nuclei and nucleoli are relatively large organelles in bio-cells, they can be easily distinguished from the other organelles. Generally, the relative sizes or characteristic shapes of organelles are not reliable markers to identify organelles in bio-cells, therefore the identification of small organelles in SX images is difficult [2].

One method to identify organelles in bio-cells is by the fluorescence labeling technique, which is generally used in biological observation to enhance the contrast of specific organelles in bio-cells. Organelles labeled with fluorophores can be identified the various organelles in a same bio-cell [3]. By the use of the fluorophores, correlative light-electron microscopy (CLEM) was applied to observe the fine structures and also to identify the locations of organelles in the bio-cells. In this method, after loading the bio-cells with a fluorophore, the locations of the organelles are obtained by fluorescence microscopy and the fine structures of the organelles are observed by electron microscopy at the exact positions of the organelle [4].

Correlative microscopy, by using an optical fluorescence microscope and SX microscope operating in the 2.3 – 4.5 nm wavelength region, will enable the observation of hydrated specimens composed of C, N and O atoms. Hence, one of the authors obtained fluorescence and SX images of Leydig cells of a mouse testis as represented in Figs. 1 [5]. The complex structures of the cellular organelles observed in the SX images were visually compared with the images obtained by fluorescence microscopy. The direct comparison between the SX and the fluorescence images showed that some structures in the SX image were identical to those observed in the fluorescence image, which suggested that the structures observed in the SX image correspond to mitochondria (MitoTracker binds well to mitochondria).

In this study, fine structures of hydrated biological specimens in an SX image, which are generally not usually identified, are identified by numerical comparison with fluorescence images that are obtained by observing the same biological specimens with a fluorescence microscope. The fine structures identified by the numerical comparison can be shown as a fine structure map.

2. Material and Methods

2.1. Experimental conditions

Poly methyl methacrylate (PMMA), with a thickness of 500 nm, was coated onto a thin glass plate and used as the X-ray photoresist film. Leydig cells from a mouse testis were placed onto the PMMA film and were cultivated for two to three days to establish good contact onto the film. The Leydig cells were loaded with MitoTracker, phalloidin and DAPI. The Leydig cells were covered with a silicon nitride (Si_3N_4) membrane with a thickness of 200 nm and enclosed into a sample holder designed to protect the bio-cells from the vacuum. The distance between the membrane and the glass substrate was controlled to be 5 μm thick by a spacer deposited on the membrane. Fluorescence images of the Leydig cells in the sample holder were observed at first by the fluorescence microscope (Eclipse Ti-U, Nikon Instruments Inc.), and then the sample holder was placed in the vacuum chamber for SX imaging. The time interval between the fluorescence- and the SX-imaging was typically 10 minutes. The spatial resolution of the fluorescence image depends on the emission wavelength of the fluorophore and was estimated to be less than 0.8 μm.

Soft X-ray images were obtained by the contact method with the use of a laser-produced plasma (LPP) light source. The X-ray pulse used for the SX imaging was generated by the high power Nd:glass laser system with the output energy of 20 J and a 600 ps pulse duration. The exposure time, which is determined by the pulse duration of the SX, is approximately 600 ps. The spatial resolution of the SX imaging was measured to be approximately 90 nm.

After exposure to soft X-rays, the PMMA was rinsed with sodium hypochlorite (NaClO) to remove the cells and was carefully observed by differential interference microscopy to make sure that no cells remained on the surface. The PMMA photoresist was then developed with a mixed solution of methyl isobutyl ketone (MIBK) and isopropyl alcohol (IPA) and the printed SX images on the surface of the PMMA were read by using an atomic force microscope.

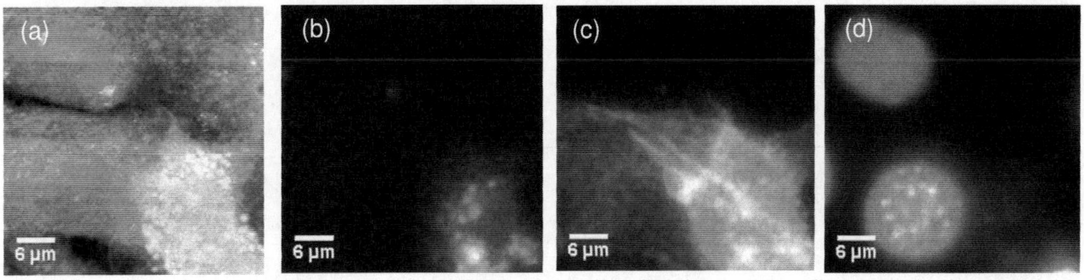

Figure 1: Experimental imaging results of Leydig cells. SX image (a), fluorescence images with MitoTracker staining (b), with phalloidin staining (c), and with DAPI staining

2.2. Application conditions for the PCA method

Principal component analysis (PCA) is a statistical procedure that converts a set of observations of possibly correlated variables into a set of values that are linearly uncorrelated variables by the use of an orthogonal transformation. As the PCA method is applied to a pair of images under a specific condition, two eigenimages are obtained. The first eigenimage represents the average of the two original images and the other represents the common structures of the two original images [6].

Image multiplication is usually used to evaluate common structures between images under the assumption that both the signal intensities and the signal dispersions of both images are roughly equal. Before applying the PCA method to the experimental images, image multiplication was applied to analyse the characteristic structures between the images. Even after the image multiplication, some common structures between the images were not identified because of the weak signal intensity of the images, which suggests we needed to try another method to analyse the experimental images.

For the investigation of the application conditions of the PCA method, an image of 300×300 pixels was expected to be a high spatial resolution image, and a low spatial resolution image was also obtained by blurring the high spatial resolution image by using a 2D Gaussian function. After several trials, the application procedure for processing the images was determined as follows. The values of the pixels were normalized to a 32-bit grayscale so that the maximum and minimum values of the pixels were 1 and 0, respectively. After the PCA analysis was applied to the pair of images [7], eigenimages were also normalized to a 32-bit grayscale by the maximum and minimum values. The first eigenimage is the same as the average image which was scaled by a grayscale normalized with maximum and minimum values. The second eigenimage reflects the common structures between the two original images, which are easily recognizable from the original images.

2.3. Position alignment of experimental images

In many SX microscopy systems, the fluorescence and SX images are obtained separately through the use of different imaging devices. Therefore, the images obtained will be different in the position of specimen, pixel size, field of view and spatial resolution. Before applying the PCA method to the experimentally obtained images, all the differentiations, except for the spatial resolution, should be adjusted between the images.

First, the alignment of an image was performed by overlapping it with another image. The accuracy of the alignment was obtained by applying the PCA method to the selected images then changing the position of the images. The simulation result suggests that the accuracy of the alignment is limited by the worst value of the spatial resolution, and that the position of the images coincide with each other within this value. Attempting to compare the two images of the same objects with each other at different wavelengths provided an easy and reliable method to check whether the structures observed in each image were correlated or not by visible inspection. Rough alignment of the images was performed by visible inspection by finding the same structures observed in each image, such as

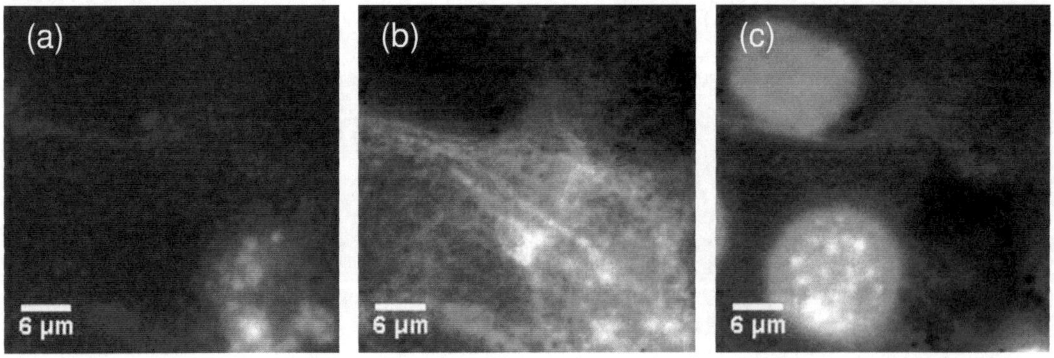

Figure 2: Second eigenimages between the SX image and the fluorescence ones of MitoTracker staining (a), of Phalloidin staining (b), and of DAPI staining (c).

some large organelles in bio-cells, and using them as landmarks in the images. The landmarks of the image were then aligned roughly to overlap one another.

The next step was to adjust the pixel sizes of the images to fit each other using the bilinear interpolation method [8]. In this method, the image with a smaller pixel size was enlarged to have the same pixel size as the other image whilst maintaining the two-dimensional intensity-distribution of the images.

The final step was to align the images numerically. In our experience, the best numerical method is largely dependent on the combination of selected images. The best results for aligning images in this study were achieved by transforming the images to be in the spatial frequency domain by the 2D-FFT method, and then the normalized correlation coefficients were obtained among the transformed images, and the position of the image was aligned to maximize the normalized correlation coefficients [9]. The images in the spatial frequency domain were retransformed to be in the spatial domain, and this was confirmed that the agreement between the images was within the accuracy criterion of the alignment suggested in Section 3. Output images with the best alignment showed subpixel accuracy.

3. Results & Discussion

Image processing, as described in Sections 2.2 and 2.3, was applied to the experimentally obtained images of Leydig cells as designated in Figs. 1 [5], and eigenimages were obtained as the result of the applications. The resulting second eigenimages are shown in Fig. 2(a), (b) and (c), which show the common structures between the SX image and the fluorescence images obtained with MitoTracker, with Phalloidin and with DAPI, respectively. The images are normalized by the minimum and maximum values of pixels and displayed by a 32-bit grayscale. Structures that were blurred in the original fluorescence images are clearly observed, especially in Fig. 2(b).

As an index for the fine structures observed in the images, the power spectra of the images are calculated. The power spectrum of a cyclically shifted signal is identical to that of the original signal. In other words, the power spectrum of a discrete periodic signal of a certain length shows no change by some offset of the signal. Therefore, the amount of fine structures in an image can be compared with that in the other image by the power spectra obtained from the images [10].

Power spectra of both the fluorescence image obtained with Mito Tracker staining and the SX image are represented in Fig. 3(a), and those of both the fluorescence image obtained with Phalloidin staining and the SX image, in Fig. 3(b). The spectra are normalized at the intensity of the lowest frequency and an eigenimage is compared with the corresponding fluorescence image. The results show that the power spectrum of the eigenimage obtained from the fluorescence image obtained with MitoTracker is the same as that of the fluorescence image, and that the power spectrum of the eigenimage obtained from the fluorescence image obtained with Phalloidin in the high frequency region is higher than that of the fluorescence image and lower than that of the SX image. In

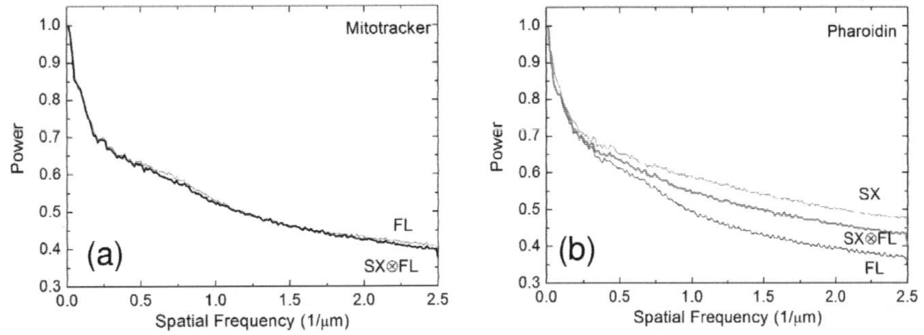

Figure 3: Power spectra of original images and second eigenimages, The results corresponding to the MitoTracker-fluorescence images (a), and to the Phalloidin-fluorescence images (b). The result corresponding to the DAPI-fluorescence images show the same tendency as in the case of Phalloidin.

the case of the images stained with DAPI, the power spectra show the same tendency as in the case of Mito Tracker.

The fluorescence images of the Leydig cells stained by Phalloidin, Mito Tracker, and DAPI reflect the distribution of the cytoskeleton, Mitochondria, and cell nucleus, respectively. The present fluorescence images show that the Leydig cells spread out widely and the Mitochondria exist around the cell nucleus. These results coincide well with that Leydig cells have a tendency to spread out with the central focus on its cell nucleus. As the result of the spreading, the thickness of the Leydig cells decrease with an increase of the distance from the cell nucleus. In addition, transmittance of the SX light increases as the decrease of the thickness of the cells. Therefore, the structures of the second eigenimages in the thick part of the cells show the same shapes of the fluorescence images, as designated in Fig. 2(a) and (c). Small structures that appeared in Fig. 2 (b) will be originated from the cytoskeleton.

4. Summary

Four pairs of images, comprising the fluorescence and SX images, were analysed by the PCA method, and common structures were found in all eigenimages. The analytical results suggest that correlations between the structures depend on the thickness of the cells. The common structures located at around the cell nucleus show the same structures of the fluorescence images, and the structures located at the edge of the cells will be originated from the cytoskeleton.

Although the images were analysed by the present method with no specific assumptions before the analysis, common structures between the two images could be extracted without visible inspection. This fact suggests that the present method can be applied to any pairs of images composed of visible and SX images. Correlations between structures in different images will be applicable for analysing microscopic images of complex systems in the SX wavelength region.

References

[1] Sayer, D., Kirz, J., Feder, R., Kim, D. M. & Spiller, E., Science **196**, (1977) pp. 1339-1340.
[2] Meyer-Ilse, W., Hamamoto, D., Nair, A., Lelievre, S. A., Denbeaux, G., Johnson, L., Pearson, A.L., Yager, D., Legros, M. A. & Larabell, C. A., J. Microscopy **201,** (2001) pp. 395-403.
[3] Alberts, B., Johnson, A., Lewis, J., Morgan, D., Raff, M., Roberts, K. & Walter, P. "*Molecular Biology of the Cell 6th ed.*", Chap. 9, (Garland Science, New York) (2014).
[4] Vicidomini, G., Gagliani, M.C., Canfora, M., Cortese, K., Frosi, F., Santangelo, C., Di Fiore, P.P., Boccacci, P., Diaspro, A. & Tacchetti, C., Traffic **9**, (2008) pp. 1828-1838.
[5] Kado, M., Kishimoto, M., Tamotsu, S., Yasuda, K. & Shinohara, K., J. Phys.: Conf. Ser. **463**, (2013) 012056.
[6] Konishi, S., "*Introduction to Multivariate Analysis*" in Japanese, (Iwanami Shoten Publishers,Tokyo), (2010).
[7] Abramoff, M. D., Kwon, Y. H., Ts'o, D. Y., Li, H., Barriga, E. S. & Kardon, R., Proc. IEEE Int. Symp. Biomed. Imaging **2**, (2004) pp. 1115-1118.
[8] Burger, W. & Burge, M.J., "*Digital Image Processing: an algorithmic introduction using Java*", (Springer, New York), (2010) p. 405.
[9] Lewis, J. P., Vision Interface '95, (1995), pp. 120-122.
[10] Burger, W. & Burge, M. J., ibid, p. 339.

IOP Conf. Series: Journal of Physics: Conf. Series **849** (2017) 012010 doi:10.1088/1742-6596/849/1/012010

Application of soft X-ray microscopy to environmental microbiology of hydrosphere

K Takemoto[1], M Yoshimura[2], T Ohigashi[3], Y Inagaki[3], H Namba[2] and H Kihara[2, 4]

[1] Kansai Medical University, 2-5-1 Shin-machi, Hirakata, Japan
[2] The SR Center, Ritsumeikan University, 1-1-1 Noji-Higashi, Kusatsu, Japan
[3] Institute for Molecular Science, 38 Nishigonaka, Myodaiji, Okazaki, Japan
[4] Himeji-Hinomoto College, 890 Koro, Kodera-cho, Himeji, Japan
E-mail: takemoto@hirakata.kmu.ac.jp

Abstract. Microstructures of unprocessed filamentous cyanobacterium, *Pseudanabaena foetida* sp., producing a musty smell were observed using soft X-ray microscopy. Carbon-enriched structures and granules as well as oxygen-enriched granules which have been already reported were observed. Except for early log growth phase, the oxygen-enriched granules were observed. However, the carbon-enriched structures were observed throughout log growth phase. The result suggests there is a relationship between the oxygen-enriched granules and 2-methylisoborneol (2-MIB) productivity, since the 2-MIB productivity of each cell is increased depending on the culture period in log growth phase.

1. Introduction

In environmental microbiology of hydrosphere, the needs for observation of microbes and the mucous matter are high.

Lake Biwa is the most important water reservoir in Kyoto-Osaka-Kobe region in Japan. The high economic growth in the 1960s caused severe environmental problems in the lake such as eutrophication, losses in biodiversity, degradation of ecosystems. Although the water quality was gradually improved by cooperation with other stakeholders such as governments, local governments, citizens, the biota began to exhibit correspondingly drastic negative changes.

In 1969, a problem of musty smell in tap water has occurred due to sudden propagation of a certain green filamentous cyanobacterium. The filamentous cyanobacterium was reported as *Phormidium tenue* (Menegh.) Gomont and 2-methylisoborneol (2-MIB) was identified as a causative substance [1]. However, results of our soft X-ray microscopy (XM) observation suggested the necessity of reidentifying of it [2, 3]. In 2016, as a result of morphological and genetical analysis, the cyanobacterium was proposed as a new planktonic species producing 2–MIB: *Pseudanabaena foetida* Niiyama, Tuji *et* Ichise sp. nov. [4]. Now, the study how the microstructure and culturing conditions are related to the 2-MIB productivity is proceeding. The aim of this study is to elucidate the microstructure related to the 2-MIB productivity of *P. foetida* sp. using the XM.

2. Materials and methods

2.1. Phytoplankton Material

IOP Conf. Series: Journal of Physics: Conf. Series **849** (2017) 012010 doi:10.1088/1742-6596/849/1/012010

P. foetida sp. was collected from Lake Biwa and subcultured. The culture was maintained on CT medium (pH 8.0) under a 12 h light/12 h dark cycle at 20 °C. The cultures were illuminated with fluorescent lamps, which provided about 85 μmol photons m^{-2} s^{-1} of photosynthetically active radiation.

2.2. *X-ray microscopy*
Phytoplankton was observed by soft XM using the BL-12 microscope at the SR Center of Ritsumeikan University [5, 6]. The microscope was equipped with Fresnel zone plate condenser and objective with 53.7 nm and 36 nm outer zone widths respectively. The images were recorded using an X-ray CCD camera (C4880-21-24WD: Hamamatsu Photonics K.K.). The achievable spatial resolution was approximately 70 nm based on the knife-edge estimation (20 - 80%) at 517 eV (see e.g. [7]).

Projection images were collected using exposure times ranging from 60 to 120 s. Background images were also collected and each image was divided by the background image. Cryogenic imaging is allowed by using the indirect contact cryogenic system which enables cooling to around -160 °C without frost [8].

2.3. *Sample preparation and observation*
For 2D imaging, the cells suspension dropped onto a 100 nm thickness silicon nitride membrane window (Silson Ltd.). After air-dried, the samples were observed by XM at 620 eV and 532 eV at room temperature.

For 3D imaging, the cell suspension was injected into a glass capillary tube with a fine tip that was created by using a puller, PC-10 (Narishige). For acquisition of whole data set, 75 X-ray projection images were acquired with rotating the sample 4 degree each (180 degree rotation). I_0 images were acquired every 5 projections to compensate decay of beam current. The cross sectional images were reconstructed with convolution back projection algorithm. The reconstructed volumetric data was visualized using VGStudio MAX 2.0 (Volume Graphics GmbH). The spatial resolution of the cross sectional image was approximately 160 nm based on the edge response (20 - 80%) of the glass capillary at 517 eV.

3. Results and discussion
Figure 1 shows XM images of air-dried *P. foetida*. Although filamentous bacteria are often surrounded with sheaths, *P. foetida* cells were not surrounded with sheaths. Any metabolite-like agar or gelatin were not also observed. Several granules were observed in the cells 4 weeks after inoculation, log growth phase (Fig.1 a). As reported previously, the granules disappear at 532 eV. This result indicates that the granules are oxygen-enriched structures such as polyphosphate granules. A cell 1 week after inoculation, early log growth phase, showed interesting structures which have been not observed under previous XM observation (Fig.1 b). In the middle of the cell, a boundary line is seen. From the boundary line, the structures extend to both ends of a cell. The structures are almost linearly symmetrical with the boundary line. It is noteworthy that the structures appear not only in 620 eV image but also in 532 eV image. This result indicates that a dividing cell has been observed and the structures are carbon-enriched structures such as nucleoid or prokaryotic cytoskeleton [9].

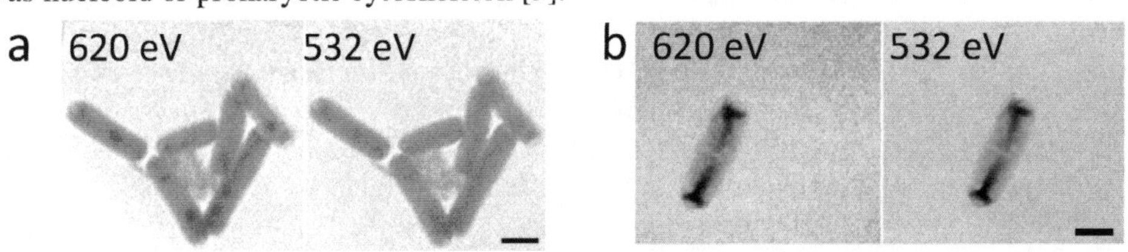

Figure 1. XM images of air-dried *P. foetida*. Images of cells 4 weeks after inoculation, log growth phase, were taken at 620 eV and at 532 eV (a). Images of a cell 1 week after inoculation, early log growth phase, were taken at 620 eV and at 532 eV (b). Exposure time was 120 s. Scale bar is 2 μm.

The carbon-enriched structures were observed throughout log growth phase. However, few oxygen-enriched granules were observed in early log growth phase. In log growth phase, 2-MIB productivity of each cell is increased depending on the period of its culture. The result suggests that the oxygen-enriched structures such as polyphosphate granule relate to 2-MIB productivity and the carbon-enriched structures such as nucleoid or prokaryotic cytoskeleton are not related to it.

To verify the oxygen-enriched structures, distribution of oxygen and oxygen K-edge XANES spectra were collected by the scanning transmission X-ray microscope (STXM) at UVSOR BL4U [10, 11]. Figure 2 shows STXM images and a distribution of oxygen in *P. foetida* grown to early stationary phase. Four oxygen-accumulated granules consisting of large and small sizes were observed. Although absorption spectra were noisy, a spectrum from the granule obviously exhibited a pre-peak which is different from that of organic carbonyl group (C=O) at about 532 eV. To identify the granules, we intend to promote a more detailed analysis.

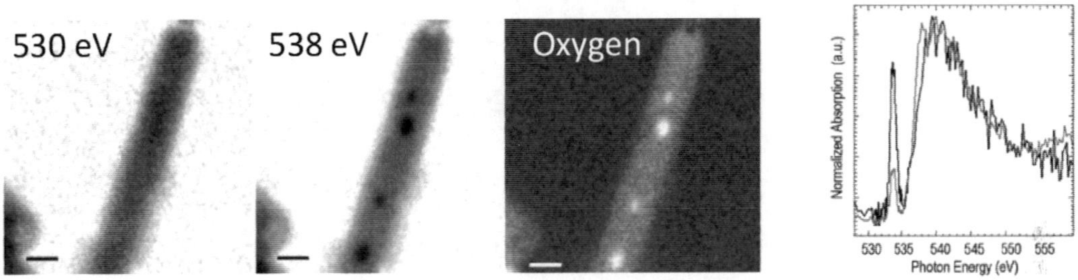

Figure 2. STXM images and oxygen K-edge XANES spectra. STXM images and distribution of oxygen in the air-dried *P. foetida* cells grown to log growth phase. Scale bar is 1 μm. Oxygen K-edge XANES spectra of cytoplasm (black line) and granule (grey line).

Figure 3 shows 3D reconstructed tomograms and 2D slice images of *P. foetida* cells grown to late log growth phase. Many rod-shaped bacteria without sheath were observed. In the cells, heterogeneous X-ray absorption was seen, whereas carbon-enriched granules were not seen. These results are the same that in air-dried 2D observation at 532 eV (Fig.1 a). However, *P. foetida* cells grown to death phase contained several carbon-enriched granules (Fig.4). The cells shrank because of radiation damage during repeated X-ray exposure. However three different granules were inside different cells. Although this result indicates that the granules are carbon-enriched structures, candidate prokaryotic organelles corresponding to the granules have not been assigned yet.

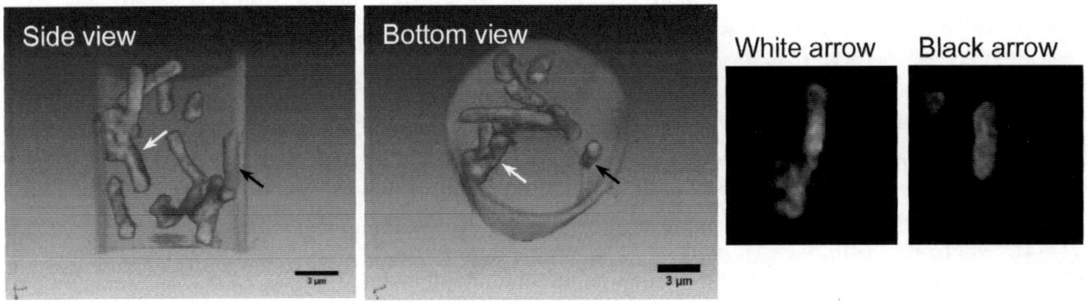

Figure 3. Cryo-XM images of *P. foetida* grown to late log growth phase in a hydrated state at -110 °C. Three-dimensional rendering images (side and bottom views) and longitudinal section images of cells (white and black arrows). Images were taken at 532 eV and exposure time was 60 s.

4. Conclusions

To investigate how the microstructure is related to the 2-MIB productivity, unprocessed *P. foetida* cells were observed by using soft XM. The *P. foetida* cells were not surrounded with sheath. Carbon-enriched structures such as nucleoid or prokaryotic cytoskeleton were observed throughout log growth phase although few oxygen-enriched granules such as polyphosphate granule were observed in early log growth phase. In addition, carbon-enriched granules were observed in cells grown to death phase although any candidate prokaryotic organelles corresponding to the granules have not been assigned yet.

During log growth phase, 2-MIB productivity of each cell is increased depending on the period of its culture. The result suggests that the oxygen-enriched structures relate to 2-MIB productivity and the carbon-enriched structures do not relate to it.

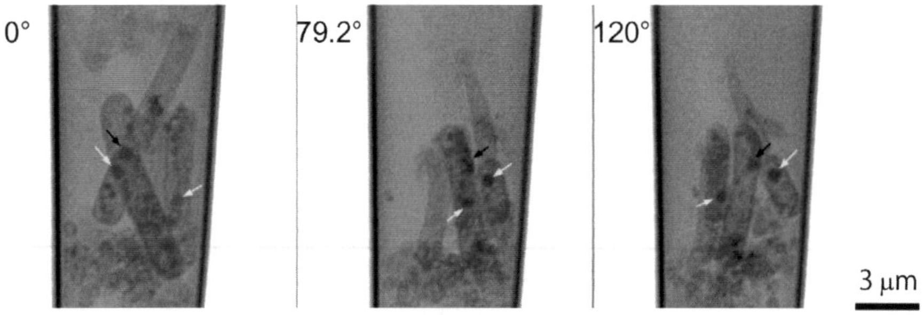

Figure 4. XM images of *P. foetida* grown to death phase in a hydrated state at room temperature. Image was taken at 517 eV and each exposure time was 120 s. Three different granules (black, white and yellow arrows) were inside different cells.

References

[1] Yagi M, Kajino M, Matsuo V, Ashitani K, Kita T and Nakamura T 1983 *Water Sci. Technol.* **15** 311–321
[2] Takemoto K, Mizuta G, Yamamoto A, Yoshimura M, Ichise S, Namba H and Kihara H 2013 *J. Phys. Conf. Series* **463** 012052
[3] Takemoto K, Yoshimura M, Yamamoto A, Ichise S, Namba H and Kihara H 2016 *AIP Conf. Proc.* **1696** 020024
[4] Niiyama Y, Tuji A, Takemoto K and Ichise S 2016 *Fottea* **16** 1–11
[5] Hirai A, Takemoto K, Nishino K, Niemann B, Hettwer M, Rudolph D, Anderson E, Attwood D, Kern D P, NakayamaY and Kihara H 1999 *Jpn. J. Appl. Phys.* **38** 274–278
[6] Ohigashi T, Namba H, Takemoto T and Kihara H 2011 *Mem. SR Center Ritsumeikan Univ.* **13** 201–202
[7] Takemoto K, Okuno K, Minemoto J, Yamamoto A, Niemann B, Hettwer M, Rudolph D, Anderson E, Attwood D, Kern DP, Okamoto T and Kihara H 2006 *IPAP Conference Series 7, Proc. 8th Int. Conf. on X-ray Microscopy* **7** 82-84
[8] Yoshimura M, Ishii T, Murakami S, Takemoto K, Kihara H , Namba H and Ohta T 2015 *Mem. SR Center Ritsumeikan Univ.* **17** 149–151
[9] Shih YL, Rothfield L 2006 *Microbiol. Mol. Biol. Rev.* **70** 729–54
[10] Ohigashi T, Arai H, Araki T, Kondo N, Shigemasa E, Ito A, Kosugi N and Katoh M 2013 *J. Phys. Conf. Series* **463** 012006.
[11] Takemoto K, Yoshimura M, Ohigashi T and Kihara H 2015 *UVSOR Act.Rep.* 2014 p.152

Acknowledgments

Samples were provided by Dr. Ichise (Lake Biwa Environmental Research Institute). The soft X-ray imaging work was performed at the SR Center of Ritsumeikan University and the STXM imaging work was performed at UVSOR under the MEXT Project for Creation of Research Platforms and Sharing of Advanced Research Infrastructure. A part of this study was supported by JSPS KAKENHI Grant Numbers 24510039 and 15K00572, and the Strategic Research Foundation at Private Universities (S1101034).

Synchrotron X-ray microscopy of marine calcifiers: how plankton record past climate change

S A T Redfern[1], O Branson[1,2], E Read[1]

[1] Department of Earth Sciences, University of Cambridge, Downing Street, Cambridge, CB2 3EQ, UK
[2] Now at: Research School of Earth Sciences, Australian National University, Building 142, Mills Road, Acton, ACT 2601, Australia

satr@cam.ac.uk

Abstract. We have used STXM and PEEM to reveal the underpinning chemistry and nanoscale structure behind palaeo-climate geochemical signatures, such as trace Mg in shells- proposed proxies for palaeo-ocean temperature. This has allowed us to test the chemical assumptions and mechanisms underpinning the use of such empirical proxies. We have determined the control on driving chemical variations in biogenic carbonates using STXM at the absorption edge of Mg, B, and Na in the shells of modern plankton. The power of these observations lies in their ability to link changes in chemistry, microstructure, and growth process in biogenic carbonate to environmental influences. We have seen that such changes occur at length scales of tens of nanometres and demonstrated that STXM provides an invaluable route to understanding chemical environment and key heterogeneity at the appropriate length scale. This new understanding provides new routes for future measurements of past climate variation in the sea floor fossil record.

1. Introduction

Earth's atmospheric CO_2 is now at a higher level than it has been at any time in the last 15 million years[1,2]. This is despite the fact that about one third of atmospheric anthropogenic CO_2 has been absorbed by the oceans, moderating climate change. Consequent decrease in seawater carbonate ion concentration may affect the ability of marine organisms to build their calcium carbonate. Past climate change, recorded in shells recovered from ocean sediment, give insights into current variation, but the methods of decoding past climate from calcified organisms remain largely empirical. One such class of organisms are the single-celled foraminifera, that live in the water column as planktonic species or at in or near the sediment as benthic species. Because they are ubiquitous and readily identified, studies of foraminiferal microfossils have formed the backbone of a huge body of work devoted to understanding past marine environments over Earth history.

There is abundant evidence in the geological record, from carbonate minerals that form the hard parts of marine organisms such as plankton (recovered from ocean floor sediment in drilled cores that represent a record back in time), that $CaCO_3$ mineral polymorphs making up those marine calcifiers have been influenced by the ratio of Mg to Ca concentrations in ancient oceanic waters, stretching back tens of millions of years[3]. The incorporation of Mg into $CaCO_3$ hard parts of marine calcifiers, as well as its incorporation into abiogenically-precipitated carbonates, is dependent on temperature, with Mg solid solubility in $CaCO_3$ increasing with increasing temperature for simple thermodynamic reasons. Although Mg/Ca ratios in biogenic and abiogenic carbonates are not identical, when grown in seawater of identical composition, the Mg/Ca ratio of foraminiferal calcite is a widely accepted and applied empirical "proxy" for ocean temperature[4,5]. The construction of Mg/Ca records from foraminifera preserved in ocean sediments has been instrumental in developing our understanding of past global climate, but the mechanisms behind this proxy have remained largely unknown. Similar proxies based on other elements have also been proposed in recent years, including the use of boron in carbonates as an indicator of variations in ocean pH, of sodium as an indicator of salinity, and of iodine as an indicator of oxygen content. In each case the application is empirical, based on observed correlations between external drivers and internal chemistry. We have applied X-ray microscopy techniques to the problems of biomineral trace element incorporation in foraminifera to reveal, for the first time, the modes of incorporation and to provide a basis of understanding of how geochemical palaeoproxies operate.

2. Synchrotron X-ray microscopy of foraminifera

Spectroscopic techniques to investigate element-specific atomic coordination within a structure have been developed over the last few decades at synchrotron light sources. Various forms of X-ray absorption spectroscopy (XAS), including Near Edge X-ray Absorption Fine Structure (NEXAFS/XANES) and Extended X-ray Absorption Fine Structure (EXAFS) spectroscopy, have been applied to investigate element coordination in bulk biomineral samples[6-9]. The development of synchrotron X-ray microscope instrumentation, providing nano-scale monochromatic X-ray probes, has added spatial discretion to these techniques, allowing the simultaneous nanoscale mapping and spectroscopy of specific elements within a structure[10-13]. Techniques such as photo-electron electron microscopy (PEEM) and scanning transmission X-ray microscopy (STXM) provide a route to developing a comprehensive picture of trace element incorporation in biominerals such as foraminifera.

2.1. PEEM studies of Mg in foraminifera

We have investigated the distribution of Mg, Na, S and Fe in foraminifera, brachiopod and ostracod specimens using the Spin-resolved Photo-Emission Electron Microscope (SPEEM) branch of the BESSY II synchrotron beamline UE49-SPEEM, Berlin. Two forms of data were collected during analysis: spectral image stacks (Mono S scan) and difference maps between specified on- and off-peak energies (X scan). Data were collected at either the K or L2-3 edges of Mg, Na, S and Fe. Here we focus on the results for Mg. A number of challenges arose in applying this technique to resin-mounted calcium carbonate specimens. The specimens considered here are electrical insulators, which poses a significant surface charging problem and consequent image distortion. The effect of ionising radiation on the specimens is largely unknown, but the samples showed no observable changes during the experiment, suggesting that there were no significant effects due to potential radiation damage imparted on the samples. The PEEM technique is extremely surface-sensitive (top 10s of nm), making the quality of the results particularly sensitive to the condition of the sample surface, and the sample preparation techniques.

Despite these difficulties PEEM images were obtained which revealed the internal heterogeneity of multiple trace elements in foraminifera, which tends to follow systematic banded patterns, normal to the growth axis of the shell[10,14-16]. Elemental maps at the Mg K-edge (Figure 1) reveal a variation in the intensity, concomitant with the bright bands seen in the off-peak image, suggesting that these bands contain elevated Mg, in agreement with previous studies[10,14,15]. The corroboration of the results with those of past studies demonstrates the ability of the PEEM instrument to map trace elements in carbonate minerals. Furthermore, the identification of Mg banding in the structure, and the shape of the Mg NEXAFS spectra identified with the PEEM instrument are in agreement with a more extensive analyses of foraminiferal Mg, using an alternative X-ray spectroscopy technique, STXM[10]. The nano-scale resolution of the PEEM instrument offers a significant advance from most previous methods, and compared to the commonly-used laser ablation mass spectrometry techniques PEEM does not alter the sample surface significantly or cause laser-pitting.

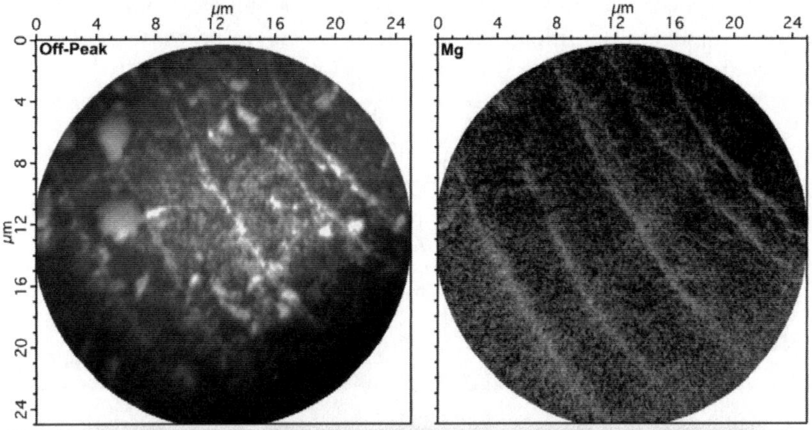

Figure 1. *(Left)* An off-peak PEEM (from 1295 eV, below the Mg K-edge) image of a section of foraminiferal shell wall, accompanied by the equivalent Mg K-edge peak intensity map *(right)* for the same region. Brightness denotes a higher value.

2.2. Scanning Transmission X-ray Microscopy (STXM) studies of Mg and B in foraminifera

We have characterised the coordination of Mg, B and Na in two disparate species of symbiont-bearing foraminifera using Near-Edge X-ray Absorption Fine Structure (NEXAFS) spectroscopy. Both the benthic *Amphistegina lessonii* and the planktic *Orbulina universa* exhibit Mg-banding[15,17-19], which we have confirmed by electron microprobe. Data were gathered at the STXM branch of beamline 11.0.2 at the Advanced Light Source (ALS, Berkeley, CA). Raster X-ray absorption images map Mg, B and Na distributions at much higher resolution than the electron microprobe (Figure 2). This high-resolution mapping reveals that the broad (~2–3 μm) bands observed in *A. lessonii* by the electron probe were in fact made up of sequential finer (<500 nm) bands in close proximity, revealing that banding is much finer-grained than previously thought. NEXAFS spectra gathered at the Mg-edge from on- and off-band regions in *A. lessonii* and *O. universa* were identical within instrumental error, as is evident from the featureless (to within instrumental noise) residuals in Figure 3.

The Mg banding could either be produced by variations in the concentration of Mg hosted in a single phase, or be the result of a multi-host-phase, where background Mg environment is augmented by an alternative phase in the bands. Our NEXAFS spectra demonstrate the sensitivity of this method to changes in the chemical environment of the Mg ion: closely related mineral phases produce disparate spectra (Figure 3). Therefore, we would expect a single-host-phase system (Mg in a single homogeneous chemical environment) to produce uniform spectra throughout the test and the spectra of

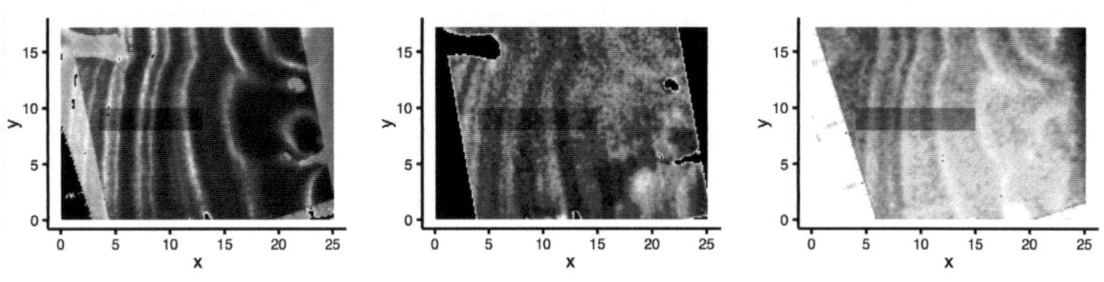

Figure 2. Mg (left), B (centre) and Na (right) distributions measured by STXM across the shell wall of *Orbulina universa*. Axes scales are labelled in μm. The heterogeneous banding is evident perpendicular to the growth direction for each element, although maxima in each elemental concentration do not coincide in position and banding is offset/out of phase between elements. The grey bar corresponds to the composition profiles show in Figure 4 below.

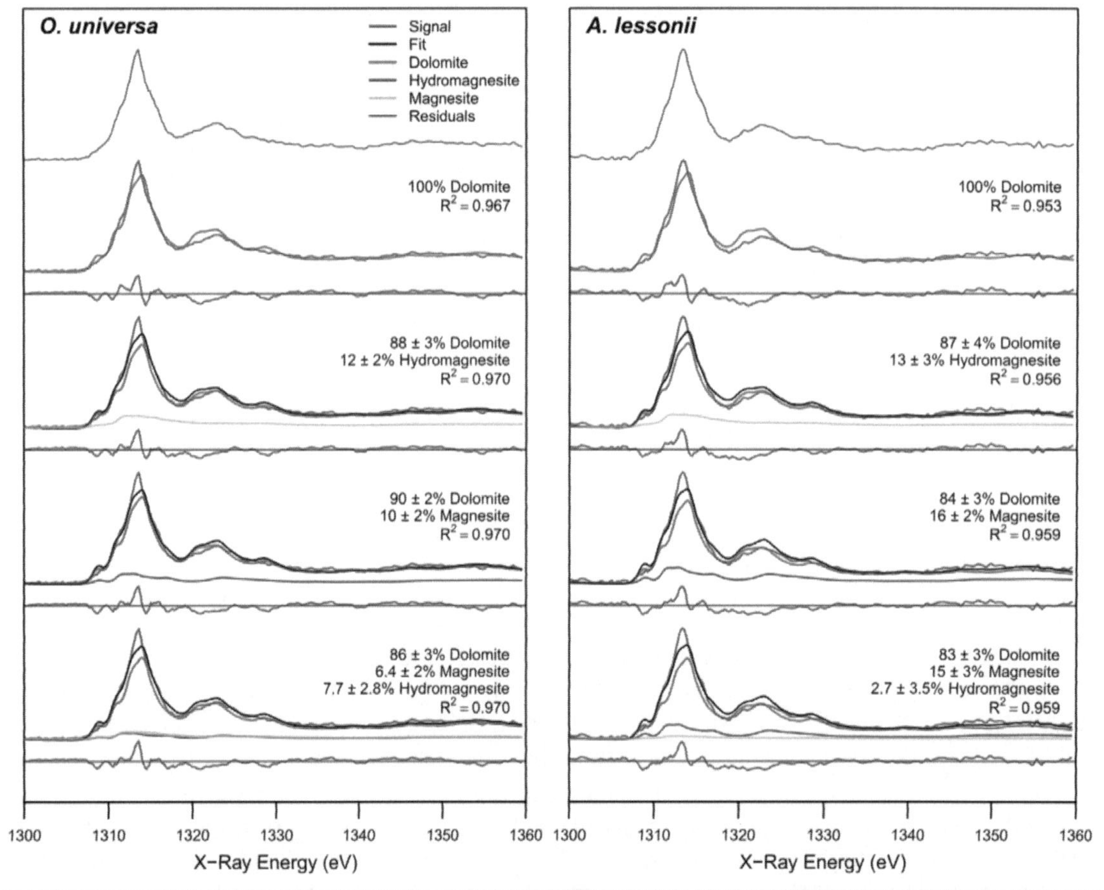

Figure 3. NEXAFS spectra from image stacks at the Mg-edge in high-Mg areas from *O. universa* and *A. lessonii*, compared to the candidate minerals dolomite, hydromagnesite and magnesite. Spectra from low-Mg areas are identical. Each fit consists of one or more candidate spectra combined with a linear background. The dolomite spectrum alone is a good fit for the foraminiferal Mg. The fit is marginally improved by adding extra phases, but the shape of the residuals (shown below each model) does not alter significantly, and the increase in goodness of fit (indicated by R^2) is insignificant.

a multi-host-phase system (with multiple Mg chemical environments) to vary between the on- and off-band regions. The presence of B and Na banding in the shell demonstrates a general chemical variation associated with such bands, which have previously been identified with daily calcification events (Figure 4). The boron NEXAFS spectra (Figure 5) measured from both high-boron and low-boron areas suggest that the boron is incorporated into the structure as BO_3, replacing the carbonate group in the rhombohedral calcite[16]. The sodium spectra are more equivocal and do not allow us to infer the chemical environment of Na in these samples.

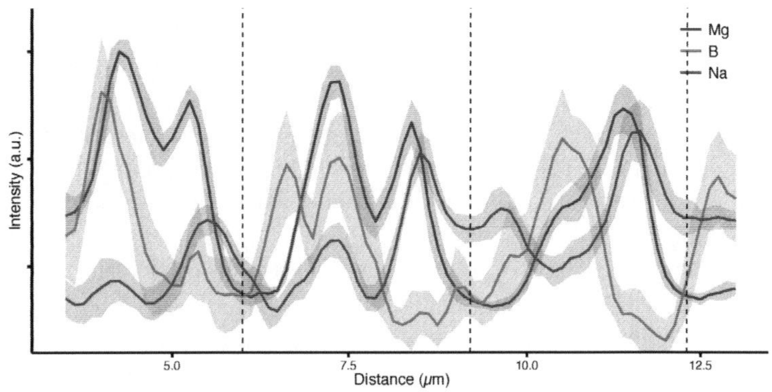

Figure 4. Relative intensities (in arbitrary units, but related to concentrations) of Mg, B and Na across the shell wall of *O. universa*, taken from the grey shaded areas indicated in Figure 2. The cyclicity of compositional banding is similar for each element, but detailed comparison demonstrates that concentrations peak at different stages in each shell growth cycle.

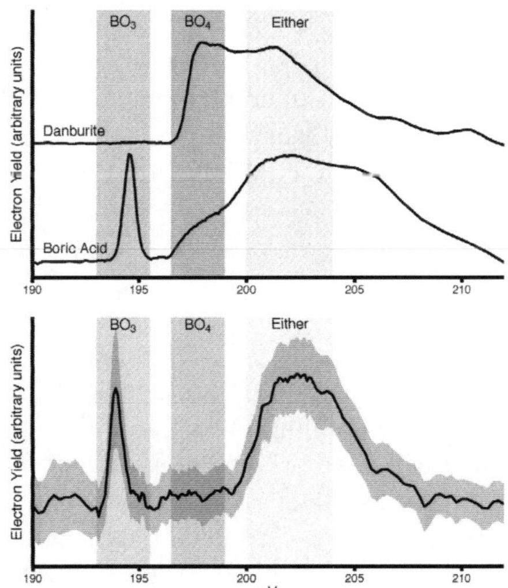

Figure 5. NEXAFS spectra at the B-edge for *O. universa (below)* compared to danburite (tetrahedral B) and boric acid (trigonal B) standards *(above)*. The peak at 190 eV in the *O. universa* sample corresponds to BO_3 within the sample, and there is no evidence for the presence of BO_4, suggesting that borate substitutes directly for carbonate in the calcite shell.

3. Discussion and conclusions

Our results provide a fundamental step forwards towards a mechanistic understanding of the Mg/Ca palaeotemperature proxy and the boron proxy for pH, as well as for the general mechanisms of biomineralisation. By conclusively demonstrating the ideal substitution of Mg into the calcite lattice we have confirmed a long-held, yet untested assumption behind the Mg/Ca proxy, supporting the presence of an inorganic thermodynamic connection between sea- water temperature and foraminiferal Mg/Ca. We demonstrate that foraminiferal Mg is not hosted in organic molecules, or in a secondary interstitial mineral or organic environment. Furthermore, the consistency of the Mg coordination throughout the shell indicates that foraminifera employ a uniform mechanism of crystal growth. This supports the links between calcite-bound trace elements (like Mg) and the external environment upon which palaeo-proxies rely. Our work on boron demonstrates that it substitutes for BO_3 in the shell and is similarly heterogeneous, showing banding at a sub-μm length scale. The B proxies (δ11B and B/Ca) have the potential to provide a record of the entire ocean carbonate system throughout Earth history. However, these proxies require the sole incorporation of tetrahedral $B(OH)_4^-$ from seawater, which has thus far been missing in the literature. Our coordination data uphold the underlying theory behind the B palaeoproxies, and give a new confidence to the interpretation of B records of past ocean carbonate chemistry. This helps validate their use as a tool to understand ocean carbonate chemistry, and the effects of ocean acidification.

We have determined the control on driving chemical variations in biogenic carbonates using STXM at the absorption edge of Mg, B, and Na in the shells of modern plankton. The power of these observations lies in their ability to link changes in chemistry, microstructure, and growth process in biogenic carbonate to environmental influences. We have seen that such changes occur at length scales of tens of nanometres and demonstrated that STXM provides an invaluable route to understanding chemical environment and key heterogeneity at the appropriate length scale. This new understanding provides new routes for future measurements of past climate variation in the sea floor fossil record.

Acknowledgements
We thank David Nicol, Iris Buisman and Martin Walker for invaluable technical assistance, and James Bryson for his help with synchrotron data collection. Jean DeMouthe (California Academy of Sciences) and Mike Rumsey (Natural History Museum, London) provided of B-containing minerals for use as references. This work was funded by NERC, Jesus College (Cambridge) and the US Department of Energy (via ALS)

References
[1] M Pagani, J C Zachos, K H Freeman, B Tipple, and S Bohaty (2005) Marked decline in atmospheric carbon dioxide concentrations during the Paleogene. *Science*, 309: 600–603.
[2] A K Tripati, C D Roberts, and R A Eagle (2009) Coupling of CO_2 and ice sheet stability over major climate transitions of the last 20 million years. *Science*, 326: 1394–1397.
[3] J B Ries (2010) Review: geological and experimental evidence for secular variation in seawater Mg/Ca (calcite-aragonite seas) and its effects on marine biological calcification. *Biogeosciences*, 7: 2795–2849.
[4] D W Hastings, A D Russell and S R Emerson (1998) Foraminiferal magnesium in Globeriginoides sacculifer as a paleotemperature proxy. *Paleoceanography*, 13: 161-169.
[5] H Elderfield and G Ganssen (2000) Reconstruction of temperature and $\delta^{18}O$ of surface ocean waters using Mg/Ca of planktonic foraminiferal calcite. *Nature*, 405: 442–445.
[6] A A Finch and N Allison (2007) Coordination of Sr and Mg in calcite and aragonite. *Mineralogical Magazine*, 71:539–552, 2007.
[7] A A Finch and N Allison (2008) Mg structural state in coral aragonite and implications for the paleoenvironmental proxy. Geophysical Research Letters, 35: L08704.
[8] L C Foster, A A Finch, N Allison, C Andersson, and L J Clarke (2008) Mg in aragonitic bivalve

shells: Seasonal variations and mode of incorporation in Arctica islandica. *Chemical Geology* 254: 113–119.

[9] L C Foster, N Allison, A A Finch, and C Andersson (2009) Strontium distribution in the shell of the aragonite bivalve Arctica islandica. Geochem Geophys Geosyst, 10, Q03003.

[10] O Branson, S A T Redfern, T Tyliszczak, A Y Sadekov, G Langer, K Kimoto and H Elderfield (2013) The coordination of Mg in foraminiferal calcite. *Earth and Planetary Science Letters*, 383: 134–141.

[11] J-P Cuif, Y Dauphin, G Nehrke, J Nouet, and A Perez-Huerta (2012) Layered growth and crystallization in calcareous biominerals: Impact of structural and chemical evidence on two major concepts in invertebrate biomineralization studies. *Minerals*, 2: 11–39.

[12] M Cusack and A Freer (2008) Biomineralization: Elemental and organic influence in carbonate systems. *Chem Rev,* 108:4433–4454.

[13] F Farges, A Meibom, A M Flank, P Lagarde, M Janousch, and J Stolarski (2009) Speciation of Mg in biogenic calcium carbonates. *Journal of Physics: Conference Series*, 190

[14] S M Eggins, P de Deckker, and J Marshall (2003) Mg/Ca variation in planktonic foraminifera tests: implications for reconstructing palaeo-seawater temperature and habitat migration. *Earth and Planetary Science Letters*, 212: 291–306.

[15] S M Eggins, A Sadekov, and P de Deckker (2004) Modulation and daily banding of Mg/Ca in *Orbulina universa* tests by symbiont photosynthesis and respiration: a complication for seawater thermometry? *Earth and Planetary Science Letters*, 225: 411–419.

[16] O Branson, K Kaczmarek, S A T Redfern, S Misra, G Langer, T Tyliszczak, J Bijma and H Elderfield (2015) The coordination and distribution of B in foraminiferal calcite. *Earth and Planetary Sciences Letters,* 416: 67-72.

[17] J Erez, J (2003) The source of ions for biomineralization in foraminifera and their implications for paleoceanographic proxies. *Reviews in Mineralogy and Geochemistry*, 54: 115–149.

[18] D Kunioka, K Shirai, N Takahata, Y Sano, T Toyofuku and Y Ujiie (2006) Microdistribution of Mg/Ca, Sr/Ca, and Ba/Ca ratios in Pulleniatina obliquiloculata test by using a NanoSIMS: Implication for the vital effect mechanism. *Geochem. Geophys. Geosyst.* 7: Q12P20.

[19] A Y Sadekov, S M Eggins, P de Deckker (2005) Characterization of Mg/Ca distributions in planktonic foraminifera species by electron microprobe mapping. *Geochem. Geophys. Geosyst.* 6: Q12P06.

Phantom creation and analysis: Improving X-Ray microtomography scanning of soft sediment cores containing volcanic ash

E E Evans[1, 2, 3], S M Davies[1], R E Johnston[2, 3]

[1] Department of Geography, College of Science, Swansea University, Singleton Campus, Swansea, Wales, UK
[2] Materials Research Centre, College of Engineering, Swansea University, Bay Campus, Swansea, Wales, UK
[3] Advanced Imaging of Materials (AIM), College of Engineering, Swansea University, Bay Campus, Swansea, Wales, UK

e.e.evans.836924@swansea.ac.uk

Abstract. We present a series of soft sediment geological phantoms constructed to simulate volcanic ash deposits preserved within sediment cores. This study aimed to systematically discern which sediment components are most effectively detected and therefore characterised by X-ray microCT (μCT). Samples with low or overlapping attenuation contrast but distinct morphologies such as coarse grained (>125 μm) ash within fine mud (<4 μm) can be manually separated using analysis software, while samples with high attenuation contrast display distinct horizons that are easily segmented using attenuation thresholds. We anticipate that this work will have broader applications within Earth Science to optimise the exploration of sediment with μCT analysis.

1. Introduction

Volcanic ash deposits, which are widely used as age-markers in soft sediment records, are prone to post-depositional displacement within sedimentary environments. These sediment cores are routinely investigated by linear and surface analyses, which only provide one and two dimensional data, respectively. With X-ray μCT a much needed third dimension is revealed to aid the detection of micro-structures that may be linked to post-depositional processes. Geological samples, however, mainly consist of similarly attenuating silicates which makes analysis using X-ray μCT challenging. To allow effective segmentation, histograms either with distinct peaks or with enough dilation to allow phase separation need to be generated. Ongoing work aims to characterise and quantify the three-dimensional structures of volcanic ash found in core samples of sedimentary sequences. However, the detection and discernment of the presence of volcanic ash in a sample remains a particular problem.

This study aims to systematically discern which combinations of volcanic ash and sediment best respond to X-ray μCT analysis. This response is measured by observing which samples show the greatest amount of attenuation contrast and therefore can be effectively segmented and the three-dimensional structures quantified. Constraining the attenuation contrast of particular combinations of ash and sediment will give clearer limitation boundaries to samples of similar composition. Previous work in this area [1] relied on additional information, such as optical analysis, to determine thresholds to separate sediment from volcanic ash rich regions.

Geochemical composition and grain-size of the volcanic ash are the key variables within these phantom samples. The success of the analyses is quantified by the level of grayscale contrast between the ash and host sediment and the ease of segmentation. The work herein informs further work where candidate specimens will be selected based on the potential for successful μCT analysis and segmentation.

2. Samples and methodology

This study uses marine mud from the Marion Dufresne cruise MD123, core MD01-2444. The sediment is continentally derived and drilled approximately 100 km west of the Beja District, Portugal. The volcanic ashes used are derived from proximal deposits on mainland Iceland. Together the sediment and ash types represent compositions found in North Atlantic marine core sequences. Three igneous rock types were selected for their contrasting chemical composition; basaltic, intermediate and rhyolitic (relatively; low, medium and high silica, respectively). The volcanic ashes were sieved into 5 size ranges, 500, 280, 125, 80 and 25 μm. The phantoms were produced by gravity settling through water within 9 mm diameter polypropylene transparent drinking straws. The layers were formed with the volcanic ash sandwiched between two layers of marine mud.

The μCT scans were performed using an Xradia Zeiss 520 Versa X-ray microscope at Swansea University's AIM (Advanced Imaging of Materials) facility. Data was reconstructed using Zeiss' proprietary XMReconstructor software (Filtered Back-Projection) along with any pre-processing Beam Hardening removal and Smoothing (see Results for settings). For all scans the centre of the sample was 86.7212 mm from the 0.4 x objective lens and 35.8028 mm from the X-ray source.

3. Results

The figures below were generated in the software package Avizo 9 (FEI 2015, Oregon, USA). The multi-coloured map for each image was generated by assigning primary colours red, yellow and blue to volcanic ash, sediment and other materials (air, straw and water) respectively, with the overlaps in attenuation being the secondary colours in between. Further exploration of this method can be found in the authors' previous work [2]. The monochrome map is the same for each sample for comparison.

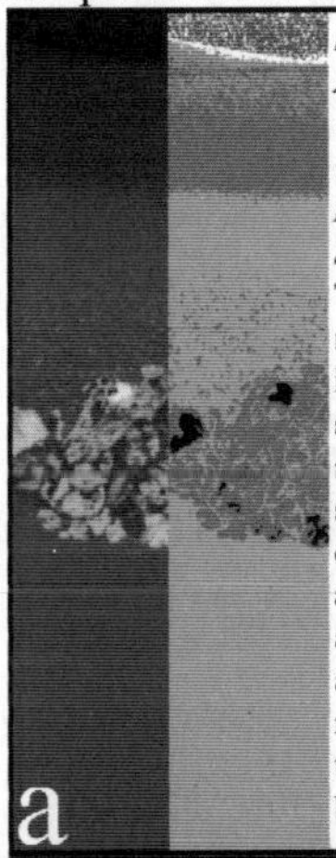

Basalt composition 280 μm
X-ray energy: 80 kV, 7W
1 sec exposure, LE3 Versa filter
11.491 μm voxel size
Beam Hardening constant: 0.19
Smoothing filter: 0.7
The grain size of the ash (centre region) is sufficiently coarse for individual grains to be observed and separated despite the overlap in attenuation between the ash and the sediment. The upper region of red is of lower attenuation than the ash and is an artefact of the lower attenuation regions seen inside the ash layer. The grain supported nature of the ash layer is probably the cause of lower density pockets between the grains.

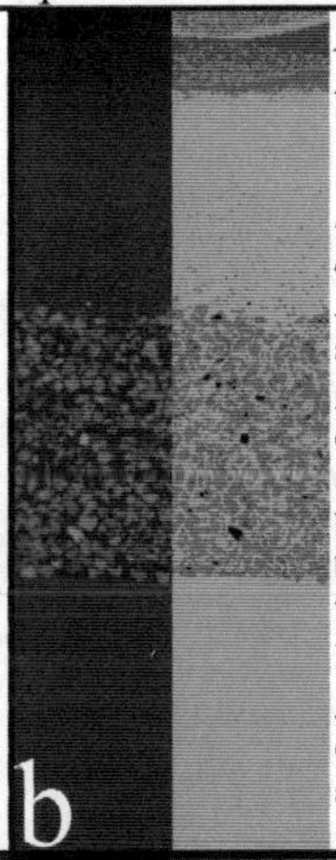

Basalt composition 125 μm
X-ray energy: 80 kV, 7W
1 sec exposure, LE3 Versa filter
11.489 μm voxel size
Beam Hardening constant: 0.2
Smoothing filter: 0.7
As with **a** the grain size is sufficiently large that individual grains can be segmented. There are clear basal and upper boundaries in the sample. The regions of purple between the grain indicate overlap between water and inter-grain space. This structure is similar to **a**.
In the case of this sample and **a** these grain sizes represent proximal deposits which, in a distal marine setting, are rarely observed.

Basalt composition 80 μm
X-ray energy: 80 kV, 7W
1 sec exposure, LE3 Versa filter
11.493 μm voxel size
Beam Hardening constant: 0.2
Smoothing filter: 0.7
As the grain size approaches that of less than eight times the voxel size (that is 2 by 2 by 2 voxels in size) the boundaries of the ash within the sample becomes more diffuse. This effect is likely a product of partial volume effects related to the scan resolution. Here the ash has to be treated as a single depositional feature instead of granular. Therefore at lower concentrations of ash discerning broad features becomes more difficult.

Basalt composition 25 μm
X-ray energy: 80 kV, 7W
1 sec exposure, LE3 Versa filter
11.495 μm voxel size
Beam Hardening constant: 0.25
Smoothing filter: 0.7
Sample **d** is the only one of the basalt samples to show regions of no attenuation overlap between the volcanic ash and the sediment, as shown by the pure red and yellow, respectively. The grain size here is barely two voxels in size however this has lead to close packing of the ash leading to a homogenous region of attenuation. Again as seen with **c** at lower concentrations individual particles may be difficult to discern from noise or similar.

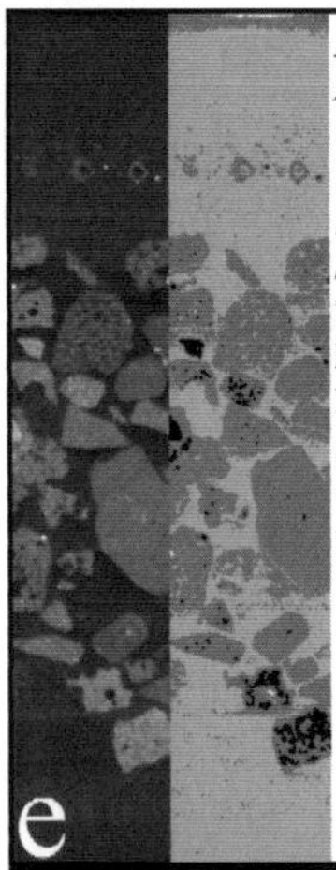

Intermediate composition 500 μm
X-ray energy: 80 kV, 7W
1 sec exposure, LE3 Versa filter
11.484 μm voxel size
Beam Hardening constant: 0.2
Smoothing filter: 0.7

The large grain size of this sample helps to clearly define each ash grain. Note the internal structure of some grains shown by variations in attenuation which may indicate the presence of lithic clasts in the case of higher attenuation or pumices in the case of lower attenuation.

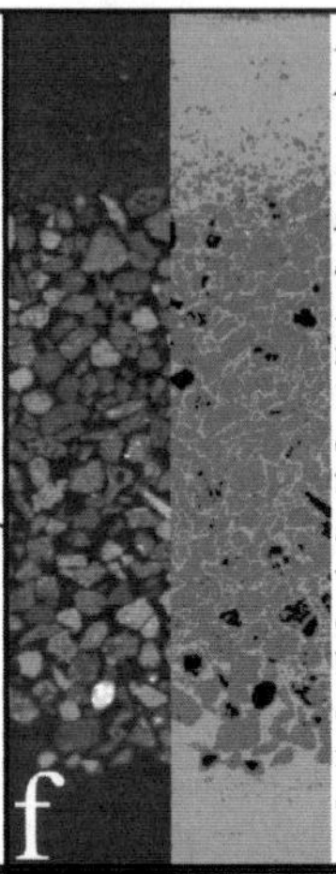

Intermediate composition 280 μm
X-ray energy: 80 kV, 7W
1 sec exposure, LE4 Versa filter
11.504 μm voxel size
Beam Hardening constant: 0.2
Smoothing filter: 0.7

As with **e** we see a heterogeneous mix of grain types. The upper and lower boundaries are somewhat diffuse. In the case of the basal boundary this may be due to additional sinking and settling. As with **a**, **b** and **e** these samples can only be used to study the contrast between ash and sediment composition and not be truly representative of sedimentary deposits seen in nature.

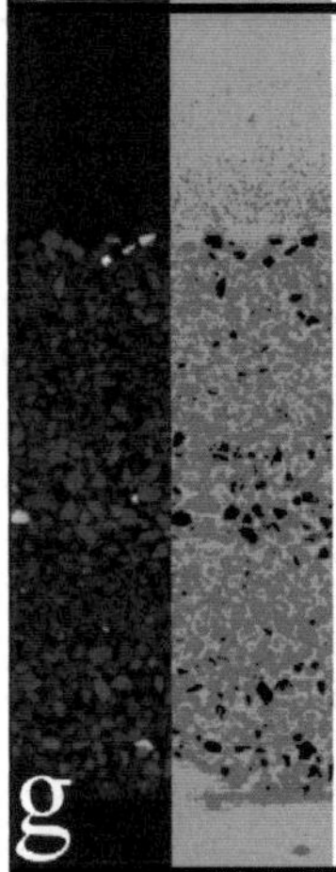

Intermediate composition 125 μm
X-ray energy: 80 kV, 7W
1.5 sec exposure, LE4 Versa filter
10.667 μm voxel size
Beam Hardening constant: 0.2
Smoothing filter: 0.7

As with the basaltic samples, the boundaries between the sediment become less distinct as the grain size decreases as seen in this sample. In this sample a noticeable fining upwards sequence can be seen within the sample. Not only is there a grain size change but those with the greater attenuation group with those of the larger grain size. This suggests a correlation between grain size and clast type.

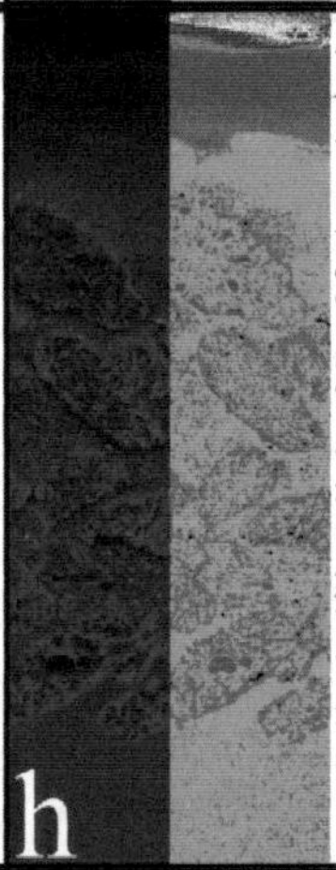

Rhyolitic composition 500 μm
X-ray energy: 60 kV, 5W
2 sec exposure, LE2 Versa filter
11.133 μm voxel size
Beam Hardening constant: 0.2
Smoothing filter: 0.7

The volcanic ash grains in this sample are highly vesicular pumice. The rims of the grains show higher attenuation than the centres. This may be due to partial volume effects from imaging such a highly pumiceous interior.

Within this sample there appears to be some buoyant movement evident from the displacement of the upper surface of the sediment.

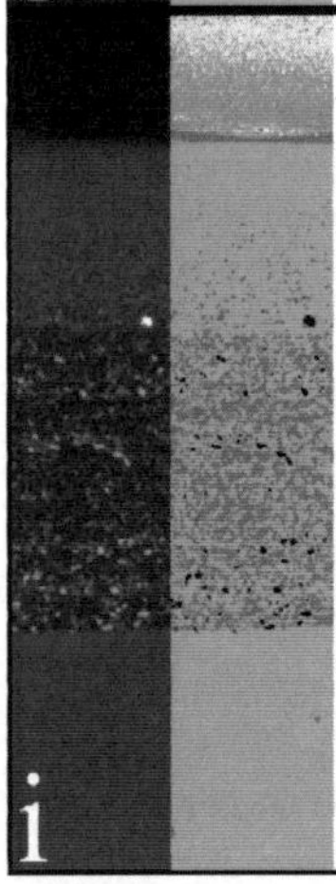

Rhyolitic composition 80 μm
X-ray energy: 80 kV, 7W
1 sec exposure, LE2 Versa filter
10.367 μm voxel size
Beam Hardening constant: 0.2
Smoothing filter: 0.7

As grain size decreases (seen here) the attenuation overlap becomes greater still for this composition. As a result if only the red regions were used as the threshold the ash layer would appear more diffuse than it actually is. Some of this may also be caused by partial volume effects from the scan resolution.

Rhyolitic composition 25 μm
X-ray energy: 70 kV, 6W
1.5 sec exposure, LE2 Versa filter
10.365 μm voxel size
Beam Hardening constant: 0.2
Smoothing filter: 0.7

For this sample, as with **d**, the fine grain structure means the ash needs to be treated as a single structure. But unlike **d** this sample has diffuse boundaries.

As with **i** the overlap and possible partial volume effects makes the ash layer appear more dilute and diffuse. This has an impact on discerning the full extent of an ash horizon.

4. Discussion

The aim of this work was to systematically study the attenuation contrast between volcanic ash of varying compositions and sizes with a host material, in this case marine mud. As SiO_2 increases within the volcanic ash the contrast between the ash and sediment becomes more overlapped. This variation appears to be independent of grain size but partially dependant on grain type (contrast the pumice of **h** with the glassy shards of **i**). The attenuation contrast between the ash and sediment is relative and so both the composition of the ash and sediment must be assessed for suitability. Exact quantification of this limit may not be possible and merely provides a priori information about the samples post-scanning. However, using effective atomic numbers and the attenuation contrasts seen in this study a semi-quantitative process of selecting suitable samples may be a viable option moving forward. Part of the aim of this study is to quantify these detection limits.

Coarser grained ash is easiest to detect in the cores regardless of composition. Although the ash may not be able to be separated using grayscale thresholds, it can be separated with manual "tagging" (user selection within analysis software). Such grain sizes are rare to encounter in distal marine deposits but common in lakes near the eruption centre. The cut off for size contributing to ease of segmentation varies with composition. Comparing figures **d** and **j**, which have the same grain size but different chemistry; **d** has regions of attenuation exclusively marked as volcanic ash (red), while in **j** the ash attenuation overlaps with that of the sediment, making it difficult to separate. For 25 μm grain sizes single ash particles may only be represented by ~8 voxels and therefore, at this resolution the ash can only be visualised as a bulk layer because the individual particles cannot be separated. These issues may be mitigated by increasing the scan resolution to produce batch or mosaic scans, which would also reduce partial volume effects such as those seen in figure **h** and the poor definition of individual grains seen in figure **j**. However the benefit of such increased scan times would have to be weighed against the likelihood of improved results. The machine maximum pre-set of 3201 projections were chosen for all scans however the authors acknowledge that this was beyond the point that additional spatial detail could be discerned and increased scan time. Similar images could result from fewer projections. Optical or physical magnification needs to be employed to improve resolution.

Additionally, samples **h**, **e**, **f** and **g** reveal interactions between coarse-grained ash deposits and the host sediment. These observations are particularly promising for future investigations of depositional processes in sedimentary environments using X-ray μCT. Whether these interactions within these phantoms are representative of natural sedimentary systems is as yet unknown.

5. Conclusions

The results of this study points to the necessity of pre-scanning knowledge of volcanic ash and sediment density/effective atomic number to determine the potential gains from μCT analysis of a sample. However, the additional information gained from the understanding of 3D structures out-weighs the additional pre-scanning data acquisition.

Further work of this nature is essential and we recommend exploring different volcanic ash compositions and sediment types to better represent the diversity of sedimentary environments. These may include marine or lake muds of differing attenuation, coarser grained sediments, peat and ice. We anticipate that this work will have broader applications for other Earth Scientists in the exploration of soft sediment with μCT analysis.

Acknowledgements

We would like to acknowledge the assistance provided by the Swansea University AIM Facility which was funded in part by the EPSRC (EP/M028267/1); technicians Gareth James, Grahame Walters and Kathryn Lacey; Dr Peter Abbott for providing samples from marine core MD01-2444. This work is funded by a Swansea University PhD scholarship.

References

The poster on which these proceedings are based: https://dx.doi.org/10.6084/m9.figshare.3988857.v1

[1] Griggs A J Davies S M Abbott P M Coleman M Palmer A P Rasmussen T L Johnston R 2015 Visualizing tephra deposits and sedimentary processes in the marine environment: The potential of X-ray microtomography *Geochem. Geophys. Geosyst.* **16** pp 4329-4343

[2] Evans E E Davies S M Johnston R E Wulf S [ONLINE] Conference Poster hosted on Figshare Available at: https://dx.doi.org/10.6084/m9.figshare.1549764

3D elemental mapping of materials and structures by laboratory scale spectroscopic X-ray tomography

C K Egan[1], S D M Jacques[1], M D Wilson[2], M C Veale[2], P Seller[2], R A D Pattrick[3], P J Withers[1], R J Cernik[1]

[1] School of Materials, University of Manchester, Manchester, UK
[2] Science and Technology Facilities Council, Rutherford Appleton Laboratory, Harwell Campus, Didcot, UK
[3] School of Earth, Atmospheric and Environmental Sciences, University of Manchester, Manchester, UK

Email: christopher.egan@manchester.ac.uk

Abstract. Using a microfocus X-ray tube and pixelated energy-resolving detector it is possible to measure the X-ray absorption spectrum of a material with high spatial resolution. Given sufficient energy resolution in the detector it is possible to detect and identify absorption edges which are characteristic to individual chemical elements. Using computed tomography the three dimensional (3D) internal elemental chemistry of an object can be reconstructed. The application of spectroscopic X-ray tomography is demonstrated by mapping distribution of heavy elements inside a mineralised ore sample. We correlate and validate this data with high resolution X-ray tomography and energy-dispersive X-ray spectroscopy data.

1. Introduction

X-ray computed tomography (XCT) can reveal the internal three-dimensional (3D) structure of an object, with most notable application to medical imaging. However it finds many more areas of application beyond this, such as in security scanning, industrial inspection, materials science, and archeology [1]. In XCT, the measured contrast is proportional to both density and atomic number variations, with high density regions appearing brighter than low density regions. However it is very difficult to obtain the exact chemical speciation due to the polychromatic nature of the X-ray beam from laboratory X-ray tubes. There are methods, such as dual energy scanning, that can be used to obtain some pseudo chemical sensitivity which are finding increasing application to the medical imaging sector [2]. But still the exact chemical speciation is unavailable. An alternative approach is to use the X-ray fluorescence (XRF) signal emitted from an object. However due to the weak intensity this method is generally limited to very bright synchrotron sources which have a major limitation from a throughput and access point of view [3]. Laboratory-based CT instruments are cheap and widely available, but due to the low intensity of radiation it is very time-consuming to undertake XRF imaging. Over recent years, a range of pixelated spectroscopic X-ray detectors have been actively researched and developed which have the ability to measure X-ray energy of individual photons with positional sensitivity. It is therefore possible use these types of detectors to measure the full attenuation spectrum and map chemical elements based on the K-edge position. Combining this with CT and using a microfocus source it is possible to identify multiple chemical elements simultaneously in 3D without prior knowledge of sample composition and with high spatial resolution [4].

2. Experimental methods

2.1 Equipment We used a microfocus tungsten target X-ray source (spot size 5μm, 160 kV, 15 μA), and HEXITEC spectroscopic X-ray detector. The detector consists of a 1 mm thick CdTe single crystal (20 × 20 mm^2) bump-bonded to a large area ASIC packaged with a high performance data acquisition system. The detector has 80 × 80 pixels on a 250 μm pitch with an energy resolution of 800 eV at 59.5 keV [5,6]. The CT scan was performed using an extended field-of-view by scanning the detector along 5 separate positions in the horizontal direction. Each sub-projection had an exposure time of 45 s (225 s for a full projection) and we performed 120 projections covering 360° (total scan time 7.5 hrs).

2.2 Tomographic reconstruction. An important consideration in spectroscopic XCT is the choice of reconstruction algorithm. Since the signal-to-noise ratio in any one energy channel is low, and with fewer projections (to compensate for longer image exposure times) this can result in poor quality reconstructions. Filtered back-projection type algorithms (such as FDK) produce poor results for sub-optimal scanning conditions (e.g. few projections). Iterative algorithms (such as SIRT) are much better noise tolerant and produce better quality reconstructions (see figure 1). In this work we perform a reconstruction for every energy channel and obtain an absorption spectrum in every voxel (producing a 4D dataset: 3 spatial and 1 spectral). It is therefore important to also consider the spectral quality. As can be seen in figure 1, the reconstructed absorption spectrum is very noisy for the FDK algorithm, with the K-edge almost lost in the noise. However, by using the flexibility of iterative algorithms, whereby the reconstructed image from the previous energy channel can be used as a starting vector for the next reconstruction (so-called channel linked), the quality of the reconstruction is improved both in the spatial and spectral domains. This approach is the algorithm of choice for this work. Due to high volumes of data processing, all reconstructions were performed using GPU accelerated code. We also used a wavelet based Fourier filter applied to the sinogram to suppress ring artefacts.

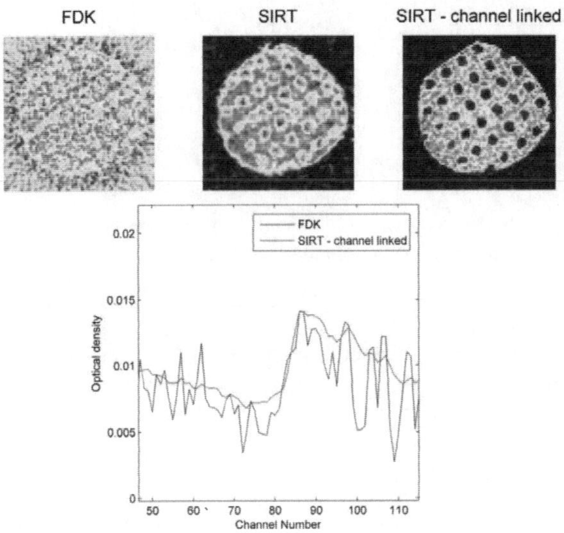

Figure 1. Testing of different reconstruction algorithms used in spectral X-ray tomography. The sample is a lattice carbon structure impregnated with palladium metal (Pd). Reconstructions using filtered back-projection (FDK), simultaneous iterative reconstruction technique (SIRT) and channel-linked SIRT methods are shown. Iterative reconstruction algorithms produce superior images, with the channel-linked method showing the best quality. The plot below shows two voxel spectra covering the Pd K-edge (step change in optical density at channel number 83) from the same voxel inside the sample, reconstructed using FDK and channel-linked SIRT methods, respectively. A clear reduction in spectral noise is evident in the latter case.

2.3 Core sample We scanned a 20 mm diameter core sample from a mineralised gold-bearing ore using spectroscopic X-ray tomography. The sample mostly comprises pyrite (FeS_2) and quartz (SiO_2), with minor amounts of gold (Au), galena (PbS), chalcopyrite ($CuFeS_2$) and bornite (Cu_5FeS_4). The Au and Pb containing phases both occur as small inclusions of about 100 μm diameter. In order to obtain a reference data set, we also scanned this sample using a high resolution X-ray CT system (Zeiss Versa XRM 520). In order to validate this data we also mechanically sliced open the core sample and performed electron probe microscopy and spectroscopy (JEOL JXA-8530F Hyperprobe).

3. Results

Figure 2 shows aligned and correlated spectral XCT and high resolution XCT reconstructions of the core sample. Small high density inclusions are evident which were theorised to be the Au and galena phases. However they were inseparable using classic gray level thresholding techniques largely because of their similar density and x-ray attenuation. In addition, due to potential alloying of gold with silver (Ag) (forming electrum), the gray levels of Au particles was found to vary wildly, making separation and identification almost impossible. However using the spectral XCT data, we can extract individual voxel spectra, where step changes in the optical density correspond to elemental K-edges. Identification of Au and Pb K-edges (at 80.7 keV and 88.0 keV respectively) and subsequent 3D mapping enables us to distinguish and separate these two phases. Unfortunately, the silver (Ag) K-edge was unidentifiable due to very high X-ray attenuation at 25.5 keV in this large sample. Using the two aligned datasets, particle inclusions were identified via spectral XCT, and 3D rendered using the higher resolution data, with blue coloured inclusions corresponding to galena (PbS) and red coloured inclusions corresponding to Au containing material. In this way a more detailed and accurate visualisation of the 3D morphology of the core sample can be obtained.

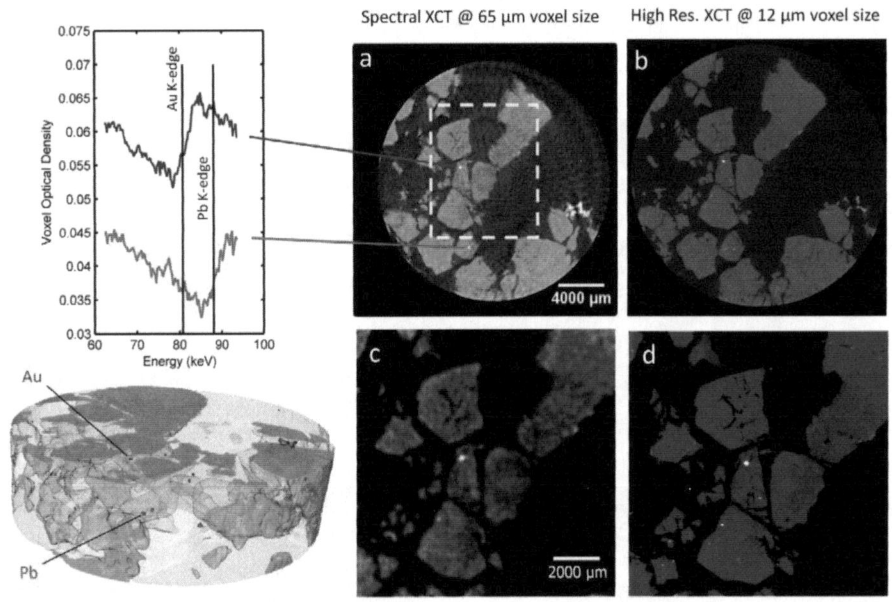

Figure 2. Correlated spectral XCT (65 µm voxel size) and high resolution XCT (12 µm voxel size) datasets. a) Slice through the spectral XCT data showing 2 voxel spectra from different particle inclusions. b) Corresponding slice through the high resolution dataset. c) Spectral XCT data zoomed region (yellow dashed box in (a)). d) High resolution XCT zoomed region. A 3D render of the whole volume with separated Au and Pb phases is shown in the bottom left of the figure.

In order to validate the data, we mechanically sectioned the core sample and performed electron probe spectroscopy mapping, which we can correlate to the XCT data, as shown in figure 3. Large area maps covering the entire surface clearly show the locations of pyrite (Fe and S maps) with low concentrations of Cu (chalcopyrite). We only observed very minor concentrations of Au and Pb in this slice. A zoomed in region in the spectral XCT data shows the location of a small inclusion which was found to have a K-edge at 88.0 keV suggesting this feature to contain Pb; the corresponding Pb WDS map confirms this. Only by high resolution electron probe mapping were we able to find concentrations of Au and Ag in very small inclusions, below the resolution of the XCT data. Some of the Au and Ag is co-located confirming the presence of electrum. It should be noted that the electron probe data originates from a very thin slice (~ 2 micron thickness) which can make it very difficult to find and quantify trace concentrations, highlighting the benefits of a 3D method like spectral XCT.

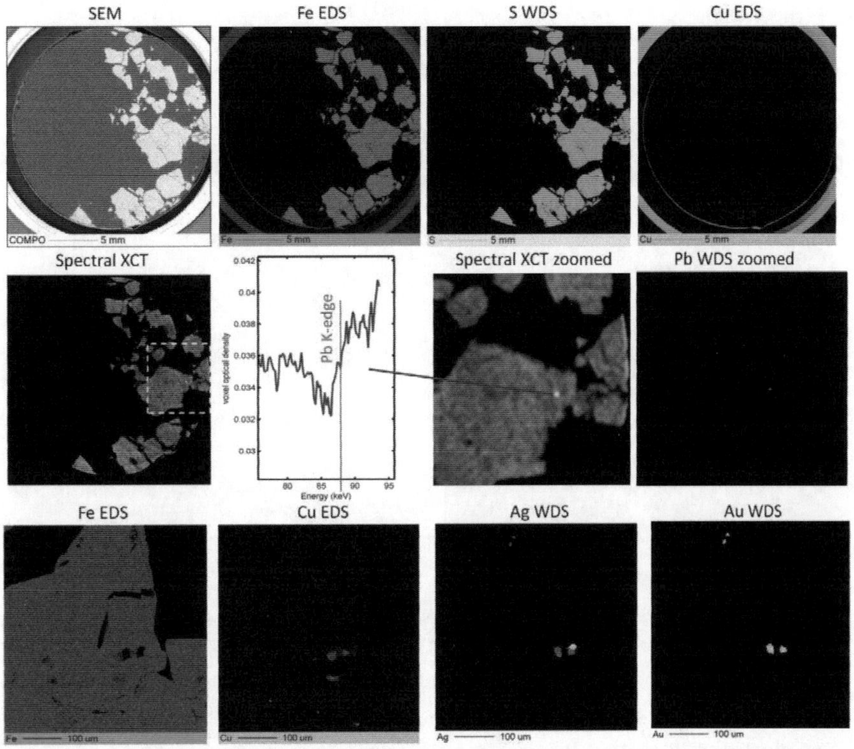

Figure 3. Electron probe microscopy and spectroscopy mapping of a polished section of the core sample. Top row: full area maps of whole section (10μm pixel size). Middle row: Correlated spectral XCT slice, with a zoomed region. Bottom row: small area, high resolution mapping (500 nm pixel size) to find small concentrations of Au and Ag. Note that Pb and S were mapped using wavelength dispersive spectroscopy (WDS) since their emission peaks are overlapping.

4. Conclusions

We have shown how laboratory-scale spectroscopic XCT can be used to non-destructively map the distribution of chemical elements in 3D. Elemental sensitivity is obtained by identifying K-edges in the reconstructed voxel attenuation spectra. This data can be combined with high resolution XCT and electron probe spectroscopy data to quantify and validate the results. This enables us to obtain a more accurate representation of an object's 3D chemical structure.

Acknowledgements

This work was funded by The Leverhulme Trust and supported by EPSRC grants: EP/F007906/1, EP/F001452/1 and EP/M010619/1. Additional thanks are due to Dr J Fellows for performing the electron probe spectroscopy work.

References

[1] Hsieh J 2009 *Computed tomography: principles, design, artifacts, and recent advances* (Bellingham, WA: SPIE)
[2] Graser A, Johnson TR, Chandarana H and Macari M. 2009 Eur. Radiol. **19** 13-23
[3] Bleuet P, Gergaud P, Lemelle L, Tucoulou R, Cloetens P, Susini J, Delette G and Simionovici A. 2010 *Trends Anal. Chem.; TrAC* **29** 518-27
[4] Egan CK, Jacques SDM, Wilson MD, Veale MC, Seller P, Beale AM, Pattrick RAD, Withers PJ and Cernik RJ 2015 *Sci. Rep.* **5** 15979
[5] Seller P *et al.* 2011 *J. Instrum.* **6** C12009
[6] Wilson MD *et al.* 2013 *IEEE Trans. Nucl. Sci.* **60** 1197-1200

Simultaneous surface and bulk sensitive XAS measurements of magnetic particle clusters

S Swaraj[1*], P M Dietrich[2] and W E S Unger[2]
[1] Synchrotron SOLEIL, L'Orme des Merisiers, Saint-Aubin - BP 48, F-91192 Gif-sur-Yvette cedex, FRANCE
[2] BAM Federal Institute of Materials Research and Testing, 12200 Berlin, Germany

*sufal.swaraj@synchrotron-soleil.fr

Abstract. Magnetic iron oxide nanoparticle clusters (mnpc) coated with organic stabilizers were investigated using scanning transmission x-ray microscopy (STXM). Simultaneous surface and bulk sensitive Fe L_3 edge absorption spectra, obtained using a photomultiplier tube and a channeltron, were used to detect subtle changes in the oxidation state in the surface and bulk of Iron Oxide mnpc. The effectiveness of this mode of STXM operation is demonstrated for these nanoparticle clusters.

1. Introduction

The application of magnetic nanoparticles of 5-20 nm diameter and their clusters are actively investigated as Magnetic Resonance (MR) imaging contrast agents due to their low toxicity, high saturation magnetization, high magnetic susceptibility, biodegradability, ease of synthesis and functionalization [1]. However, small particle size also means a decrease in the saturation magnetization [2] and formation of ferromagnetic agglomerates. Stabilizers in a controlled manner are used to overcome the magnetic attraction and to form superparamagnetic clusters.

While the shape, size, agglomeration, coatings and other synthesis parameters are well understood and controlled [3], the interplay and transformation between different iron oxide phases in these nanoparticles clusters is largely uncontrolled and not fully understood. The existence of Fe^{2+} cations in these Fe_3O_4 (magnetite) magnetic nanoparticle clusters (mnpc's) can lead to oxidation and conversion to γ-Fe_2O_3 (maghemite) and finally to α-Fe_2O_3 (hematite) that has no magnetic properties [4]. The iron oxide phase inside the bulk and at the surface of these clusters can have a significant effect on the overall magnetic properties. However, the similar spectroscopic and crystallographic signature of different iron oxides has withheld researchers from correlating magnetic properties with synthesis techniques.

In this paper we present the bulk sensitive and surface sensitive x-ray absorption based characterization of iron oxide (Fe_xO_y) mnpc's coated with dopamine stabilizer. This detection methodology and experimental configuration (Figure 1e) has already been successfully applied for organic materials [5]. Bulk sensitivity was achieved by detection of photons transmitted through the whole samples using a photomultiplier tube. Simultaneous detection of emitted photoelectrons from the sample's surface was achieved using a channeltron. Since the escape depth of Fe 2p photoelectrons at the Fe L-edge is about 4.5 nm for iron oxide [6] the information depth corresponds to the top

IOP Conf. Series: Journal of Physics: Conf. Series **849** (2017) 012014 doi:10.1088/1742-6596/849/1/012014

surface of the investigated region. We have focused our studies on the Fe L-edge as it is highly sensitive to oxidation state of iron.

2. Experimental

STXM and X-ray absorption spectroscopy (XAS) measurements were performed at the Pollux beamline at the Swiss Light Source (SLS), Villigen, Switzerland [7]. A 35 nm zone plate (ZP) is used to focus the x-rays through an order sorter aperture (OSA) on to the sample. The mnpc's were dropcasted on a 100 nm thick silicone nitride window for measurements. A scintillator coupled with a photomultiplier tube (indicated "PMT") is used to detect the bulk sensitive signal in the form of transmitted photons. Surface sensitivity was made possible by detecting surface photoelectrons using a channeltron (indicated "CT") biased at 2.5 kV (model no. KBL10RS, Dr. Sjuts Optotechnik Gmbh). The CT was mounted adjacent to the PMT and the sample was mounted facing the detectors. A wet-cell sample was also prepared for dopamine coated mnpc's and was measured only using the PMT. This particular sample was prepared by sealing a drop of mnpc solution in a sandwich of two silicone nitride windows with epoxy. The chamber was pumped down to 10^{-6} mbar. The normalization of the CT signal was done by dividing it with the measured photon flux on the PMT. Measurements used in the present work are not intended to be quantitative in nature hence the photon energy dependence of the detector efficiencies was ignored. The PMT signal was normalized using signal from a blank silicone nitride window. The XAS spectra were extracted through line scans after choosing a sample region representative of the whole sample. Plasma treatment was done in an Atto plasma cleaner (Diener electronics GmbH) for 30 min using a H_2/He gas mixture at 0.6 mbar of H_2 and 100% power. Samples were transferred to the STXM experiment station with minimum exposure to ambient air.

3. Results and discussion

3.1. STXM imaging

STXM images obtained at 711 eV from the PMT and the CT are shown in figure 1 for the uncoated, and dopamine coated mnpc's. The thickness of these coatings was calculated in a separated study on core shell nanoparticles to be ca. 1nm [8]. In general large clusters of several hundred nms in size and about 150 nm in thickness was observed in all the cases. Since these samples were drop-cast and air-dried, the analysed clusters are not perfectly dispersed. Several spectra were obtained through a line scan to ensure the chemical homogeneity of the sample before choosing representative spectra.

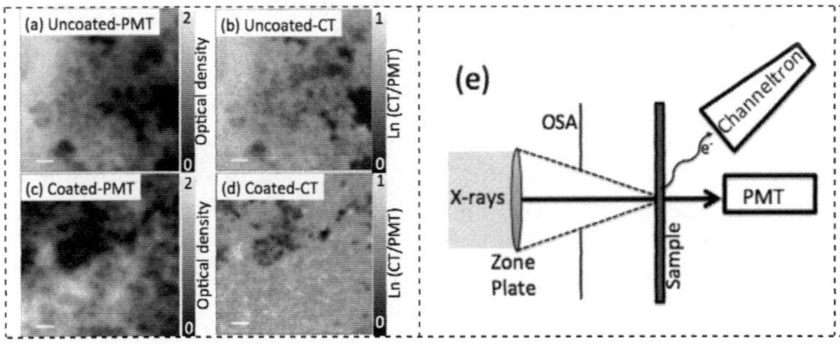

Figure 1 : (a)-(d) STXM optical density images at 711 eV for uncoated and coated mnpc's using PMT and CT. (Scale bar: 500nms) (e) Schematics of the STXM experimental set-up

3.2. Fe L_3 edge Absorption spectra

Figure 2a shows the Fe L_{23} XAS spectra of uncoated and dopamine coated Fe_xO_y mnpc's. Two regions corresponding to the $L_3(2p_{3/2})$ peak around 710 eV and $L_2(2p_{1/2})$ peak around 722 eV can be observed. The ligand field splitting and other atomic interactions form the basis of the fine structure within these regions [9]. We have focussed our discussion on the L_3 region of the spectra as there are significant differences in the PMT and CT signals in this region of the uncoated and dopamine coated mnpc's. The **A**, **B** and **C** energy positions (Figure 2a,b) marked in the L_3 region correspond to site occupancies of Fe cations in the spinel structure. **A** corresponds to Fe^{2+} species at octahedral sites (Fe^{2+} O_h), **B**

IOP Conf. Series: Journal of Physics: Conf. Series **849** (2017) 012014 doi:10.1088/1742-6596/849/1/012014

corresponds to Fe^{3+} species at tetrahedral sites (Fe^{3+} T_d) and **C** corresponds to Fe^{3+} species at octahedral sites (Fe^{3+} O_h). Relative intensity differences can be observed around the feature **A**. The features **B** and **C** are not clearly resolved. In order to obtain a relative quantification of the oxidation state of iron in the mnpc samples we have used the linear combination of simulated Fe^{2+} and Fe^{3+} spectra using the CTM4XAS software [10].

A linear combination of the simulated spectra of only two species, $Fe^{2+}(O_h)$ and Fe^{3+} (O_h), were used to obtain a consistent fit. The simulations parameters were used according to ref. [9, 10]. A representative fitted spectrum is shown in figure 2b. The integrated intensity of each component is used to estimate the Fe^{2+}/Fe^{3+} ratio. Figure 2c shows the Fe^{2+}/Fe^{3+} ratios as obtained from the fits for all the spectra. These include the measurements on a reference maghemite sample (Fe_2O_3) and from a wet-cell sample wherein the dopamine-coated mnpc's were dispersed in water. Firstly, it is evident that although the reference maghemite (Fe_2O_3) sample, measured in TEY (Total Electron Yield) mode, should be purely Fe^{3+}, it was necessary to introduce the Fe^{2+} component to obtain a reasonable fit. This could be due to reduced sites below the top surface of the reference sample. As expected, the Fe^{3+} phase is the dominant phase in the reference sample. Among the mnpc's, the uncoated clusters show the maximum amount of Fe^{2+} species (Fe^{2+}/Fe^{3+} ~0.9). The bulk and the surface content of the two phases are not very different, with Fe^{3+} content slightly higher in the surface. This observation is expected since the surface is inevitably oxidised. Reduction by H_2 plasma results in an overall decrease in the Fe^{3+} content at the surface as well as in the bulk as illustrated by strong differences in the L_3 region. Figure 2b shows exemplarily the spectra of dopamine coated mnpc's treated with H_2 plasma for 30 minutes.

In the case of dopamine coated mnpc's, it is clear that the bulk content of Fe^{2+} is lower than the surface content of Fe^{2+} indicating a more oxidized bulk compared to the surface. Such behaviour is consistent with initial oxidation behaviour of magnetite as observed by McCarty et al. [11]. In general the transformation of iron oxide starting from magnetite (Fe_3O_4) to maghemite (γ-Fe_2O_3) can be described by the progressive oxidation of octahedral [Fe^{2+}] to [Fe^{3+}] by losing some structural iron and creating cation vacancies [12] according to the following substitutional series $[Fe^{3+}]\{Fe^{3+}_{1+2\delta}\,Fe^{2+}_{1-3\delta}\,{}_{\delta}\}O_4$ where [] indicates tetrahedral sites, {} indicates octahedral sites, represents cation vacancies in the spinel structure and $0 \leq \delta \leq 1/3$ [13]. This implies

$$\frac{Fe^{2+}_{(Oh)}}{Fe^{3+}_{(Oh)} + Fe^{3+}_{(Td)}} = \frac{1-3\delta}{2+2\delta} \tag{1}$$

δ values calculated for different samples are presented in figure 2c along with the theoretical values for pure magnetite and pure maghemite. Values of δ greater than 1/3 and less than 0 indicate the presence of mixed phases not consistent with either magnetite or maghemite phases. The dopamine coated dry samples clearly show a magnetite like composition. The dopamine coated wet-cell sample shows the most maghemite like composition that is very different from dry dopamine mnpc sample. This could be due to the fact that the wet cell sample was significantly older than the dry sample. This could have allowed for sufficient oxidation time for magnetite like mnpc's. The uncoated sample's composition indicates the abundance of Fe^{2+} sites originating from neither magnetite nor maghemite phases but most probably from mixed iron oxide phases. It has to be noted that we have ignored the presence of metallic Fe^0 species in our fitting routine. The energy position of Fe^{2+} is closer to Fe^0 species and some of the contributions of Fe^{2+} species account also for the Fe^0 species, especially in the case of broad L_3 peaks obtained for uncoated mnpcs.

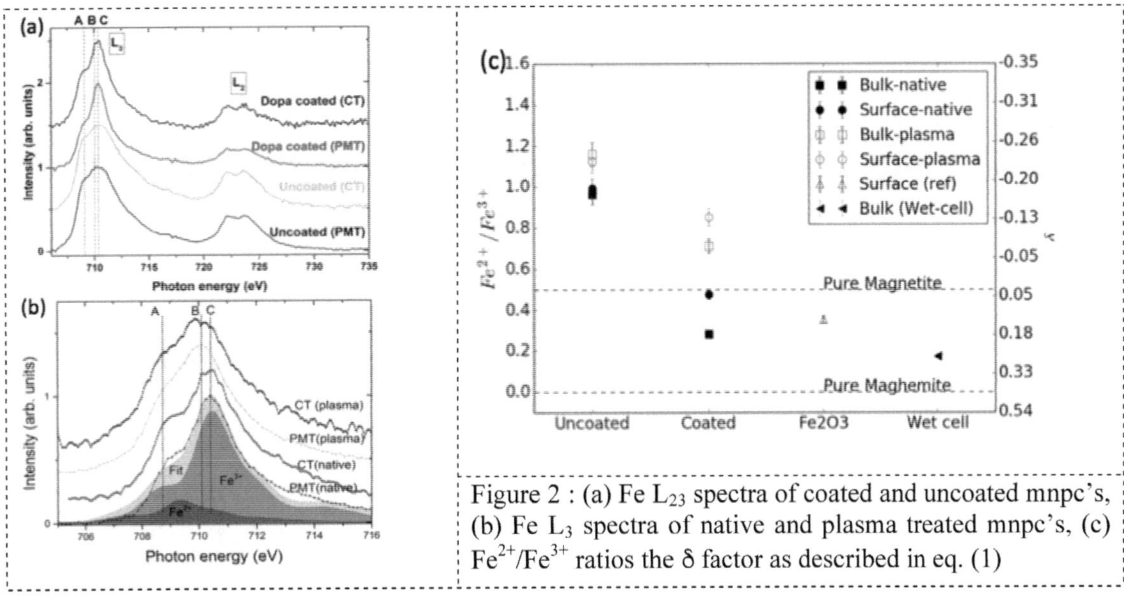

Figure 2 : (a) Fe L_{23} spectra of coated and uncoated mnpc's, (b) Fe L_3 spectra of native and plasma treated mnpc's, (c) Fe^{2+}/Fe^{3+} ratios the δ factor as described in eq. (1)

4. Conclusion

We have shown that using bulk and surface sensitive measurements in STXM, it is possible to disentangle subtle changes in oxidation state of iron oxide mnpc's. Furthermore the applied characterization strategy of iron oxide based nanoparticles combined with computational simulations of Fe L_3 spectra allows distinguishing differently coated iron oxide np's in terms of iron oxide nature, especially of magnetite or maghemite characteristics. Dopamine coatings were found to introduce oxidation state changes leading to increase of Fe^{3+} species. Also mixed Fe_xO_y species were identified by simultaneous bulk and surface sensitive STXM/XAS measurements as demonstrated for the "naked" Fe_xO_y core particles.

References

[1] Y. Sahoo et al., 2005, *J. Phys. Chem. B* **109** 3879; W. Ryoo et al. 2003, *Ind. Eng. Chem. Res.* **42** 6348; B.-S Kim et al.2005, *Nano Lett.* **5** 1987; W. Zheng et al. 2005, *J. Magn. Magn. Mater.* **288** 403; G. Fritz et al. 2002, *Langmuir* **18** 6381; D. Reith et al. 2002, *J. Chem. Phys.* **116** 9100.
[2] G. Vaidyanathan et al. 2007, *J. Magn. Magn. Mater.* **313** 293.
[3] J. Park et al. 2005, *Angew. Chem.* **117** 2932; J. Park et al. 2004, *Nat. Mater.* **3** 891; B. H. Kim et al. 2011, *J. Am. Chem. Soc.* **133** 12624; L. M. Bronstein et al. 2011, *Langmuir* **27** 3044.
[4] L. Signorini et al. 2003, *Phys. Rev. B* **68** 195423.
[5] B. Watts et al. 2010, *Macromol. Rapid Commun.* **31** 1706.
[6] S. Gota et al. 2000, *Phys. Rev. B* **62** 4187.
[7] J. Raabe et al. 2008, *Rev. Sci. Instrum.***79** 113704.
[8] H. Kalbe et al. 2016, *J. Electron Spectrosc. Relat. Phenom.* **212** 34.
[9] P. Kuiper et al. 1997, *J. Electron Spectrosc. Relat. Phenom.* **86** 107; R.A. Pattrick et al. 2002, *Eur. J. Mineral.* **14** 1095;V.N. Antonov et al. 2003, *Phys. Rev. B* **67** 024417.
[10] E. Stavitski et al. 2010, *Micron* **41** 687.
[11] T. Fujii T et al. 1999 *Phys. Rev.* B **59** 3195.
[12] K. F. McCarty et al. 2014 *J. Phys. Chem. C* 2014 **118** 19768.
[13] G. P. Santana et al. 2001, *Cio Solo* **25** 33;N. Mahmed et al. 2014, *J. Magn. Magn. Mater* **353** 15.

Chemical state mapping of heterogeneous reduction of iron ore sinter

M. Kimura[1,2], Y. Takeichi[1,2], R. Murao[3], I. Obayashi[4], Y. Hiraoka[4], Y. Liu[5]
[1]Photon Factory, Institute of Materials Structure Science, High Energy Accelerator Research Organization, Tsukuba, Ibaraki 305-0801, Japan
[2]Department of Materials Structure Science, School of High Energy Accelerator Science, Graduate University for Advanced Studies (SOKENDAI), Tsukuba, Ibaraki 305-0801, Japan
[3]Advanced Technology Research Laboratories, Nippon Steel & Sumitomo Metal Co., Chiba 293-8511, Japan
[4]Advanced Institute for Materials Research (AIMR), Tohoku University, 2-1-1 Katahira, Aoba-ku, Sendai 980-8577, Japan
[5]Stanford Synchrotron Radiation Lightsource, SLAC, 2575 Sand Hill Rd., Menlo Park, CA 94025, USA
E-mail: masao.kimura@kek.jp

Abstract. Iron ore sinter constitutes the major component of the iron-bearing burden in blast furnaces, and its reduction mechanism is one of the keys to improving the productivity of ironmaking. Iron ore sinter is composed of multiple iron oxide phases and calcium ferrites (CFs), and their heterogeneous reduction was investigated in terms of changes in iron chemical state: Fe^{III}, Fe^{II}, and Fe^{0} were examined macroscopically by 2D X-ray absorption and microscopically by 3D transmission X-ray microscopy (TXM). It was shown that the reduction starts at iron oxide grains rather than at calcium ferrite (CF) grains, especially those located near micropores. The heterogeneous reduction causes crack formation and deteriorates the mechanical strength of the sinter. These results help us to understand the fundamental aspects of heterogeneous reduction schemes in iron ore sinter.

1. Introduction

Iron ore sinter constitutes the major component of the iron-bearing burden in blast furnaces in most Asian-Pacific countries. Iron ore sinter is formed by liquid sintering at $T > 1500$ K, and is composed of iron oxide phases (mainly α-Fe_2O_3 and partially Fe_3O_4) and calcium ferrites (CFs) [1,2]. Iron oxide grains with a diameter of several tens to hundreds of μm are bonded by CFs such as $CaFe_2O_4$, $Ca_2Fe_2O_5$, and SFCA: $Ca_2(Ca,Fe,Al)_6(Fe,Al,Si)_6O_{20}$, which contain many pores with a diameter from μm to mm. In industrial processes, it is necessary to optimize the speed of reduction and minimize the degradation of mechanical strength during reduction, because the iron ore sinter can pile up as high as 30–40 m in the blast furnace. Therefore, it is important to understand where the reduction starts and how it progresses. We attempted to reveal the reduction process in terms of changes in the iron chemical state (Fe^{III}, Fe^{II}, and Fe^{0}) and crack formation in the sinter by X-ray absorption and TXM on both a macro and microscopic scale.
The reduction reaction rates for Ca-Fe-O phases possibly found in iron ore sinter were measured by in situ QXAFS[3,4], and the results are summarized in Figure 1. Reduction of the CF phases takes longer

IOP Conf. Series: Journal of Physics: Conf. Series **849** (2017) 012015 doi:10.1088/1742-6596/849/1/012015

than that of the Fe-O system, because the decomposition of Ca-Fe-O into Ca-O and Fe-O phases precedes the reduction[3,4].

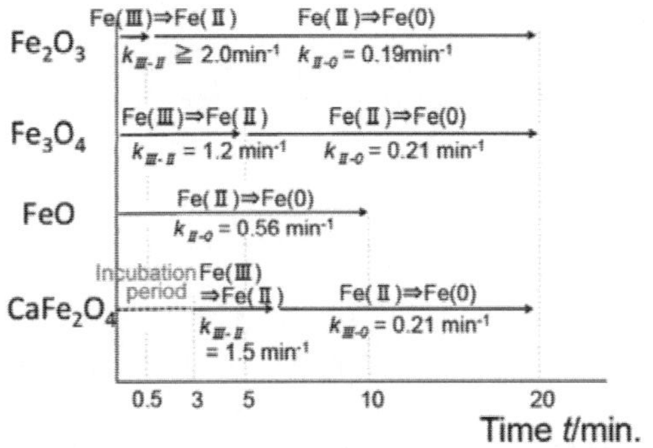

Figure 1. Reduction reaction rates for Ca-Fe-O systems [3,4].

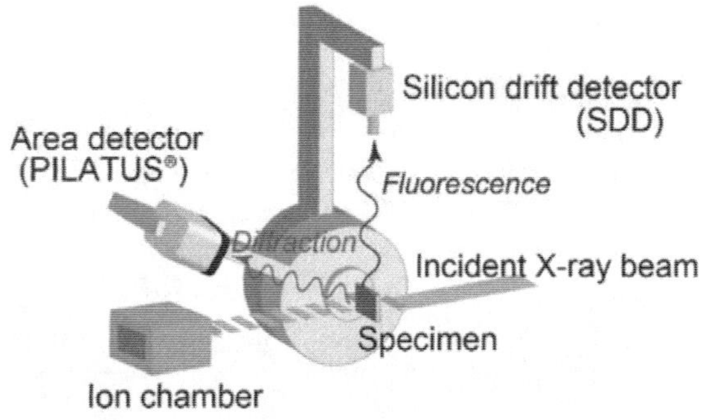

Figure 2. Schematic diagram of the experimental setup for chemical mapping at BL-15A1 in PF.

2. Experimental

Iron oxide ores were mixed with limestone flux and coke breeze, and heated by the combustion of coke breeze, resulting in heating up to temperatures of 1450–1600 K (above the eutectic temperature of $CaO–Fe_2O_3$: 1478 K). The sinter was heated up to T = 1173 K in a reducing gas atmosphere of $CO:CO_2$ = 0.5:0.5 ($\times 10^{-1}$ MPa).

Macroscopic chemical mapping in 2D was carried out based on X-ray absorption measurements obtained by scanning the specimen (Figure 2). The sinter was embedded in resin and sliced into specimens with a thickness of 20–30 μm. X-ray absorption near-edge structure (XANES) spectra were measured using ion chambers before and after absorption by the specimen. The XANES spectrum at each point is assumed to be expressed as the linear combination of those of $Fe^{II}O$ and $Fe^{III}{}_2O_3$ and the ratios were determined by least squares fitting. X-ray fluorescence was also measured using a silicon drift detector (SDD) in order to determine the chemical composition of iron and calcium. Experiments were carried out at the synchrotron undulator beamline BL-15A1 [5] of the Photon Factory, IMSS, KEK in Japan. The typical time for chemical mapping for 30 photon energies is a few hours for a 1-mm square in "on the fly" mode.

Microscopic chemical mapping in 3D was carried out by X-ray microscopy (TXM) at the synchrotron wiggler beamline 6-2C of Stanford Synchrotron Radiation Lightsource in the USA, where 3D chemical mapping can be achieved with the capability of imaging a 32-micron field-of-view at 30-nm resolution using a capillary condenser and a zone plate [6]. A cylindrical pillar measuring approximately 20 um (dia.) × 30 μm (length) was cut from the iron ore sinter by FIB.

Crack formation was investigated using X-CT with an in-house X-ray source. A rectangular specimen of 2×2×10 mm^3 was cut out from the iron ore sinter. A CCD camera with a scintillator was used as a detector. The transmission image was measured in steps of 1° or 2° in a 360° rotation with an energy of 70 keV. The minimum spatial resolution was as small as 0.7 μm.

3. Results and discussion

Figure 3 shows the results of iron chemical state mapping in 2D and compositional mapping of iron ore sinter before and after reduction. Before the reduction, most areas show a chemical state of Fe^{III} (red) and Fe^{III}+Fe^{II} (pink) corresponding to α-Fe_2O_3 and Fe_3O_4, respectively. Some areas with a high calcium concentration show a chemical state between Fe^{III} and Fe^{II} (white) corresponding to CFs. These results are consistent with the phase fraction obtained by X-ray diffraction: α-Fe_2O_3 (approx. 55 vol.%), Fe_3O_4 (approx. 20 vol.%), $CaFe_2O_4$ (approx. 20 vol.%) and $Ca_2Fe_2O_5$ (approx. 5 vol.%). After the reduction, most areas shifted to a reduced chemical state, but this change is heterogeneous rather than homogeneous. Areas with a low calcium concentration changed to Fe^{III} (blue; typical areas are shown by blue arrows in Figure 3), while areas with a high calcium concentration remained as Fe^{III} (red) or between Fe^{III} and Fe^{II} (white; shown by red arrows in Figure 3). This is also the case in areas near large pores where the flow of reductive gas is high enough for easy reduction. Thus the heterogeneous reduction is caused not by the heterogeneous microstructures but by the different reaction rates intrinsic to iron oxides and CFs, which can be explained by the different reduction rates measured for single-phase powder specimens (Figure 1).

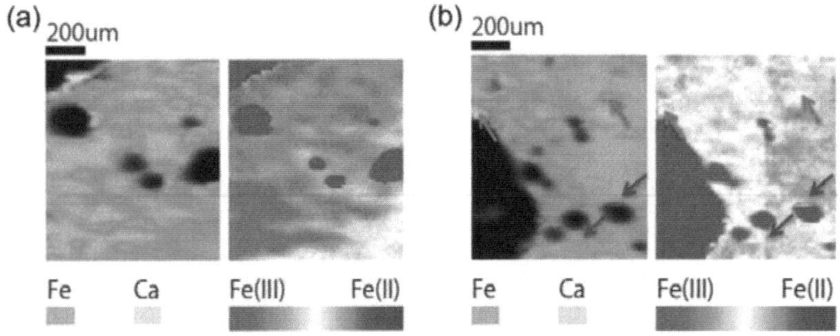

Figure 3. Results of iron chemical state mapping (right) and compositional mapping (left) of iron ore sinter (a) before and (b) after reduction.

This kind of heterogeneous reduction of bulk sinter is expected to easily cause microcracks, because the reduction leads to a large volume decrease[7]. Figure 4 shows (a) an internal slice taken from the volume X-CT dataset of the reduced sinter, and (b)-(e) extracted images from (a): (b) large pores, (c) cracks, (d) Fe-O oxides, and (e) CFs. Extraction of each component was carried out by setting threshold values for three regions: pores/cracks, Fe-O oxides, and CFs, and image processing considering the contrast difference at the boundaries. Crack formation was found to be highly correlated with the areas containing CFs as shown by the cross marks in (e). This suggests that possible sites for cracking are the interface between the reduction-advanced and the reduction-delayed area and/or the interface between the reduction-advanced area and the pre-existing pore area. The heterogeneous progress of reduction may trigger cracking; further investigation of the correlation between the heterogeneity and crack initiation is in progress.

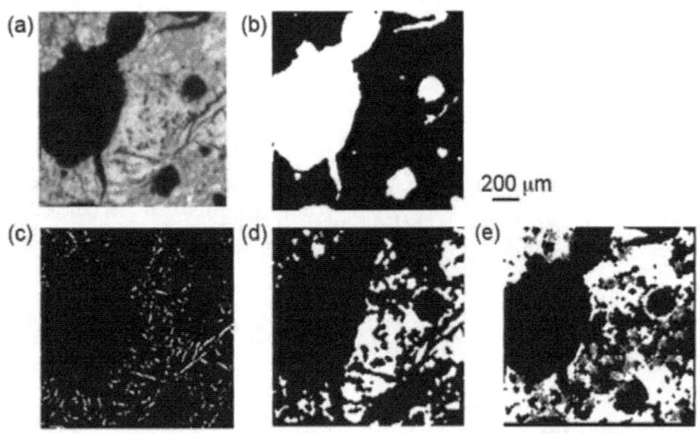

Figure 4. (a) A slice image from the volume X-CT dataset of the reduced sinter, and (b)-(e) extracted images from (a): (b) large pores, (c) cracks, (d) Fe-O oxides, and (e) CFs. Cross marks in (e) indicate areas where the CF phases show a high correlation

Figure 5 shows the microscopic results of iron chemical state mapping in 3D obtained by TXM. Chemical states in areas indicated by pink broken lines corresponding to Fe-O oxides with high density and large iron concentrations, showed larger chemical shifts (red), suggesting the progress of reduction. In comparison, areas indicated by white and blue lines have small densities and edge jumps, corresponding to fewer iron phases, i.e., CFs. These areas showed smaller chemical shifts (orange or green), showing less progress in reduction. These results showed that the heterogeneous reduction progresses on a scale of μm or less and that this is caused by the different reaction rates intrinsic to iron oxides and CFs.

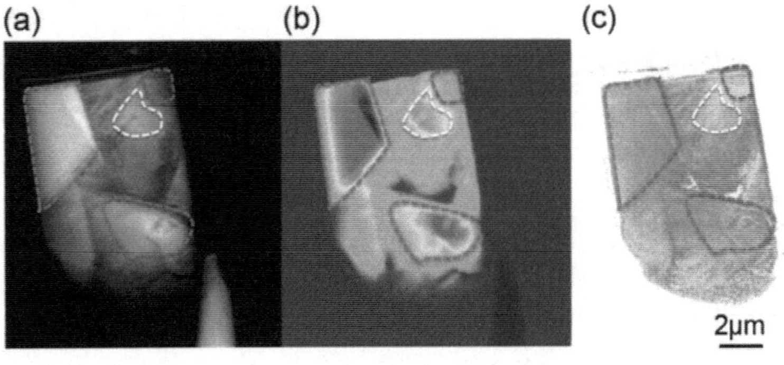

Figure 5. Microscopic results of iron chemical state mapping in 3D obtained by TXM: (a) Transmission absorption, (b) edge jump at the Fe K-edge, and (c) edge shift. In (b) and (c), the amounts are shown by color: blue (small) – green – red (large).

4. Conclusions

Heterogeneous reduction reaction of iron ore sinter was investigated macroscopically by XANES in 2D and microscopically by TXM in 3D. It was shown that reduction starts at iron oxide grains rather than at calcium ferrite (CF) grains, because the reduction rate of CFs is smaller than that of Fe-O oxides. This heterogeneous reduction starts on a scale of μm or less, and progresses on a scale of a few tens of μm. Crack formation is shown to be highly correlated with this heterogeneity, and will deteriorate the mechanical strength of the sinter. Chemical state mapping both microscopically and macroscopically helps us to understand the fundamental aspects of heterogeneous reduction, which is essential for optimizing microstructures of iron ore sinter in order to achieve a more highly efficient reduction process in ironmaking.

Acknowledgements

Part of this work carried out by M. Kimura, Y. Takeichi, I. Obayashi, and H. Hiraoka was supported by the Cross-Ministerial Strategic Innovation Promotion Program (SIP, units D66) operated by the Cabinet Office of Japan. Experiments at the PF were performed under the approval of the PF Program Advisory Committee (Proposal Nos. 2014G707, 2015S2-002 and 2016S2-001). Experiments at the SSRL were carried out under the approval of SSRL (LOI and Proposal Nos. 4468, 2016). The cooperative research between Nippon Steel & Sumitomo Metal Co. and KEK is highly appreciated. We thank Ms. M. Ito and Mr. T. Ishii for experimental support.

References

[1] Ahsan S N, Mukherjee T and J.A.Whiteman 1983 *Ironmaking and Steelmaking* **10** 54.
[2] Mumme W G, Clout J M F and Gable R W 1998 *Neues Jahrbuch für Mineralogie - Abhandlungen* **173** 93.
[3] Kimura M, Uemura K, Nagai T, Niwa Y, Inada Y and Nomura M 2009 *Journal of Physics: Conference Series* **190** 012163.
[4] Kimura M, Uemura Y, Takayama T, Murao R, Asakura K and Nomura M 2013 *Journal of Physics: Conference Series* **430** 012074.
[5] Igarashi N, Shimizu N, Koyama A, Mori T, Ohta H, Niwa Y, Nitani H, Abe H, Nomura M and Shioya T 2013 *Journal of Physics: Conference Series* **425** 072016.
[6] Liu Y, Meirer F, Wang J, Requena G, Williams P, Nelson J, Mehta A, Andrews J C and Pianetta P 2012 *Analytical and bioanalytical chemistry* **404** 1297.
[7] Turkdogan E, Olsson R and Vinters J 1971 *Metallurgical and Materials Transactions B* **2** 3189.

IOP Conf. Series: Journal of Physics: Conf. Series **849** (2017) 012016 doi:10.1088/1742-6596/849/1/012016

In-operando studies of Ag-TCNQ nanocrystals using Raman and soft x-ray microspectroscopy

Benedikt Rösner,[1,3*] Ute Schmidt,[2] Rainer H. Fink[1]

[1] Department Chemie und Pharmazie & ICMM, Friedrich-Alexander-Universität Erlangen-Nürnberg (FAU), Egerlandstr. 3, 91058 Erlangen, Germany.

[2] WITec GmbH, Lise-Meitner-Str. 6, 89081 Ulm, Germany

[3] Laboratory for Micro- and Nanotechnology, Paul Scherrer Institut, 5232 Villigen PSI, Switzerland (current affiliation)

*benedikt.roesner@psi.ch

Abstract. We characterize individual Ag-TCNQ nanocrystals during switching their resistivity state *in operando*. Raman and soft X-ray absorption microspectroscopy are employed to disclose the electronic state of the organic component in dependency of applied voltage. Whereas Raman microspectroscopy offers qualitative insight into the conversion of negatively charged TCNQ molecules to their neutral counterpart, quantification of the neutral fraction can be achieved using X-ray absorption spectroscopy. These results allow a detailed investigation of resistivity switching in electrically bistable Ag-TCNQ nanocrystals.

1. Introduction

Electrically bistable Ag-TCNQ (TCNQ = 7,7',8,8'-tetracyanoquinodimethane) nanowires have been under intense investigation for more than a decade with respect to their promising capabilities to undergo reversible transitions between two resistance states [1]. Although reverse charge transfer has been proposed being the reason for changes of the conductivity of metal-organic TCNQ-based salts [2], unambiguous proof of this concept has not been demonstrated in individual nanometre-sized crystals so far.

Microspectroscopic methods represent powerful tools for characterizing electronic properties of metal-organic and organic materials. Herein, we report on the potential of photon-based methods to conduct spectroscopy on these functional electronic materials in high spatial resolution down to the nanoscale [3]. We employed confocal Raman microscopy and scanning transmission X-ray microscopy (STXM) to keep track of the electronic changes, which occur in individual Ag-TCNQ crystals upon switching their resistivity state.

2. Experimental

Ag-TCNQ crystallites have been synthesized in wet-chemical environment as described elsewhere [4]. Commercial 100 nm thick silicon nitride membranes (Silson Ltd.) or silicon wafers (with a top 100 nm silicon oxide layer) were used as substrates. Individual wires suited for *in operando* experiments have been identified under the light microscope and contacted by thermal evaporation of 30 nm thick gold contacts.

Raman spectromicroscopy was conducted with a WITec alpha 500 R confocal Raman microscope. After mounting the contacted Ag-TCNQ wires on the microscope, the two gold electrodes were contacted with two gold-coated needles using micromanipulators. A constant negative voltage was applied to the Ag-TCNQ wire. The whole sample was then scanned together with the micromanipulator to ensure a stable electric contact. Arrays of Raman spectra were recorded while the current flow was monitored by a Keithley 4200-SCS parameter analyser. For different sets of spectra, the voltage was increased in several steps. *In operando* STXM experiments were performed using contacted samples mounted on a dedicated holder for *in-situ* experiments in the STXM at the PolLux beamline [3] at the Swiss Light Source (Figure 1) in a similar way. The X-ray beam was focused using a Fresnel zone plate with 35 nm outermost zone width and an order-sorting aperture with 70 µm diameter, and was detected using a photomultiplier tube.

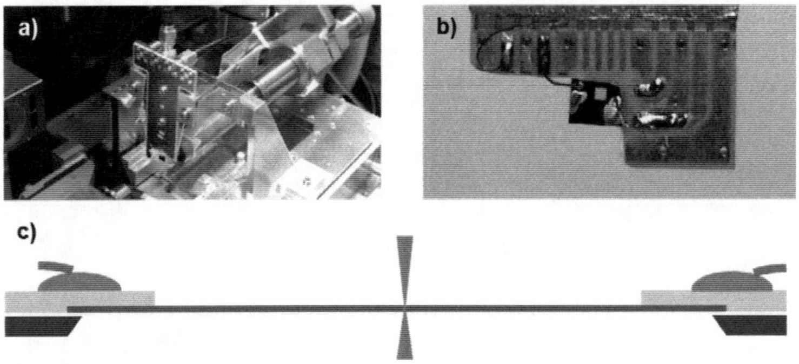

Figure 1: Experimental for *in-operando* STXM experiments. a) Photograph of the contacted and mounted sample inside the PolLux-STXM, and b) photograph of the sample holder with the membrane and the contacted gold electrodes. The lower left part of the holder has been cut off to provide enough space for the order-sorting aperture as the working distance becomes extremely small at the carbon edge. c) Schematic illustration of the contacted Ag-TCNQ wire on the Si_3N_4 membrane being probed with X-rays.

3. Results and Discussion

In order to confirm the presence of electronic switching in Ag-TCNQ nanocrystals, we conducted a survey study probing several dozen wires with diameters between 30 nm and 380 nm inside a transmission electron microscope [5]. Quite important, the Ag-TCNQ crystals exhibit similar electrical switching characteristics over the whole diameter range, enabling to use wires with diameters between 100 nm and 400 nm diameter as model material for microspectroscopic *in-situ* experiments.

3.1. Raman spectroscopy

After the proposed reverse charge transfer [2] from the negatively charged TCNQ anion to the silver cations, the material provides a significantly different vibrational structure which can be probed using Raman spectromicroscopy. Experimental spectra of TCNQ, Na-TCNQ and Cu-TCNQ allow for distinguishing the neutral and the negatively charged TCNQ molecule as the latter shows a clear shift of the vibration band at 1457 rel. cm^{-1} to 1388 rel. cm^{-1}, due to antibonding effects at a specific C=C double bond [6, 7].

Whereas Raman spectra recorded without voltage and with -1 V applied show clearly the clean spectrum of negatively charged TCNQ. The vibrational peak of the neutral molecule at 1457 rel. cm^{-1} starts to rise when the applied voltage is driven to negative values (Figure 2). This change proves qualitatively that reversed charge transfer takes place upon switching the resistivity state of an individual Ag-TCNQ wire. The pronounced evolution of the peak at -15 V furthermore demonstrates that a remarkable part of the material is converted. However, an absolute quantification of affected molecules based on Raman spectroscopy is rather than straightforward due to dissimilar polarizability tensors of neutral and negatively charged molecules [8].

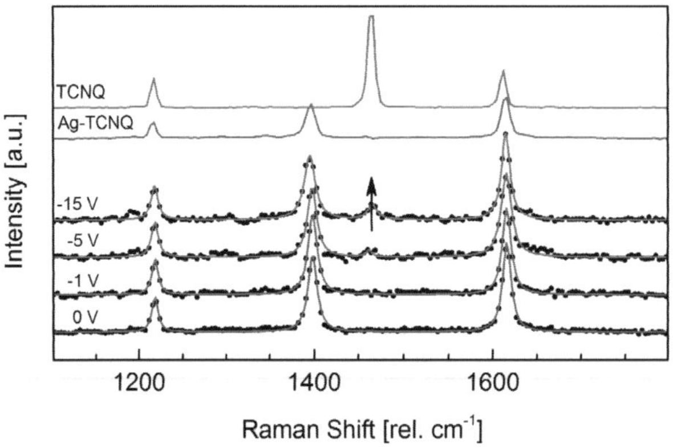

Figure 2: Raman spectra of an individual Ag-TCNQ wire with different voltages applied. Reference spectra are shown for comparison.

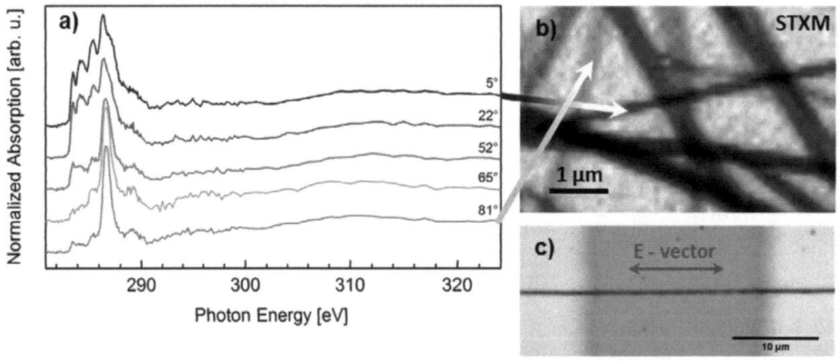

Figure 3: Linear dichroism in Ag-TCNQ. a) NEXAFS spectra of Ag-TCNQ oriented differently with respect to the polarization vector of the X-rays. b) Scanning transmission X-ray micrograph showing different absorption contrast with respect to the orientation of Ag-TCNQ crystallites. c) Defined geometry for *in operando* STXM experiments.

3.2. Scanning transmission X-ray microspectroscopy

Near-edge X-ray absorption fine structure (NEXAFS) spectroscopy provides a complementary tool for disclosing the electronic structure of TCNQ. Whereas Raman spectroscopy shows a clear shift of one vibrational band, the NEXAFS spectra of TCNQ and $TCNQ^-$ both show several overlapping C 1s to π^* absorption peaks between 283 and 289 eV. However, their intensities are much lower for the negatively charged species where an additional electron is already present in the unoccupied states. In addition, the peaks at 288.0 eV and 289.6 eV are distinctive for neutral TCNQ [9, 10].

For quantitative comparability, the absorption probability of differently oriented wires in NEXAFS spectroscopy has to be taken into account. Due to their very ordered crystal structure [11], Ag-TCNQ needles show a pronounced linear dichroism (Figure 3 a,b). We thus conducted all experiments with the E-vector of the X-rays oriented parallel to the crystals (see Figure 3 c) to ensure maximum absorption in the π^* region.

Applying various negative voltage steps, the contacted Ag-TCNQ is switched and its X-ray absorption spectrum is recorded *in operando*. The fraction of neutral TCNQ can then be derived expressing the experimental spectra as linear combination of the reference spectra of TCNQ and Ag-TCNQ (Figure 4). The experiments show remarkable fractions of the neutral TCNQ molecule up to 22.3% when its resistivity is switched, enhancing the conductivity of the material remarkably [5]. More important, the switched state seems to be maintained when the voltage is switched of for several minutes (red square in Figure 4 b).

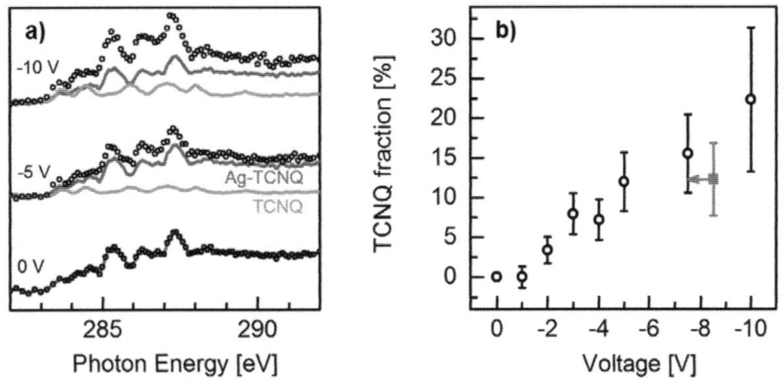

Figure 4: a) NEXAFS spectra of an individual Ag-TCNQ nanowire with 0, -5 and -10 V applied. The spectra can be expressed by a linear combination of the reference spectra of TCNQ (orange) and Ag-TCNQ (green), giving the percentage of neutral TCNQ contained in the material (b). The neutral TCNQ fraction stays present switching off the voltage for 5 min after measuring the spectrum at -7.5 V (red square).

4. Conclusions

Employing microspectroscopic probes in high lateral resolution offers exciting possibilities to investigate electronic organic building blocks *in operando*. We are thus able to confirm and to quantify the effect of reverse charge transfer in individual Ag-TCNQ nanocrystals upon resistivity switching. These experiments enable important insights into charge-induced processes on the nanoscale, which is crucial in ongoing miniaturization of electronic devices.

References

[1] Fan Z, Mo X-L, Lou C, Yao Y, Wang D, Chen G-R and Lu J G 2005 *IEEE Trans Nanotech* **4** 238-41
[2] Potember R S, Poehler T O and Cowan D O 1979 *Appl Phys Lett* **34** 405-7
[3] Raabe J, Tzetkov G, Flechsig U, Böge M, Jaggi A, Sarafimov B, Vernooij M G C, Huthwelker T, Ade H, Kilcoyne D, Tyliszczak T, Fink R H and Quitmann C 2008 *Rev Sci Instrum* **79** 113704
[4] Rösner B, Späth A and Fink R H 2013 *J Cryst Growth* **380** 34-8
[5] Ran K, Rösner B, Butz B, Fink R H and Spiecker E 2016 *Nanotechnology* **accepted**
[6] Khatkale M S and Devlin J P 1979 *J Chem Phys* **70** 1851-9
[7] Umbach T R, Fernández-Torrente I, Ruby M, Schulz F, Lotze C, Rurali R, Persson M, Pascual J I and Franke K J 2013 *New J Phys* **15** 083048
[8] Machado H J S and Hinchliffe A 1997 *Electron J Theor Chem* **2** 49-55
[9] Bässler M, Fink R H, Buchberger C, Väterlein P, Jung M and Umbach E 2000 *Langmuir* **16** 6674-81
[10] Koshino M, Kurata H and Isoda S 2004 *J Electr Spectr Rel Phen* **135** 191-200
[11] Shields L 1985 *J Chem Soc, Faraday Trans 2* **81** 1-9

X-ray attenuation properties of commonly employed solid oxide fuel cell materials

T M M Heenan, D J L Brett and P R Shearing*

Electrochemical Innovation Lab, Department of Chemical Engineering, UCL, London, UK, WC1E 7JE

*E-mail: p.shearing@ucl.ac.uk

Abstract. X-ray nano CT has been vastly applied to study the microstructure of solid oxide fuel cell (SOFC) electrodes. One widely accepted indicator of electrochemical performance is the triple phase boundary (TPB): a location where the three materials responsible for ionic, electronic and gas-phase reactant transport are in contact. X-ray absorption tomography has been used extensively in the characterisation of these TPBs, utilising the different attenuation properties of the constituent materials. Here we present a quantitative comparison of the attenuation properties for elements commonly employed in solid oxide fuel cell materials.

1. Introduction

Offering high efficiencies and fuel versatility, solid oxide fuel cells (SOFCs) offer great promise in the future of energy conversion technology. In their most simple form, a SOFC is comprised of two electrically conducting electrodes separated by an electrically insulating, ionically conducting electrolyte. Reduction and oxidation reactions occur at the cathode and anode electrodes respectively, with ion transport occurring through the electrolyte layer. It is widely accepted that the reaction sites in composite electrodes are located at the boundary of the three transport phases (electronic, ionic and reactant transport), the so-called triple phase boundaries (TPBs). Characterisation methods including X-ray nano CT and FIB-SEM slice and view allow for the calculation of TPB densities [1].

Commonly employed anode materials include Ni cermets, with a ceramic such as yttria-stabilised zirconia (YSZ) or gadolinium-doped ceria (GDC), whereas cathodes commonly contain a lanthanum strontium compound containing either manganite (LSM) or cobalt ferrite (LSFC) which can also be coupled with the YSZ or GDC ceramics [1 – 3].

Presented here, is an analysis of the key X-ray properties which dictate the reliability of image segmentation of SOFC materials via X-ray nano CT techniques. This includes consideration of the influence of dopant concentrations on the attenuation of composite metal-ceramic cermets and the location of the optimum imaging energies to achieve three-phase segmentation for such cermet materials.

2. Properties of solid oxide fuel cell materials: elements, compounds and cermets.

X-ray absorption tomography has been widely employed in the characterization of SOFC microstructures [1]. Triple phase boundaries (TPBs) within SOFC electrodes have become a widely accepted metric as an indicator of electrochemical performance. In order to locate such boundaries, multi-phase segmentation is required, which can only be achieved with knowledge of several attenuation characteristics of the constituent materials. One method to achieve such segmentation is X-ray absorption spectroscopy (XAS) imaging which has been employed to achieve solid-solid segmentation (typically metal and ceramic) [4, 5].

Nomenclature

L	Thickness	μm
I	X-ray intensity	counts
w	Mass Fraction	no-units
AC	Attenuation Coefficient	no-units
x	Dopant concentration	mol %
Greek Letters		
μ	Linear *AC*	μm^{-1}
μ_m	Mass *AC*	$\mu m^2 g^{-1}$
λ	Attenuation Length	μm
ρ_m	Mass Density	$g \mu m^{-3}$
η	Contrast Coefficient	no-units
Subscripts		
i	Material i	
k	Compound k	
0	Incident	

Three metrics of the ability to visualise the difference in attenuation of the constituent materials, and thus the ability for image segmentation, are the mass attenuation coefficient, μ_m, linear attenuation coefficient, μ, and the attenuation length, λ. These three properties can be derived from the exponential decay of X-ray intensity when passing through a material described by Equation 1 for material i [5].

Comprehensive lists of the mass attenuation coefficients and attenuation lengths are presented elsewhere [6, 7, 8] and the linear attenuation coefficient can be calculated from the mass attenuation coefficient using Equation 2.

The attenuation length is defined as the distance, L, perpendicular to the material surface through which the X-ray intensity, I, has decayed to 1/e that of the intensity at the material surface, I_o. Inserting these conditions into Equation 1 results in the relationship between attenuation length and the linear attenuation coefficient displayed in Equation 3.

$$\frac{I}{I_0} = \exp\left[-\mu_{m_i} L \rho_{m_i}\right] \quad (1)$$

$$\mu_i = \mu_{m_i} \rho_{m_i} \quad (2)$$

$$\lambda_i = \frac{1}{\mu_i} \quad (3)$$

Although elemental data for the attenuation properties of solid oxide fuel cells is readily available, many materials contain several elements in order to achieve the desired thermo-mechanical and electrochemical properties [9]. Examples being yttria – stabilised zirconia (YSZ), gadolinium – doped ceria (GDC), lanthanum strontium manganite (LSM) and lanthanum strontium cobalt ferrite (LSCF). Equation 4 can be employed in order to obtain the mass attenuation coefficients for compound materials via summation of the i constituent elements weighted according to their respective mass fractions (w_i) in compound k [7].

$$\mu_{m_k} = \sum_i w_i \left(\frac{\mu_i}{\rho_{m_i}}\right) \quad (4)$$

Common laboratory X-ray sources using copper, Cu, or chromium, Cr, for X-ray generation provide peak energies at 8.0 and 5.4 keV respectively. Hence, when inspecting the variation of attenuation properties using Equation 4 across the commonly employed range of dopant concentrations used during xYSZ fabrication ($3 \leq x \leq 8$ mol% yttria addition to zirconia) a fixed energy of 5.4 keV was chosen [10]. It was found that the compound mass attenuation coefficient for xYSZ deviated insignificant amounts through the entire dopant range; the xYSZ mass attenuation coefficient reduced by less than 0.5 $cm^2 g^{-1}$ from 3 to 8 mol %. The elemental contribution to attenuation properties was then inspected by considering the difference in mass attenuation coefficient between zirconium, the primary element within xYSZ, and Ni; $\mu_{m_{Zr}}$ is ca. 166% larger than $\mu_{m_{Ni}}$ at 5.4 keV. Concluding that at 5.4 keV the ability to obtain triple phase segmentation is greatly dependent upon the elemental contribution whereas doping level has negligible effect for xYSZ.

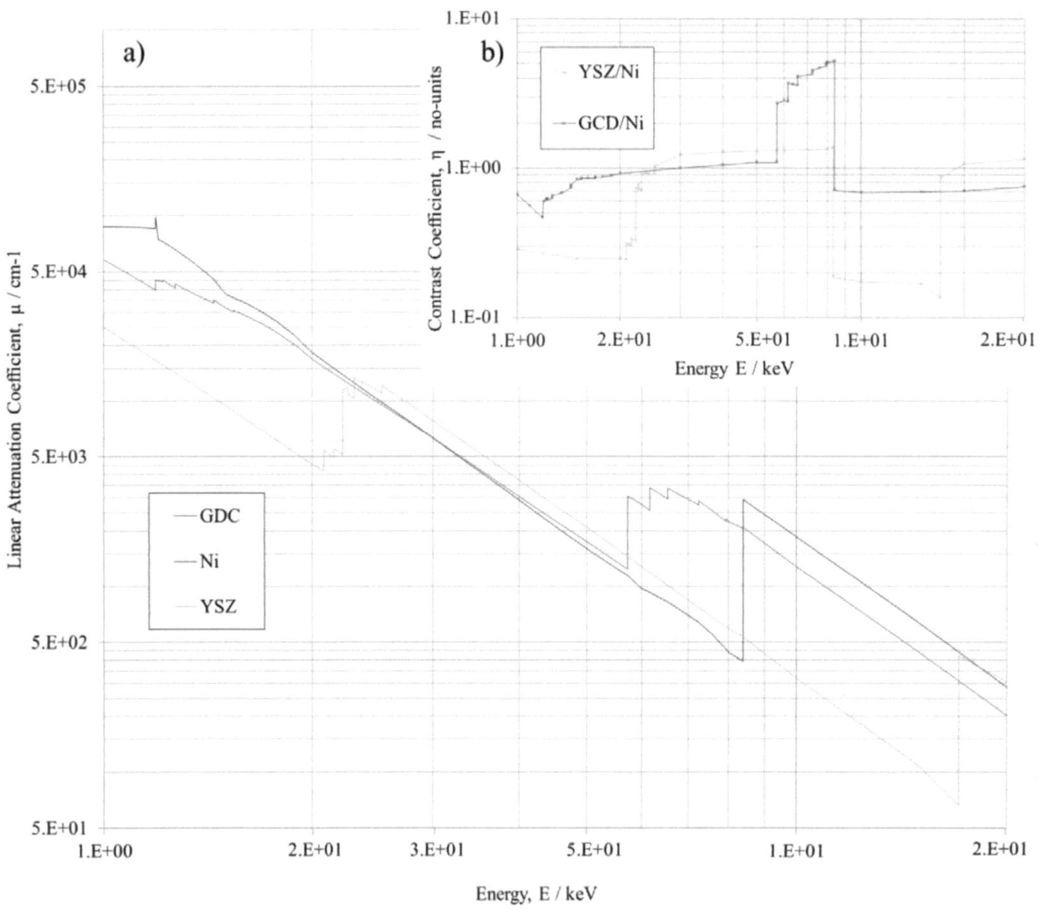

Figure 1. Locating the optimal segmentation energies between 1 and 20 keV using XAS techniques for Ni-8YSZ and Ni-3GDC electrodes: a) the linear attenuation coefficients for commonly employed solid oxide fuel cell materials: pure nickel, zirconia stabilised with an addition of 8 mol % yttrium oxide and ceria doped with 3% gadolinium, and b) the contrast quality through the same energy range quantified by the contrast coefficients (the ratio of linear attenuation coefficients) for the cermet materials.

Monochromatic X-ray tomography is conducted at a fixed energy chosen specific to the experiment in order to optimise segmentation of the microstructure. Therefore, with use of Equation 4 and the elemental data for nickel, yttrium, zirconium, gadolinium, cerium and oxygen from [7], Figure 1 a. was produced displaying the variation of linear attenuation coefficient for Ni, 8-YSZ and 3-GDC with X-ray energies between 2 and 30 keV. The linear attenuation coefficients for 8YSZ and 3GDC were then compared to the linear attenuation coefficient for Ni using Equation 5, by defining a variable which quantifies contrast: the contrast coefficient, η.

$$\eta = \frac{\mu_{Ceramic}}{\mu_{Ni}} \quad (5)$$

Segmentation of the ceramic and metal is optimal at contrast coefficients most distant from unity i.e. optimal segmentation is achieved when $\eta \gg 1 \; or \; \eta \ll 1$. The contrast coefficients for both Ni-8YSZ and Ni-3GDC are displayed in Figure 1 b. From the figure it is evident that, given a choice of x-ray energies such as at a synchrotron, XAS conducted at energies just below the Ni K-edge will provide greatest segmentation possibility for the Ni-3GDC cermet while XAS conducted just above the Ni K-edge will produce the best results for the Ni-8YSZ cermet, 8 and 9 keV respectively. It is also seen that, due to the highly attenuating properties of 3GDC below the Ni K-edge, the contrast coefficient for Ni-3GDC is potentially far greater than Ni-8YSZ if the correct energy is chosen. For the commonly used laboratory X-ray sources Cr and Cu the contrast coefficients for Ni-3GDC and Ni-8YSZ is quantified and presented in Table 1.

Table 1. Contrast coefficients for two commonly employed lab-based X-ray sources Cr and Cu for two cermet materials Ni-3GDC and Ni-8YSZ.

X-ray source	Peak Energy / keV	$\eta_{Ni-3GDC}$ / no-units	$\eta_{Ni-8YSZ}$/ no-units
Chromium, Cr	5.4	1.09	1.31
Copper, Cu	8.0	5.08	1.34

3. Conclusion

In order to achieve TPB data for common SOFC materials from X-ray techniques, the different attenuation properties of the constituent materials must be understood.

Presented here, are the attenuation properties, namely the mass attenuation coefficient, linear attenuation coefficient and attenuation length, for commonly employed solid oxide fuel cell materials. The linear attenuation coefficient for two ceramic compounds, YSZ and GDC is calculated for a large range of energies (2 – 30 keV) and the theoretical phase contrast quality is quantified via definition of a contrast coefficient for both the Ni-8YSZ and Ni-GDC cermets. It is concluded that optimal contrast can be obtained at ca. 8 keV for Ni-3GDC and ca. 9 keV for Ni-8YSZ, i.e. just below and above the Ni K-edge for Ni-GDC and Ni-YSZ respectively. Furthermore, given the choice of either chromium or copper for a fixed energy lab source, copper would present the greatest solid-phase contrast for both Ni-YSZ and Ni-GDC.

Improved application of X-ray attenuation properties for electrochemical devices such as solid oxide fuel cells will aid in the understanding of cell degradation thus lead to improved life times via reduced performance losses leading to greater competitiveness in the energy market.

References

[1] Shearing P R, Brett D J L and Brandon N P, 2013, *International Materials Reviews*, **55**.6, 347 – 63.
[2] Chunwen S, Hui R, and Roller J, 2009, *J Solid State Electrochem*, **14**.7, 1125 – 44.
[3] Zhu W Z, and Deevi S C, 2003, *Materials Science and Engineering: A*, **362**.1, 228 – 9.
[4] Rehr J J and Albers R C, 2000, *Reviews of modern physics*, **72**.3, 621.
[5] Shearing P R, Gleb J, Yi J, Lee W K, Drakopolous M and Brandon N P, 2010, *Electrochemistry Communications*, **12**.8, 1021-1024.
[6] Centre for X-ray Optics, *CXRO, 13/09/2016*, http://henke.lbl.gov/optical_constants/.
[7] Hubbell J H and Sheltzer S M, 12/09/2016, *National Institute of Standard and Technology, NIST*, http://physics.nist.gov/PhysRefData/XrayMassCoef/chap2.html.
[8] Henke B L, Gullikson E M, and Davis J C, 1993, X-ray interactions: photoabsorption, scattering, transmission, and reflection at E=50-30000 eV, Z=1-92, *Atomic Data and Nuclear Data Tables*, **54**.2, 181-342.
[9] Hayashi H, Saitou T, Maruyama N, Indaba H, Kawamura K and Mori M, 2005, *Solid State Ionics*, **176**.5, 613-619.
[10] Zeiss Xradia 810 Ultra 'Nanoscale X-ray Imaging: Explore at the speed of science', 2016, *http://pages.microscopy.zeiss.com/rs/zeiss/images/EN_42_011_080_Product_810Ultra.pdf*.

Acknowledgments

The authors would like to acknowledge the EPSRC (EP/M014045/1), the Centre for Doctoral training in Fuel Cells and their Fuels and the Royal Academy for Engineering for financial support.

IOP Conf. Series: Journal of Physics: Conf. Series **849** (2017) 012018 doi:10.1088/1742-6596/849/1/012018

Understanding transport phenomena in electrochemical energy devices via X-ray nano CT

B Tjaden[1], J Lane[2], D J L Brett[1], P R Shearing[1,*]

[1]Electrochemical Innovation Lab, Department of Chemical Engineering, University College London, WC1E 7JE, UK
[2]Praxair Inc., 175 E. Park Dr., Tonawanda, NY 14150, USA
*Corresponding author: e-mail: p.shearing@ucl.ac.uk

Abstract. Porous support layers in electrochemical devices ensure mechanical stability of membrane assemblies such as solid oxide fuel cells and oxygen transport membranes (OTMs). At the same time, porous layers affect diffusive mass transport of gaseous reactants and contribute to performance losses at high fuel utilisation and conversion ratios. Microstructural characteristics are vital to calculate mass transport phenomena, where tortuosity remains notoriously difficult to determine. Here, the tortuosity of tubular porous support layers of OTMs is evaluated via high resolution X-ray nano computed tomography. The high resolution reveals the complex microstructure of the samples to then execute a selection of image-based tortuosity calculation algorithms. Visible differences between geometric and flux-based algorithms are observed and have thus to be applied with caution.

1. Introduction

Electrochemical energy devices have the potential to revolutionize distributed low-carbon power supply. However, in order to optimise their performance and enhance durability, a much improved understanding of the underlying material microstructure is required. In particular, the rate limiting step of fuel cells and oxygen transport membranes (OTMs) is governed by mass transport resistance in porous media, which, in turn, is a function of microstructural characteristics such as tortuosity, porosity and mean pore diameter [1]. These parameters are interrelated in a complicated manner [2].

Here, we utilise high resolution X-ray nano computed tomography (X-ray nano CT) to reveal the complex microstructure and resolve geometric features affecting the mass transport within tubular porous support layers of OTMs. Among the aforementioned parameters, tortuosity remains notoriously difficult to calculate [3]. Geometrically, tortuosity (τ) is calculated by dividing the shortest path length through a structure (Δl) by the Euclidean distance between the two endpoints of that path (Δx). However, in transport problems, the tortuosity in combination with the porosity (ε), is also applied to relate the effective transport property (D_{eff}) within a structure to its bulk transport property (D_{bulk}). Eq. (1) and Eq. (2) present both expressions, respectively.

$$\tau = \frac{\Delta l}{\Delta x} \quad \text{Eq. (1)} \qquad\qquad D_{eff} = \frac{\varepsilon}{\tau^2} D_{bulk} \quad \text{Eq. (2)}$$

IOP Conf. Series: Journal of Physics: Conf. Series **849** (2017) 012018 doi:10.1088/1742-6596/849/1/012018

This work compares a wide range of frequently used image-based tortuosity calculation approaches. These approaches are subdivided into geometric-based and flux-based algorithms, according to the differentiation shown in the above equations. Their suitability for quantifying diffusive mass transport in porous structures is then assessed.

2. Methodology

Four tubular yttria partially-stabilized zirconia porous support membrane samples for the application in OTMs were prepared for X-ray tomography. The samples differed in sintering temperature and wall thickness as listed in Table 1.

Table 1: Description of analysed samples.

Sample ID	Wall thickness	Sintering temperature
PS 1400 1.3	1.3 mm	1,400 °C
PS 1400 1.0	1.0 mm	1,400 °C
PS 1450 1.3	1.3 mm	1,450 °C
PS 1450 1.0	1.0 mm	1,450 °C

Table 2: Imaging specifications for X-ray nano CT.

Field of view	65 µm
Camera binning	1
Pixel size	63.1 nm
Exposure time	90 s
X-ray energy	5.4 keV
Number of images	901

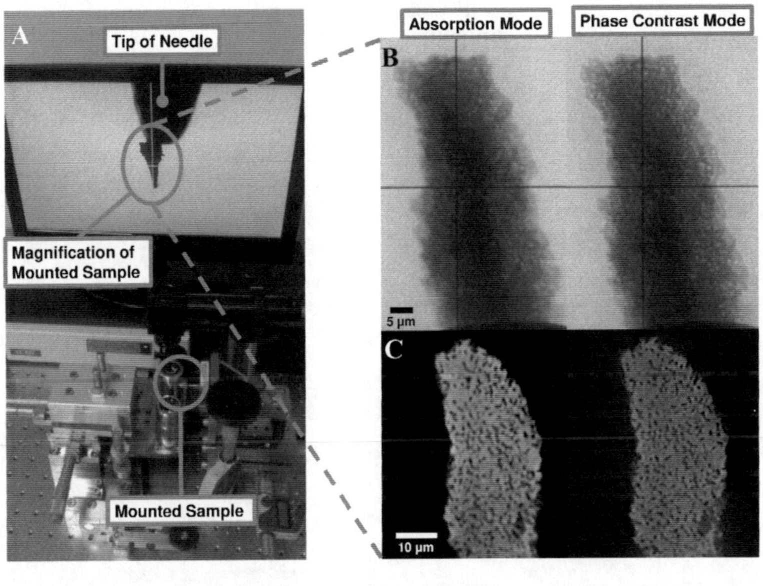

Figure 1: Mounted sample ready for X ray CT (A); radiographs (B) and image slices (C) comparing the phase and absorption imaging mode.

X-ray nano CT using the Zeiss Xradia 810 Ultra was applied to reconstruct the samples in 3D, where the imaging specifications for each sample were identical, as presented in Table 2. The Zeiss Xradia 810 Ultra achieves a quasi-monochromatic X-ray beam of 5.4 keV by the use of X-ray optics: the combination of a capillary condenser and a Fresnel zone plate filter out X-rays with energies above and below the K-edge of the chromium target and focus the beam onto the sample and detector. Figure 1A shows the small sample size of < 65 µm necessary to avoid artefacts during imaging. The Zeiss Xradia Ultra 810 is capable of imaging in absorption as well as in Zernike phase contrast mode [4]. Both modes were evaluated for their achieved image quality as shown in Figure 1B and C. It is visible that the edges and boundaries of the solid phase are finer and better defined in phase contrast mode. Hence, phase contrast imaging was chosen over absorption imaging despite the highly attenuating sample materials.

After reconstruction using a standard filtered back projection algorithm (XRM Reconstructor), the image datasets were segmented using threshold segmentation and two volumes of different dimensions were cropped for each sample to verify the consistency of results: a cube with a side length of 6.31 µm and a cuboid with the side lengths of 6.31 µm × 6.31 µm × 12.62 µm. The tortuosity was then calculated via a range of computational algorithms executed across these volumes: the fast marching method (FMM) [5], the built-in Avizo (FEI) module "Centriod Path Tortuosity" [6], by simulating the heat flux in StarCCM+ (CD-adapco) [7–9] and mass flux in COMSOL Multiphysics (COMSOL Inc.) as well as via a MATLAB based Laplace equation solver called "TauFactor" [10, 11]. While the first two calculation algorithms calculate tortuosity geometrically by applying Eq. (1), the latter three approaches simulate a flux and correlate the effective transport properties to the bulk transport properties via Eq. (2). These algorithms arrive at distinct tortuosity values along each axis of the volume and, hence, the characteristic tortuosity (τ_C) [7] was calculated for each sample. Moreover, well-known porosity-tortuosity relationships, including the Bruggeman relationship [12] and the Maxwell relationship [13], were used for comparative reasons.

3. Results and Discussion

Table 3 compares microstructural characteristics for each of the eight extracted sample volumes. The porosity was calculated via the pixel counting method while the pore diameter was achieved using the continuous pore size distribution code from [14]. It is evident that samples with lower sintering temperature feature larger pore diameter and porosity, favouring diffusive mass transport.

Table 3: Microstructural characteristics of each sample volume.

Sample ID		d_P [µm]	ε [-]
PS 1400 1.3	Cube	0.73	0.31
	Cuboid	0.74	0.32
PS 1400 1.0	Cube	0.71	0.34
	Cuboid	0.70	0.32
PS 1450 1.0	Cube	0.66	0.25
	Cuboid	0.65	0.24
PS 1450 1.3	Cube	0.63	0.23
	Cuboid	0.63	0.21

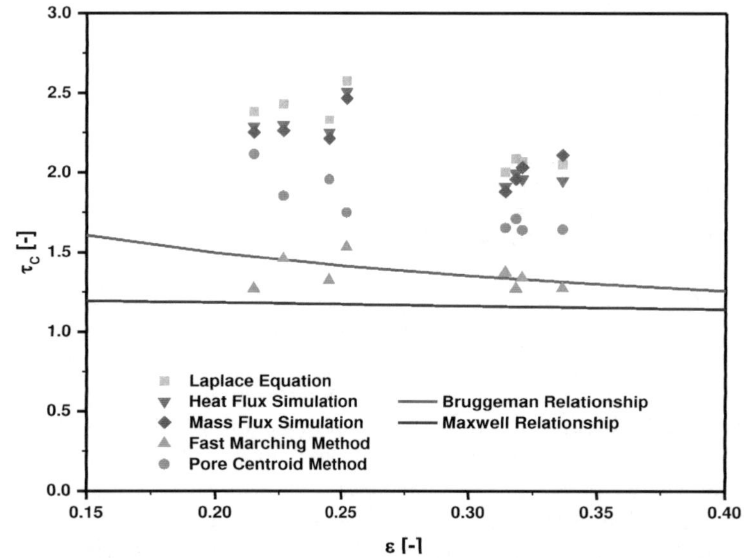

Figure 2: Comparison of tortuosity calculation algorithms which show the difference in geometric and flux-based algorithms. Empirical relationships such as the Bruggeman and Maxwell relation are unfit for the analysed microstructure treated here.

Figure 2 shows the characteristic tortuosity values of both geometric-based algorithms and of all three flux-based algorithms, similar as in [7]. The FMM achieved the lowest tortuosity values among all algorithms with the pore centroid method results slightly above. However, the former approach has the least significance in analysing the tortuosity of the porous phase. The algorithm locates the pore centroid of each slice and follows it in the in-plane direction. Hence, this algorithm is considered only as an indicator for the homogeneity of the porous phase within the sample: the centroid of a homogeneous sample will be located closer to the centre of each image slice and thus, the resulting tortuosity will be closer to unity. The low porosity samples feature a higher pore centroid tortuosity, suggesting a higher heterogeneity in the microstructure. Both empirical relationships form the lower boundary of tortuosity, where the FMM achieved values around the Bruggeman correlation. All tortuosity algorithms broadly followed the trend of decreasing tortuosity with increasing porosity.

The results of the three flux-based algorithms were very close to each other, with values around 2.3 for the low porosity samples and around 1.9 for the high porosity samples. The Laplace equation solver achieved slightly higher values compared to the other simulation-based algorithms, which might be caused by some minor smoothing effects during mesh generation for the flux simulations.

The visible differences between the FMM and the flux-based algorithms were caused by the lack of the FMM to account for constrictions and bottlenecks while only considering the minimal Euclidean distance between two planes. Hence, any pore connection, even a single pixel, was included in the calculation and thus, the geometric definition of tortuosity is closely followed. However, a migrating flux is affected by the variation of pore diameters within the sample and choses the path of least resistance and constriction. As a consequence, flux-based tortuosity calculation algorithms always result in a higher tortuosity value. In the presented study, this difference amounts to almost a factor of two. Thus, when calculating the effective transport property of a transport phenomenon following Eq. (2), a flux-based tortuosity computation algorithm has to be applied to correctly cater for the effect of the microstructure on the transport process. Among the included flux-based tortuosity calculation methods, the Laplace equation solver produced consistent results and is easiest to execute, as the binarized image sequence is sufficient as input file, making the generation of a volume mesh unnecessary.

4. Conclusions

X-ray nano CT was used to reveal the complex microstructure of porous support membranes and evaluate the effect of microstructural characteristics on transport phenomena. The high resolution allowed the execution of a wide range of tortuosity calculation approaches, where the high number of different algorithms is testimony of the importance of tomography techniques in analysing transport processes in porous structures. Significant differences between geometric and flux-based algorithms were observed. The reason for this is that geometric approaches do not take constrictions and bottlenecks into account, which affect the migration path of a flux. Consequently, care must be taken when comparing tortuosity values calculated via algorithms following different concepts.

References

[1] Shearing P R, Brett D J L and Brandon N P 2013 *Int. Mater. Rev.* **55** 347–63
[2] Shearing P R, Howard L E, Jørgensen P S, Brandon N P and Harris S J 2010 *Electrochem. Commun.* **12** 374–7
[3] Tjaden B, Brett D J L and Shearing P R 2016 *Int. Mater. Rev.*
[4] Holzner C, Feser M, Vogt S, Hornberger B, Baines S B and Jacobsen C 2010 *Nat Phys* **6** 883–7
[5] Jørgensen P S, Hansen K V, Larsen R and Bowen J R 2011 *J. Microsc.* **244** 45–58
[6] Gostovic D, Smith J R, Kundinger D, Jones K S and Wachsman E D 2007 *Electrochem. Solid-State Lett.* **10** B214-7
[7] Cooper S J, Kishimoto M, Tariq F, Bradley R S, Marquis A J, Brandon N P, Kilner J A and Shearing P R 2013 *ECS Trans.* **57** 2671–8
[8] Trogadas P, Taiwo O O, Tjaden B, Neville T P, Yun S, Parrondo J, Ramani V, Coppens M-C, Brett D J L and Shearing P R 2014 *Electrochem. Commun.* **48** 155–9
[9] Tjaden B, Lane J, Withers P J, Bradley R S, Brett D J L and Shearing P R 2016 *Solid State Ionics* **288** 315–21
[10] Cooper S J, Bertei A, Shearing P R, Kilner J A and Brandon N P 2016 *SoftwareX.*

[11] Finegan D P, Cooper S J, Tjaden B, Taiwo O O, Gelb J, Hinds G, Brett D J L and Shearing P R 2016 *J. Power Sources* **333** 184–92.
[12] Tjaden B, Cooper S J, Brett D J L, Kramer D and Shearing P R 2016 *Curr. Opin. Chem. Eng.* **12** 44–51
[13] Chueh C C, Bertei A, Pharoah J G and Nicolella C *Int. J. Heat Mass Transfer 2014* **71** 183–8
[14] Münch B and Holzer L *J. Am. Ceram. Soc.* 2008 **91** 4059–67

IOP Conf. Series: Journal of Physics: Conf. Series **849** (2017) 012019 doi:10.1088/1742-6596/849/1/012019

Electronic structure of polycrystalline CVD-graphene revealed by Nano-ARPES

Chaoyu Chen, José Avila & Maria C. Asensio*

ANTARES Beamline, Synchrotron SOLEIL, Université Paris-Saclay, L'Orme des Merisiers, Saint Aubin-BP 48, 91192 Gif sur Yvette Cedex, France.

E-mail: `maria-carmen.asensio@synchrotron-soleil.fr`

Abstract. The ability to explore electronic structure and their role in determining material's macroscopic behaviour is essential to explain and engineer functions of material and device. Since its debut in 2004, graphene has attracted global research interest due to its unique properties. Chemical vapor deposition (CVD) has emerged as an important method for the massive preparation and production of graphene for various applications. Here by employing angle-resolved photoemission spectroscopy with nanoscale spatial resolution ~ 100 nm (Nano-ARPES), we describe the approach to measure the electronic structure of polycrystalline graphene on copper foils, demonstrating the power of Nano-ARPES to detect the electronic structure of microscopic single crystalline domains, being fully compatible with conventional ARPES. Similar analysis could be employed to other microscopic materials

1. Introduction

Graphene has attracted intensive research attentions in the past decade due to its excellent structural and electronic properties as well as great potentials in applications[1, 2, 3, 4, 5]. CVD[6, 7, 8, 9, 10] is, among others to date, the dominant technology that yields to scale and performance for graphene growth on metals[11, 12]. Recently, commercially available cooper foils have become an efficient and inexpensive substrate for scalable CVD growth of large-area graphene films[12, 13]. However, in most cases the CVD-grown graphene films on Cu foils are polycrystalline with microscopic single crystal domains. This naturally lowers the possibility for traditional ARPES to characterise the detailed electronic structure of graphene. The recent conceptional and technical progress represented by Nano-ARPES[14, 15, 16] has opened a new path to characterise the electronic structure of microscopic crystal. Here we will discuss the general approach for ARPES and Nano-ARPES to measure the comprehensive band structure and Fermi surface of CVD-grown graphene on Cu foils. We observe clear features for the Fermi surface of both graphene and Cu, with that the relative orientation between graphene and Cu surface can be discussed. We then analyse in detail the approach for Nano-ARPES to study the electronic structure of microscopic graphene grains. These comprehensive electronic structures of polycrystalline graphene/Cu provides valuable information to understand and further to optimise the properties of CVD-graphene. We also expect that these general approaches combining ARPES and Nano-ARPES techniques would be employed to many other different types of nanomaterials.

IOP Conf. Series: Journal of Physics: Conf. Series **849** (2017) 012019 doi:10.1088/1742-6596/849/1/012019

2. General Fermi surface and orientation of graphene/Cu revealed by ARPES

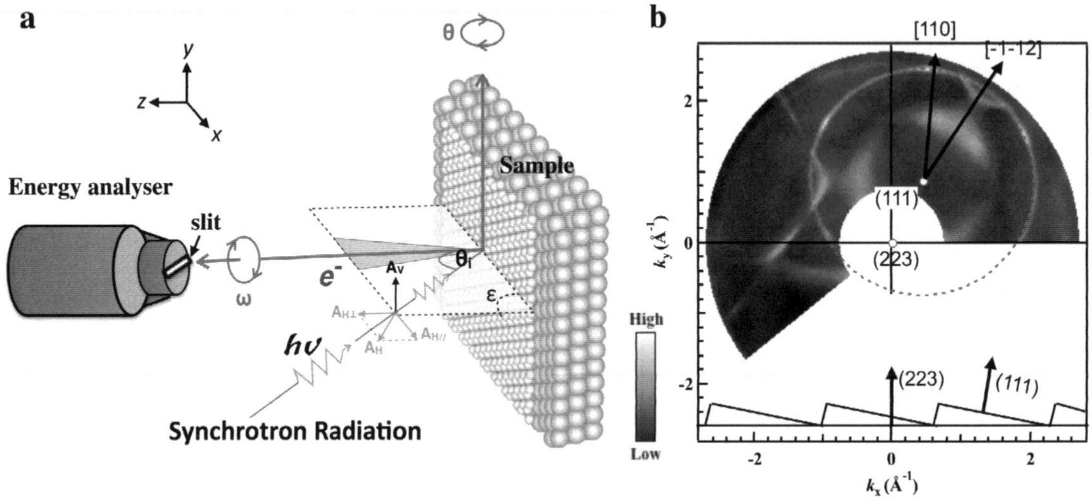

Figure 1. General ARPES measurement geometry and Fermi surface of polycrystalline graphene/Cu. a, Schematic for Fermi surface mapping by ARPES. **b,** Fermi surface of polycrystalline graphene on Cu. The schematic orientations of graphene and Cu terraces are shown, Red dashed circle indicates the Fermi surface of graphene.

Fig. 1a showns the experimental geometry for conventional ARPES measurements. $\mathbf{A_H}$ ($\mathbf{A_V}$) is the vector potential of the incident light with linear horizontal (vertical) polarisation, θ and ω denote the polar and azimuthal degree of freedom and θ_i is the polar angle of incidence. The analyser slit horizontally oriented permits photoelectron detection from one cut in angular (momentum) space along the x axis. For all different θ, the incident photon momentum, the photoelectron momentum and the surface normal lie on the same plane (ε). The Fermi surface of the sample is mapped by rotating the sample ω axis while taking spectra along x axis for each ω angle. Fig. 1b shows the examplary Fermi surface mapping of polycrystalline graphene/Cu. First we rotate the sample to the right polar angle θ, in order to cover the Dirac cone at K points. Then we rotate the azimuthal angle ω to make individual spectra and integrate them into Fermi surface.

As analysed in detail in Fig. 1b, the Fermi surface for graphene consists of a circle with radius being $\sim$1.7 $\AA^{-1}$, which corresponds to the Brillouin zone corner **K** point. This circle, rather than isolated points for the Brillouin zone corner of single crystalline graphene, results from the polycrystalline nature of graphene. As for traditional ARPES, the beam size on the sample is around 100 μm, this continuous circle means that single crystalline microscopic domains of graphene take their orientations randomly. This is in complete contrast with graphene grown on nickel, iridium and cobalt[7], in which cases graphene can only have single or few orientations respect to the substrate surface lattice.

It is quite interesting to note that, the substrate surface is dominated by the (111) crystal plane of Cu single crystal, which can be judged by the Fermi surface shape of Cu *sp* bands. However, the (111) orientation is not normal to the substrate surface. As shown in Fig. 1b, the Brillouin zone centres of both graphene and Cu (111) locate at (k_x, k_y) = (0.46$\pm$0.01, 0.86$\pm$0.01)$\AA^{-1}$. This coordinate corresponds in angular space an offset of 11.4$\pm$0.1 $^\circ$ for the (111) crystal plane, indicating that graphene grows along the (111) plane terraces of Cu single crystal, which is not the normal direction of Cu substrate surface. As analysed in detail in

Fig.1b, the normal surface of Cu substrate actually represents the (223) crystal plane of Cu single crystal. Also in Fig. 2b, the crystal orientations [-1,-1,2] and [110] are indicated by black arrows, with the former corresponding to the $\Gamma - M$ direction for Cu (111) hexagonal lattice.

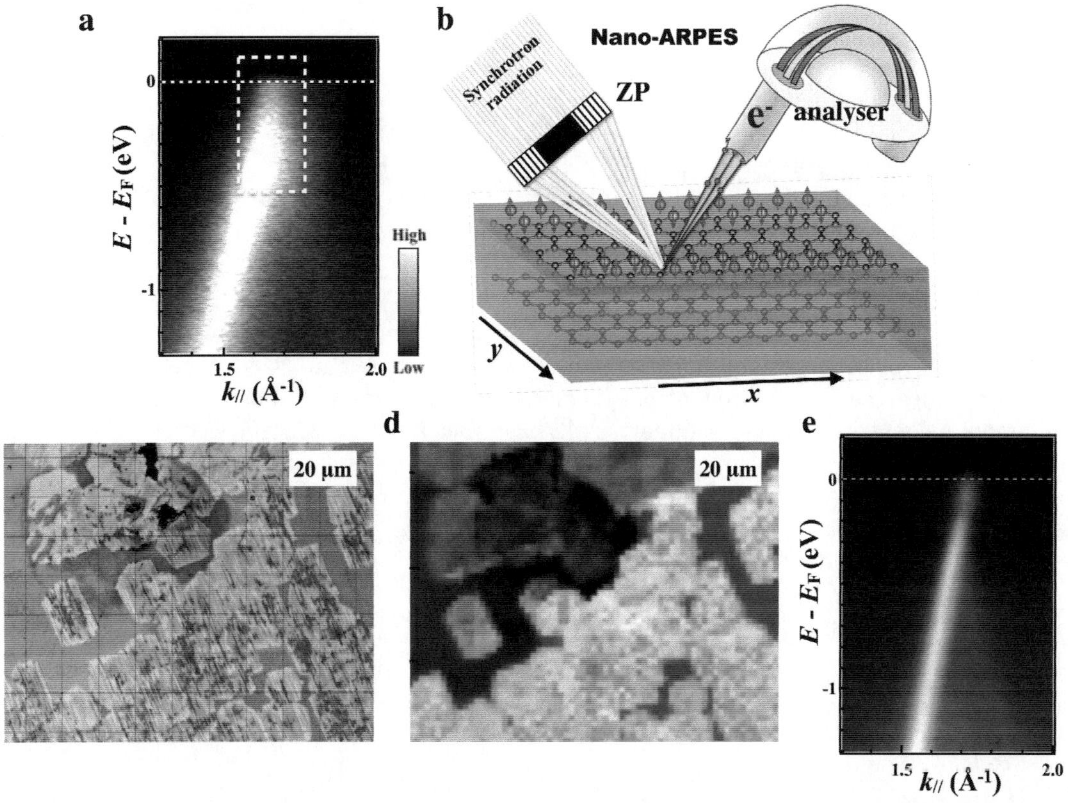

Figure 2. Nano-ARPES working principle and its study on the electronic structure of microscopic graphene grains. a, Graphene Dirac one measured by ARPES. White dashed frame indicates the energy-momentum window integrated to generate the valence band intensity distribution image (measured by Nano-ARPES) shown in **d**. **b,** Schematic for working principle of Nano-ARPES. **c**, Optical image of CVD-grown graphene grains on Cu. **d**, Valence band intensity distribution image mapped by Nano-ARPES.**e**, typical Dirac cone electronic structure from one single graphene domain measured by Nano-ARPES.

3. Microscopic characterisation of graphene grains by Nano-ARPES

Having established the polycrystalline nature with random orientation of graphene, we now use Nano-ARPES to study the electronics structure for single crystalline domains. Fig. 2b shows the schematic of our Nano-ARPES system. Synchrotron radiation (beam) is focused by Zone Plate (ZP). The focused spot size on sample is around 100 nm. The manipulation of sample is realised by a XY scanning controller with interferometric feedback to avoid thermal perturbation. Fig. 2a shows one spectra of Dirac cone measured by ARPES as discussed in Fig. 1. The broad bandwidth and not well-defined band gap result from the fact that many graphene grains with different orientation are measured for this spectra. By integrating the spectral intensity within one particular energy-momentum window while scanning the sample in real space along x and

y directions, the angle-resolved spectral intensity distribution image can be obtained, as shown in Fig. 2d.

It is clear that in this area the CVD-graphene actually consists of many grains no larger than few tens of micrometers, with some of them merging together. The optical image shown in Fig. 2c proves the existence of these graphene grains. The perfect match between the Nano-ARPES spectral intensity image and optical image undoubtedly proves the ability of Nano-ARPES to locate the microscopic sample by photoemission spectral weight.

Fig. 2e shows one typical Nano-ARPES spectra of graphene Dirac cone from one single graphene grain. Compared with ARPES spectra in Fig. 2a, the bandwidth of Dirac cone is much less and the Dirac gap is better resolved. This is naturally due to the single crystalline nature of the studied graphene grain. Together with the perfect match between Nano-ARPES image and optical image, these results demonstrated the full capability of Nano-ARPES to study the electronic structure of microscopic materials.

4. Discussion

Taking polycrystalline CVD-graphene/Cu as an example, we have fully shown the capability of Nano-ARPES to study the electronic structure of microscopic materials. Electronic structure, which describes the distribution of electronic states in reciprocal space, is one of the most fundamental concepts in condensed matter physics, since it determines the electrical, optical and magnetic behaviours of materials. The rapid development of nanotechnology and nanomaterial requires electronic structure probes with nanometer resolution. In this context, Nano-ARPES, although still being at its very early stage, has a promising future for the electronic structure study of various nanomaterials.

5. Acknowledgments

The Synchrotron SOLEIL is supported by the Centre National de la Recherche Scientifique (CNRS) and the Commissariat à l'Energie Atomique et aux Energies Alternatives (CEA), France.

6. References

[1] Castro Neto, A. H. et al. 2009. *Reviews of Modern Physics* **81**109-162
[2] Geim, A. K. & Novoselov, K. S. 2007 *Nature Materials* 6 183-191
[3] Geim, A. K. 2009 *Science* **324**, 1530-1534
[4] Han, W., Kawakami, R. K., Gmitra, M. & Fabian, J. 2014 *Nat Nanotechnol* **9** 794-807
[5] Bonaccorso, F., Sun, Z., Hasan, T. & Ferrari, A. C. 2010 *Nat. Photonics* **4** 611-622
[6] Voloshina, E. & Dedkov, Y. 2012 *Physical chemistry chemical physics : PCCP* **14** 13502-13514
[7] Batzill, M. 2012 *Surface Science Reports* **67**, 83-115
[8] Wintterlin, J. & Bocquet, M. L. 2009 *Surface Science* **603** 1841-1852
[9] Zhang, Y., Zhang, L. & Zhou, C. 2013 *Accounts of Chemical Research* **46** 2329-2339
[10] Mattevi, C., Kim, H. & Chhowalla, M. 2011 *J. Mater. Chem.* **21**, 3324-3334
[11] Kim, K. S. et al. 2009 *Nature 457* 706-710
[12] Li, X. et al. 2009 *Science* **324** 1312-1314
[13] Bae, S. et al. 2010 *Nat Nanotechnol* **5** 574-578
[14] Avila J and Asensio M C. 2014 *Synchrotron Radiation News* **27** 24-30
[15] Bostwick, A. et al. 2012 *Synchrotron Radiation News* **25** 19-25
[16] Jose, A. et al. 2013 *Journal of Physics: Conference Series* **425** 192023

IOP Conf. Series: Journal of Physics: Conf. Series **849** (2017) 012020 doi:10.1088/1742-6596/849/1/012020

Study of the tortuosity factors at multi-scale for a novel-structured SOFC anode

X Lu[1], T Li[2], O O Taiwo[1], J Bailey[1], T Heenan[1], K Li[2], D J L Brett[1] and P R Shearing[1]

[1]Department of Chemical Engineering, University College London, London, WC1E 7JE, UK
[2]Department of Chemical Engineering, Imperial College London, London, SW7 2AZ, UK
Email: xuekun.lu@ucl.ac.uk

Abstract. Gas transport properties are closely related to the tortuosity of the pore network within porous materials. For the first time, this study explores a multi-scale imaging and modelling method to measure the tortuosity of an Solid Oxide Fuel Cell (SOFC) electrode material with pore sizes spanning over hundreds of orders of magnitude. This analysis is normally challenging using image-based techniques, as pores of different sizes may not be easily resolved at the same time using X-ray computed tomography (CT). In this study, a tubular SOFC anode, fabricated by a phase inversion technique, is used to illustrate this approach. A heat flux analogy is used to simulate mass transport and the results show that the embedded large-scale finger-like pores can significantly improve mass transport by providing less tortuous pathways.

1. Introduction

Gas transport through porous anodes is critical to the electrochemical performance of solid oxide fuel cells (SOFC) as the partial pressures of the reactants and products are closely related to the anode polarization, which significantly contributes to the loss of operating voltage. The porosity (ε) and tortuosity of the gas transport path each play a significant role in determining the diffusivity of the fuel gas in the porous anode [1]. The tortuosity factor (τ) is a material parameter used to characterise gas diffusion resistance due to the tortuous pore volume [2, 3]. This effective transport parameter can be obtained by the modelling of mass/heat flux through a 3D porous phase using computational fluid dynamics (CFD) methods [4].

A newly developed tubular SOFC fabricated by phase inversion technique has attracted considerable attention [5]. Apart from the controllable pore size in the spongy layer, the exchange of solvent and non-solvent in the hollow fiber generates micro-channels which grow in the radial direction, with a diameter of approximately 20 μm. This design can significantly improve gas transport in the anode with little sacrifice of mechanical robustness. In this study, non-destructive X-ray CT is used to image the SOFC anode at micrometer and nanometer resolution for the whole anode and the porous phase in the spongy layer, respectively. The tortuosity factor extracted from the high resolution scan was subsequently used for the simulation of the entire anode.

2. Methodology

2.1 Tortuosity Factor Measurement

The analogue between heat and mass transfer is well established [6] and is regularly used to measure the tortuosity factor. The heat flow driven by a temperature gradient in a porous material is described using modified Fourier's law as

$$Q_p = -A\frac{\varepsilon}{\tau}k\frac{\Delta T}{L} \tag{1}$$

Where A is the cross sectional area of the flow volume, k is the thermal conductivity, ΔT is the temperature difference and L is the length of flow volume, ε is the porosity and τ is the tortuosity factor. As ε can be measured by image analysis, by dividing Q_p by the heat flow in a fully porous volume Q (i.e. $\varepsilon = 1$, $\tau = 1$), τ can be obtained. For simplicity, the thermal conductivity of the micro-channels are set to unity (i.e. $k = 1$), so that the thermal conductivity is equivalent to the effective transport parameter ε_s/τ_s for the spongy layer.

2.2 X-ray Computed Tomography

The entire tubular anode and the spongy layer were imaged using a Versa 520 and Ultra 810 X-ray microscope (Zeiss Xradia, Carl Zeiss, CA, USA) respectively [7]. Detailed scanning parameters are shown in Table 1. FDK and filtered-back projection algorithms were used for the reconstruction of the full anode and the spongy layer scans respectively [5, 8]. The reconstructed sample volumes were segmented using the Avizo V9.0 software (VSG, Bordeaux) package.

Table 1. Scanning parameters for the spongy layer and entire anode

	Tube target	Voltage (kV)	Voxel size (μm)	Field of view (μm^2)	Projections	Exposure time (s)	Camera binning
Spongy	Cr	35	0.032 (binned)	16×16	1201	60	2
Anode	W	140	1.07	2000×2000	2001	18	1

2.3 Computed Dynamics Fluid (CFD) Simulation

The segmented porous phase of the spongy layer was imported into the CFD software Star-CCM+ (CD-adapco Inc.) for meshing followed by heat flux simulation. ε_s/τ_s was then used to define the effective thermal conductivity of spongy layer in the simulation of the entire anode on a control volume with full radial thickness, an axial depth of 700 μm and 90° arc angle (see Figure 1). The detailed simulation parameters are shown in Table 2.

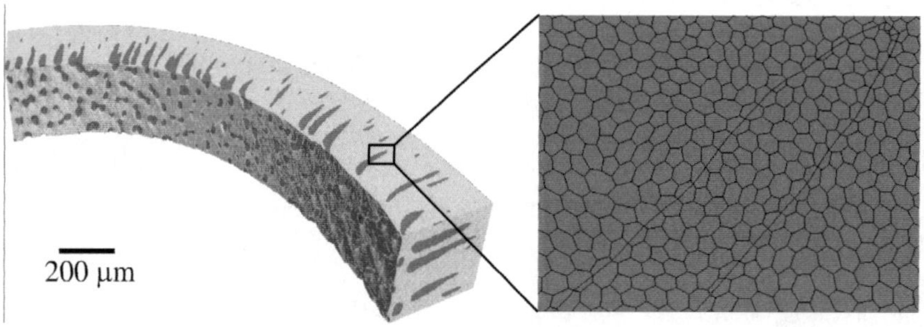

Figure 1. The effective thermal conductivities of the micro-channels (red) and spongy layer (green) are assigned as unity "1" and ε_s/τ_s respectively. Conformal meshes are generated between the two regions.

Table 2. CFD simulation parameters for spongy layer and entire anode

	Spongy layer	Full anode
Number of cells	1.2 million	1.6 million
Inlet temperature (°C)	1000	1000
Outlet temperature (°C)	300	300
Thermal conductivity	ε_s/τ_s	1
Sample dimension (μm^3)	9×6×4	800×800×800

3. Results and discussion

A virtual slice of the anode is shown in Figure 2a. The radially aligned micro-channels are clearly resolved. The porous phase in the spongy layer, which appears as solid phase in Figure 2a, is segmented from the CT data and imported to Star-CCM+ to simulate the heat transfer driven by the temperature difference from 1000 K (LHS) to 300 K (RHS) (Figure 2b). The porosity and the mean diameter of the porous phase in the spongy layer are measured to be 0.19 and 200 nm respectively by image analysis. The heat flux in the 3D microstructure is shown in Figure 2c. It is observed that the tortuosity and constrictivity of the pore network generate a heterogeneous distribution of the flux, and the local flux maxima appear at the position where the transition of the pore thickness is sharp in the pathway. By integrating the heat flux on the cross-sectional plane, and dividing it by the heat flow on the cross-sectional plane of the empty volume, the material parameter ε_s/τ_s is equal to 0.015, which is subsequently used as the thermal conductivity of spongy layer in the anode simulation.

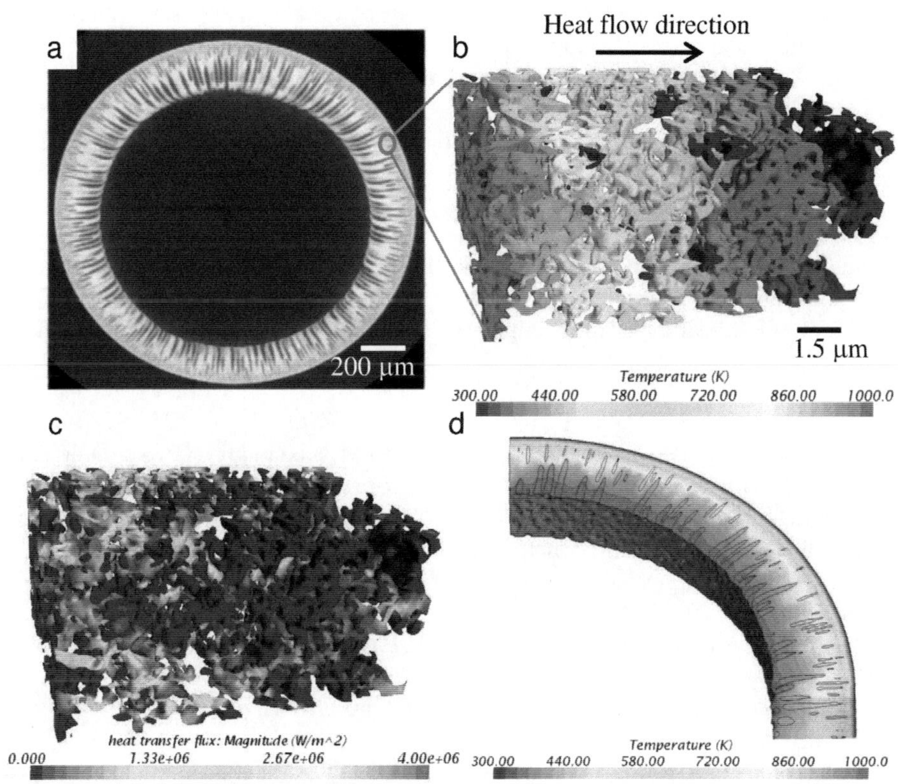

Figure 2. (a) Virtual ortho slice of the anode; (b) simulation of heat transfer in the spongy layer; (c) heat flux in the spongy layer; (d) Heat transfer simulation on the anode with different thermal conductivity defined in the micro-channel and spongy layer.

The result of the heat transfer simulation for the whole anode is shown in Figure 2d. The temperature contour displays a non-uniform distribution with a ragged edge where there are isolated micro-channels. These are caused by the discrepancies of the effective thermal conductivity between the micro-channels and the spongy layer. As the heat flows from one micro-channel to the other, temperature gradient occurs at the spongy layer between them due to the low thermal conductivity.

The microstructure parameters extracted from the anode heat transfer simulation are compared with those in the spongy layer in Table 3. With the micro-channels embedded in the anode, the macroscopic material parameter for the whole anode ε_a/τ_a is approximately 4 times as large as ε_s/τ_s, indicating a 300 % increase in terms of heat flow.

Table 3. Comparison of the material parameters measured in spongy layer and full anode

	ε	τ	ε/τ
Spongy	0.19	12.1	0.015
Anode	0.33	5.2	0.063

4. Conclusion
Multi-scale imaging and hierarchical simulation is an effective method to extract the material parameters (i.e. porosity and tortuosity) in the samples with distinctly different pore sizes spanning over several orders of magnitude, under which circumstances the porous phases cannot be resolved at the same time. These geometrical parameters are critical for effective diffusivity estimation in macroscopic electrochemical simulations. Results show that mass transfer in novel-structured tubular SOFC anodes embedded with radially aligned micro-channels is significantly improved four-fold due to the decrease of global tortuosity compared to the traditional homogeneous tubular SOFC anode with the pure spongy layer.

Acknowledgement
The authors acknowledge the sample provided by Dr. Tao Li from Kang Li's research group in Imperial College London, and also appreciate support from the EPSRC under grants EP/N032888/1 and EP/M014045/1. PR Shearing acknowledges funding from the Royal Academy of Engineering.

References
[1] Virkar A V, Chen J, Tanner C W and Kim J-W, 2000 *Solid State Ionics* **131** 189-198
[2] Iwai H, Shikazono N, Matsui T, Teshima H, Kishimoto M, Kishida R, Hayashi D, Matsuzaki K, Kanno D and Saito M, 2010 *Journal of Power Sources* **195** 955-961
[3] Wilson J R, Kobsiriphat W, Mendoza R, Chen H-Y, Hiller J M, Miller D J, Thornton K, Voorhees P W, Adler S B and Barnett S A, 2006 *Nature materials* **5** 541-544
[4] Shearing P, Howard L, Jørgensen P S, Brandon N and Harris S, 2010 *Electrochemistry communications* **12** 374-377
[5] Droushiotis N, Doraswami U, Ivey D, Othman M H D, Li K and Kelsall G, 2010 *Electrochemistry Communications* **12** 792-795
[6] Tjaden B, Lane J, Withers P J, Bradley R S, Brett D J and Shearing P R, 2016 *Solid State Ionics* **288** 315-321
[7] Schurch R, Rowland S M, Bradley R S and Withers P J, 2015 *IEEE Transactions on Dielectrics and Electrical Insulation* **22** 709-719
[8] Scherl H, Koerner M, Hofmann H, Eckert W, Kowarschik M and Hornegger J, 2007 *Medical Imaging* 651058-651058-10

IOP Conf. Series: Journal of Physics: Conf. Series **849** (2017) 012021 doi:10.1088/1742-6596/849/1/012021

Simultaneous heating and compression of irradiated graphite during synchrotron microtomographic imaging

A J Bodey[1], Z Mileeva[2], T Lowe[2], E Williamson-Brown[3], D S Eastwood[2], C Simpson[2], V Titarenko[2], A N Jones[2], C Rau[1] and P M Mummery[2]

[1] Diamond Light Source, Oxfordshire, OX11 0DE, UK
[2] University of Manchester, Manchester, M13 9PL, UK
[3] Deben UK Ltd., Bury St. Edmunds, Suffolk, IP30 9QS, UK

andrew.bodey@diamond.ac.uk

Abstract. Nuclear graphite is used as a neutron moderator in fission power stations. To investigate the microstructural changes that occur during such use, it has been studied for the first time by X-ray microtomography with in situ heating and compression. This experiment was the first to involve simultaneous heating and mechanical loading of radioactive samples at Diamond Light Source, and represented the first study of radioactive materials at the Diamond-Manchester Imaging Branchline I13-2. Engineering methods and safety protocols were developed to ensure the safe containment of irradiated graphite as it was simultaneously compressed to 450N in a Deben 10kN Open-Frame Rig and heated to 300°C with dual focused infrared lamps. Central to safe containment was a double containment vessel which prevented escape of airborne particulates while enabling compression via a moveable ram and the transmission of infrared light to the sample. Temperature measurements were made in situ via thermocouple readout. During heating and compression, samples were simultaneously rotated and imaged with polychromatic X-rays. The resulting microtomograms are being studied via digital volume correlation to provide insights into how thermal expansion coefficients and microstructure are affected by irradiation history, load and heat. Such information will be key to improving the accuracy of graphite degradation models which inform safety margins at power stations.

1. Introduction

The Diamond-Manchester Imaging Branchline I13-2 at Diamond Light Source (DLS) provides high-flux, partially-coherent 5-35keV X-rays which are used primarily for tomographic imaging experiments. Such experiments are often highly bespoke and involve complex manipulation of sample environments.

Nuclear graphite bricks function as both moderators and structural components in fission power stations. Over time, radiation and oxidation degrade the graphite, leading to mass loss, distortion and eventual cracking. When distortion and cracking become too severe, it can become difficult to insert and remove control rods, thereby affecting the risk profile of power station operation. Owing to uncertainties about the mechanisms and impacts of graphite degradation, UK power stations operate with very conservative safety margins. Computational models of degradation (eg. EDF Energy's FEAT-DIFFUSE6 and [1]) inform these safety margins; such models rely upon data relating to coefficients of thermal expansion (CTEs) and the manner in which these vary as a function of

IOP Conf. Series: Journal of Physics: Conf. Series **849** (2017) 012021 doi:10.1088/1742-6596/849/1/012021

temperature and pressure. To generate CTE data and investigate how microstructure is also affected by irradiation history, load and heat, we performed X-ray microtomography on nuclear graphite in a custom sample environment and are now analysing the results via digital volume correlation (DVC) [2]. Development and use of the sample environment posed a number of technical challenges relating to the safe containment of samples undergoing simultaneous heating and compression.

2. Experimental design

Cylindrical samples, of 5mm diameter and 6mm height, were trepanned from both unirradiated graphite bricks and irradiated bricks at a UK power station. Samples were pretreated offsite to boil-off radioactive gases.

Samples were mounted in double containment vessels which were produced for the experiment as a collaborative design between University of Manchester and Deben UK Ltd. Double containment was provided by two cylindrical glass walls which were sealed against upper and lower plates. The sample was mounted on a static platform and compressed from above by a ram. Ram movement was controlled externally, with seals maintained by bellows.

The containment vessels did not require a very high degree of air tightness, but did need to contain any airborne particulates that could be generated during the course of an experiment. Helium leak testing was conducted jointly by Deben and the DLS Vacuum Group.

Samples arrived at DLS mounted in individual containment vessels and were handled on site in accordance with DLS's internal procedures. Since Deben Open-Frame Rigs had not been used previously for synchrotron experiments and the containment vessels had not been used with these rigs before, dry runs were performed with unirradiated samples to test various elements of the setup. Vessels were mounted inside the Open-Frame Rig, which was equipped with two Optron 150W IR Spot (Optron GmbH, Germany) focused infrared lamps (figure 1). The whole vessel was free to rotate inside the rig. To prevent sample cracking and associated particulate generation, the rig had been programmed with a compression limit and the option for torsion disabled.

Figure 1. A double containment vessel in the Open-Frame Rig, with focused infrared lamps radiating orthogonally to the direction of the X-ray beam. A nuclear graphite sample can be seen undergoing compression in the vessel.

3. Microtomography

Samples were imaged with a partially-coherent, polychromatic 'pink' beam (5-35 keV) of parallel geometry generated by an undulator of 5mm gap. The beam was reflected from the platinum stripe of a grazing-incidence focusing mirror and filtered with 950μm pyrolytic graphite and 2mm aluminium. 2,001 projection images, each of 75ms exposure time, were acquired at equally-spaced angles over 180° of stepped rotation by a pco.edge 5.5 (PCO AG, Germany) detector. The image recorded at 180°

was compared to the image recorded at 0° to check for various experimental problems [3]. The detector was coupled to a 500μm $CdWO_4$ scintillator and a visual light microscope, providing 2.5× total magnification, a field of view of 6.7 × 5.6 mm (2560 × 2160 pixels) and an effective pixel size of 2.6μm. A propagation distance of approximately 70mm was used, providing a moderate level of in-line phase contrast.

4. Preliminary results

Tomographic slices of unirradiated and irradiated graphite are shown in figure 2. These images demonstrate widespread changes in microstructure caused by irradiation and radiolytic oxidation, including increased porosity and filler-particle compression. CTEs were calculated from displacements of features as measured by DVC. For the unirradiated samples, CTEs were found to increase with increasing compressive load. Other results, as well as image analysis and DVC methods, will be presented in an upcoming publication.

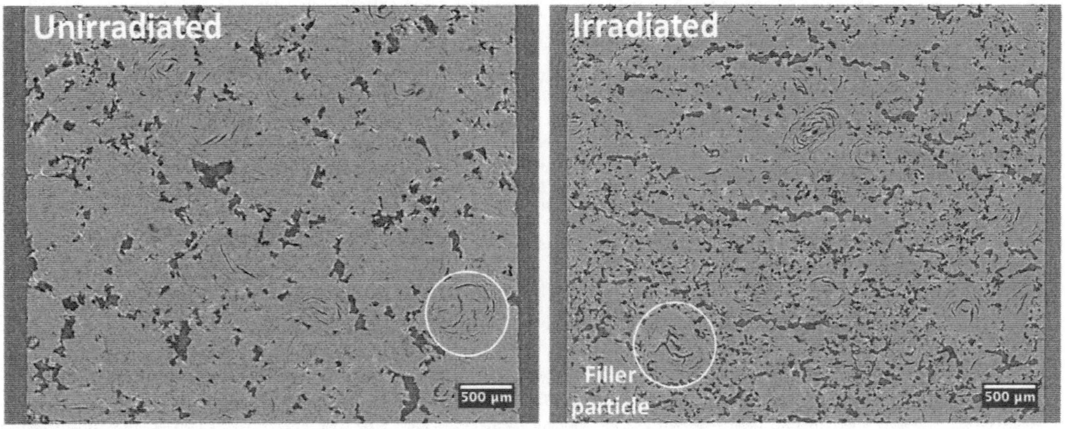

Figure 2. Comparison of unirradiated and irradiated graphite. Both samples had been raised to a temperature of 300°C and loaded to 450N. Note the increased porosity and compression of filler particles associated with irradiation.

5. Acknowledgements

We acknowledge financial support from the Innovate UK project *The Influence of Graphite Irradiation Creep on Plant Life Optimisation* (101437) and thank EDF Energy for the graphite samples. We thank DLS and the Diamond-Manchester Collaboration for beamtime at I13-2 under proposal MT12200. We would like to express our gratitude to K Wanelik and A A Wilson for helping to commission the Open-Frame Rig and integrate it into the DLS data acquisition systems, L Lea and colleagues in the DLS Vacuum Group for containment vessel testing, S Logan for technician support, and R Doull of the DLS Health Physics Group for assistance with health and safety matters.

References

[1] McNally K, Fahad M, Tan E, Warren N, Hall G N and Marsden B J 2016 A numerical study of internal brick stresses in AGR moderator bricks *Nuclear Engineering and Design* **309** 277-93

[2] Bay B K, Smith T S, Fyhrie D P and Saad M 1999 Digital volume correlation: Three-dimensional strain mapping using X-ray tomography *Experimental Mechanics* **39** 217-26

[3] Atwood R C, Bodey A J, Price S W T, Basham M and Drakopoulos M 2015 A high-throughput system for high-quality tomographic reconstruction of large datasets at Diamond Light Source *Philosophical Transactions of the Royal Society A: Mathematical, Physical and Engineering Sciences* **373**

IOP Conf. Series: Journal of Physics: Conf. Series **849** (2017) 012022 doi:10.1088/1742-6596/849/1/012022

In-situ X-ray nano-CT System for Polymer Electrolyte Fuel Cells under Operating Conditions

O Sekizawa[1], T Uruga[1,2], N Ishiguro[3], H Matsui[4], K Higashi[1], T Sakata[1], Y Iwasawa[1] and M Tada[3,4]

[1] Innovation Research Center for Fuel cells, The University of Electro-Communications, Chofu, Tokyo 182-8585, Japan

[2] Japan Synchrotron Radiation Research Institute, SPring-8, Sayo, Hyogo 679-5198, Japan

[3] RIKEN SPring-8 Center, Sayo, Hyogo 679-5148, Japan

[4] Research Center for Materials Science, Nagoya University, Nagoya, Aichi 464-8602, Japan

E-mail: iwasawa@pc.uec.ac.jp

Abstract. We developed two types of in-situ three-dimensional imaging systems on the basis of full-field transmission X-ray computed tomography (XCT) methods for polymer electrolyte fuel cells (PEFCs) under operating conditions at beamline BL36XU at SPring-8. One was for a wide field of view (more than 500 μm) to obtain the whole membrane electrode assembly (MEA) images, and the other was for nano spatial resolution (less than 100 nm) using a Fresnel zone plate as objective optics. We succeeded in in-situ three-dimensional visualization of an MEA in PEFC using both XCT measurement systems and show preliminary results.

1. Introduction

A crucial issue for developing next generation polymer electrolyte fuel cells (PEFCs) is spatiotemporal information on the active factor and reaction mechanism, especially the degradation factor and mechanism in PEFC cathode catalysts. A PEFC has a multi-layered structure consisting of a membrane electrode assembly (MEA), gas diffusion layer (GDL), bipolar plate, and separator. An MEA consists of an anode catalyst layer, polymer electrolyte membrane (PEM), and cathode catalyst layer. The electrochemical reactions occur spatially inhomogeneously under potential operating conditions. To visualize the degradation mechanisms for the electrocatalysis, the internal structures of MEA in PEFC require in-situ non-destructive three-dimensional (3D) imaging under operating conditions.

We succeeded in 3D imaging of the chemical state distribution of electrode catalysts in an MEA by using an X-ray computed laminography (XCL) XAFS technique [1-3]. XCL is a 3D imaging technique suitable for laterally extended planar objects, such as MEAs. The geometric restriction of XCL measurement, however, imposed a specific design on the PEFC, which made it difficult to maintain the operating conditions of the PEFC due to heat dissipation from a large X-ray window. Also, X-ray scattering from support of PEFC disturbed transmission X-ray images of PEFCs. To improve this situation, we developed two types of in-situ 3D full-field transmission X-ray computed tomography (XCT) systems consisting of a newly designed PEFC at BL36XU at SPring-8 [3, 4]

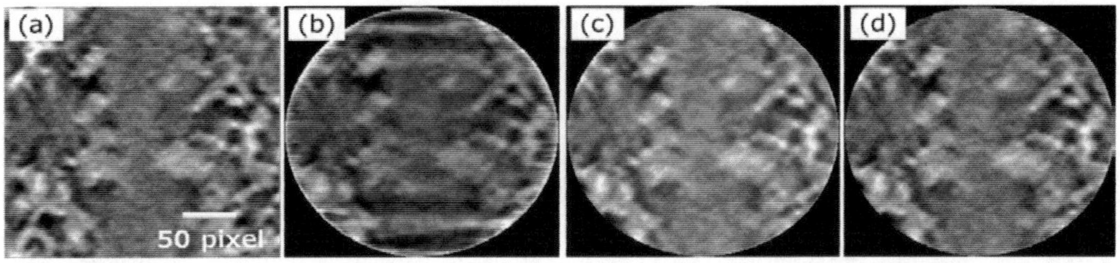

Figure 1. Simulation of reconstruction calculation with and without approximate correction during iterative reconstruction cycles. Original image (a), reconstructed image after 50 cycles without correction (b), after 10 cycles with correction (c), after 50 cycles with correction (d).

2. Methods and results

3D reconstruction ideally requires two-dimensional (2D) projection images measured at a projection angle from 0 to 180 deg. In the case of measurements of real PEFCs, 2D projection images are difficult to measure at all angles. To measure images at the widest possible angle, we designed a new PEFC single cell to cover a measurement angle from -80 to 80 deg for the XCT measurement. The newly designed PEFC single cell was based on the Japan Automobile Research Institute (JARI) standard cell [5], which made it possible to conduct XCT measurements under experimental conditions almost the same as usual laboratory conditions. The MEA consisted of an anode catalyst layer (10~15 μm thick), a Nafion membrane (50 μm thick) and a cathode catalyst layer (10~15 μm thick). The MEA was sandwiched with GDL (190 μm thick) in the PEFC cell [6]. The MEA was larger than an X-ray beam, so X-ray absorption of the area outside of the reconstruction produced artifacts on reconstructed images. We used a simple correction method to reduce the artifacts by approximating absorption of the area outside of the reconstruction by using the average value of the reconstruction area. Figure 1 shows the simulation results with and without the correction. The original image used was a cross-sectional image of PEFC measured by XCL. The number of projection was 1600 images (-80 to 80 deg) and reconstructed by OS-EM method [7]. The original image was well reconstructed using the approximate correction.

We developed the following two types of in-situ full-field XCT measurement system having different fields of view (FOVs) and spatial resolutions to conduct the nondestructive microscopic imaging of an MEA in a PEFC under operating conditions. One is a wide full-field transmission XCT system for obtaining microscopic images of the overall MEA between the cathode and anode GDLs, and the other is a full-field high-spatial resolution imaging nano-XCT system using a Fresnel zone plate (FZP) as objective optics to achieve a detailed imaging of the cathode catalyst layer and PEM.

2.1. In-situ wide full-field transmission XCT measurement system

Figure 2 shows the layout of a wide full-field transmission XCT measurement system. Gas tubes and heater cables were connected to PEFCs through a rotary joint and slip ring to rotate the PEFCs smoothly and freely during full-field imaging measurement. The transmission X-ray images of PEFCs were measured by an indirect high-resolution X-ray imaging system, which is an imaging unit (Hamamatsu Photonics K.K., AA50) combined with a low noise sCMOS camera (Hamamatsu photonics K.K., ORCA-Flash 4.0). The transmitted X-ray

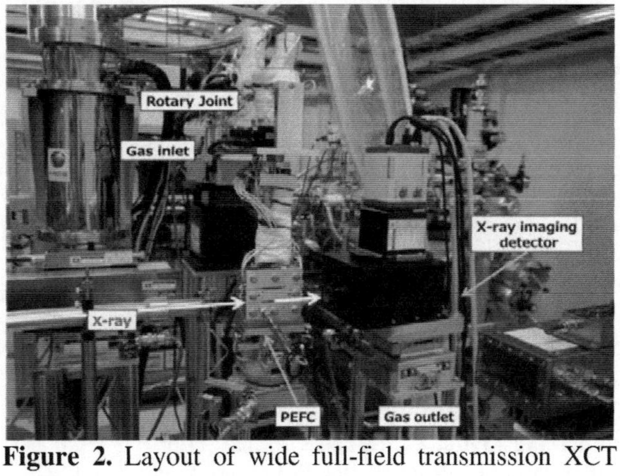

Figure 2. Layout of wide full-field transmission XCT system.

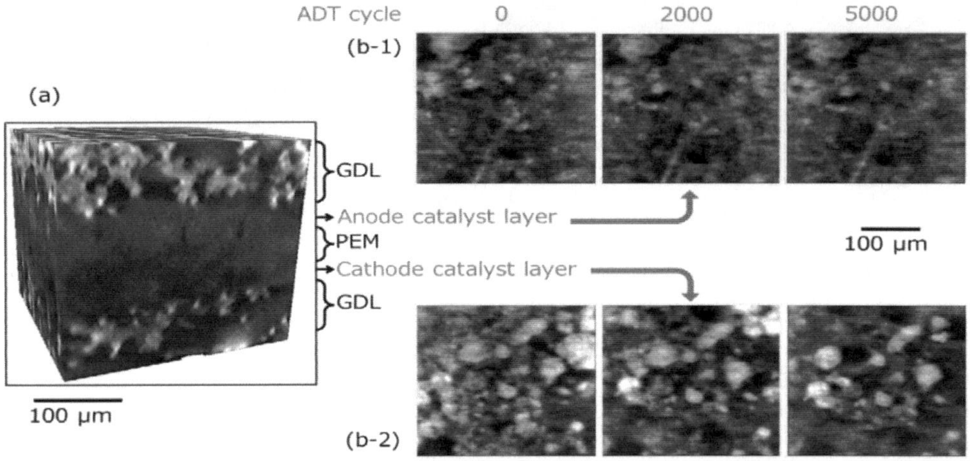

Figure 3. Reconstructed 3D image of MEA in PEFC (a) and its cross-sectional images of anode (b-1) cathode catalyst layer (b-2) during ADT. Catalyst: Pt_3Ni/C (cathode) and Ru/C (anode), gas: N_2 (cathode) and H_2 (anode), cell temperature: 80 °C, operation cell voltage during ADTs: 0.6 to 1.0 V Measurement time of each image: 3 s.

image was converted into a visible light image by a single crystal scintillator ($Ce:Gd_3Al_2Ga_3O_{12}$) and magnified by 20 on the camera by an objective lens in the imaging unit. The FOV and spatial resolution of the system were ca. 500×500 μm^2 and 1 μm, respectively.

Figure 3(a) shows preliminary results of 3D image reconstruction of an MEA in a PEFC under the operation conditions. The bright area shows high absorption area, such as metal catalyst-rich area. The whole the MEA in the PEFC was clearly visualized from anode to cathode GDL. Strong X-ray absorption of the area outside of the reconstruction, especially that of the GDLs, at high incident angles produced the artifact on the reconstructed images. Our simple approximate correction in this study could not fully reduce the artifact, which lowered the image resolution slightly. Figures 3(b-1) and (b-2) show reconstructed cross-sectional images at the cathode and anode catalyst layer during accelerated durability tests (ADTs). The anode catalyst layer had relatively uniform distribution of the catalyst particles before the ADTs and no significant change occurred during the ADTs. In contrast, in the cathode catalyst layer, the catalyst particles were ununiformly distributed before the ADTs, and some catalyst particle-rich areas disappeared during the ADTs. The location-dependent degradation process of the MEA was successfully observed under operation conditions.

2.2. In-situ full-field high-resolution imaging nano-XCT measurement system

The wide full-field transmission XCT system mentioned in chapter 2.1 had FOV to observe the overall MEA, but its spatial resolution (ca. 1 μm) was insufficient to achieve detailed imaging of inside of the Pt catalyst layer (10~15 μm thick). To achieve high-resolution imaging, we developed an in-situ full-field nano-XCT measurement system using a FZP as objective optics. Figure 4 shows the layout of the system. The optics of the system was based on that

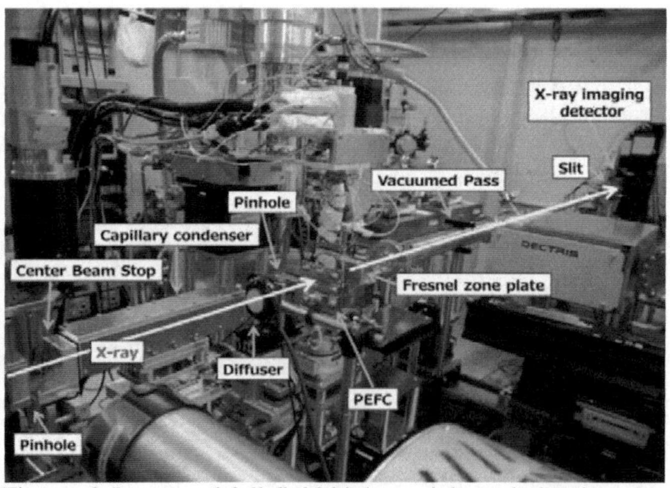

Figure 4. Layout of full-field high-spatial resolution imaging nano-XCT system.

designed by Suzuki et al. at beamline BL47XU at SPring-8 [8]. An optical system for illuminating objects was a beam condenser consisting of a single-bounce conical-shape mono-capillary made of Pyrex glass. The objective FZP was a commercially available one (NTT-AT Corp., FZP-155/50). The zone material is 0.5-µm-thick tantalum. The outermost zone width of FZP is 50 nm, and the diameter is 155 µm. The focal length of the FZP is 75 mm at 12 keV. A detector was placed 2400 mm from FZP, and the magnification by FZP was ca. 30 at 12 keV. The magnification of the X-ray imaging detector was set to 5, so the total magnification of the imaging system was 150. By 2D imaging measurement of X-ray chart pattern (NTT-AT Corp., XRESO-50HC, Ta: 500 nmt), the minimum pattern (50 nm) was clearly distinguished. The sample cell and its connected equipment, such as a rotary joint and slip ring, were the same as those used for the wide full-field transmission XCT measurement system.

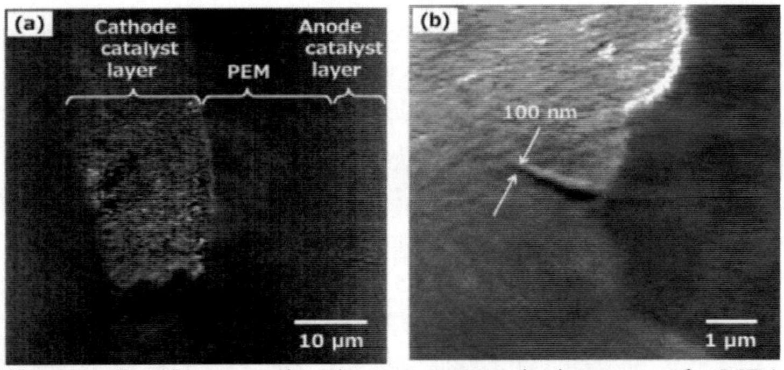

Figure 5. Cross-sectional reconstructed images of MEA perpendicular (a) and parallel (b) to MEA plane. Catalyst: Pt/C (cathode) and Pd/C (anode) [6], gas: N_2 (cathode and anode), cell temperature: 25 °C.

Figure 5 shows a preliminary result of a reconstructed cross-sectional image of an MEA measured using the nano-XCT measurement system. The FOV of the system was 60×60 µm^2. The fine image of the inside of the cathode catalyst layer, such as 100 nm width cracks, and its boundary with the PEM was clearly obtained, which indicates the spatial resolution of nano-XCT system was less than 100 nm.

3. Conclusion

We developed two types of in-situ full-field XCT measurement systems for PEFCs: a wide full-field transmission XCT system (FOV: 500 µm, spatial resolution: 1 µm) and a high-resolution imaging nano-XCT system (FOV: 60 µm, spatial resolution: < 100 nm). We successfully observed 3D structural changes in MEA under operating conditions during ADT. We plan to develop an automatic switching system for both XCT measurement systems to achieve sequential observations with wide FOV and with high spatial resolution on the same area in a PEFC.

4. Acknowledgement

We thank Drs. Kentaro Uesugi and Masato Hoshino for their helpful discussions with us about constructing full-field XCT measurement systems. We also thank Mr. Yoshiki Miura for his help in designing parts of XCT measurement system. XCT measurements were performed with the approval of SPring-8 proposals (2014A7821, 2014B7821, 2015A7840, 2015B7840, 2016A7840). This work was supported by the New Energy and Industrial Technology Development Organization (NEDO), Japan.

References

[1] Saida T, Sekizawa O, Uruga T, Tada M, *et al.* 2012 *Angew. Chem. Int. Ed.* **51** 10311-14
[2] Sekizawa O, *et al.* 2014 *X-Ray Microscopy 2014 Conference Program Handbook* p 287
[3] Sekizawa O, Uruga T, Takagi Y, Nitta K, Kato K, *et al.* 2016 *J. Phys.: Conf. Ser.* **712** 012142
[4] Sekizawa O, Uruga T, Tada M, Nitta K, Kato K, *et al.* 2013 *J. Phys.: Conf. Ser.* **430** 012020
[5] Hashimasa Y, *et al.*, 2006 *J. Power sources* **155** 182
[6] Ishiguro N, Sekizawa O, Uruga T, Matsui H, *et al.* 2016 *J. Phys. Chem. C.* **120** 19642-51
[7] Hudson HM, Larkin RS, 1994 *IEEE Trans Med Imag* **13** 601-09
[8] Suzuki Y, Takeuchi A, Uesugi K, Terada Y, *et al.* 2013 *J. Phys.: Conf. Ser.* **463** 012028

IOP Conf. Series: Journal of Physics: Conf. Series **849** (2017) 012023 doi:10.1088/1742-6596/849/1/012023

Observation of the Interface between Resin and Carbon Fiber by Scanning Transmission X-ray Microscopy

T. Harano[1], R. Murao[1], Y. Takeichi[2], M. Kimura[2], Y. Takahashi[3]

[1] Advanced Technology Research Laboratories, Research and Development, Nippon Steel & Sumitomo Metal Corporation, Futtsu, Chiba 293-8511, Japan
[2] Photon Factory, Institute of Materials Structure Science, High Energy Accelerator Research Organization, Tsukuba, Ibaraki 305-0801, Japan
[3] Department of Earth and Planetary Science, The University of Tokyo, Bunkyo-ku, Tokyo 113-0033, Japan
E-mail: harano.ya6.takayuki@jp.nssmc.com

Abstract. The sub-micron spatial distribution of chemical states of carbon fiber (CF) and the interface between CF and resin must affect physical properties of carbon-fiber-reinforced plastic (CFRP). In order to evaluate scanning transmission X-ray microscopy (STXM) techniques for application to the chemical-state analysis of CFRP, we performed STXM measurements near C *K*-edge energies. The results of the spectral deconvolution analysis suggest the presence of another phase at the interface of CF and resin, which may be a coating layer. In addition, the preferred orientation of the stuck of graphene sheets to the fiber axis direction of CF was observed by using linear polarized X-ray beams.

1. Introduction
It is essential to investigate the interface between materials as well as their microstructures to reveal the origins of the physical properties of composite materials. For example, chemical states (valence, functional groups, *etc.*) around the interface between carbon fiber (CF) and resin in carbon-fiber-reinforced plastic (CFRP) gives helpful information in clarifying the enhancement mechanisms of tensile strength and elastic modulus. Conventional techniques for these studies include X-ray photoelectron spectroscopy and X-ray absorption near-edge structure (XANES) spectroscopy. However, their spatial resolutions are in the order from mm to μm. This makes it difficult to observe the interface structures in detail. The spatial resolution of electron energy-loss spectrometry (EELS) is much higher (less than 1 nm) than those of conventional spectroscopy techniques. However, EELS might not be appropriate for organic compounds because the strong electron beams sometimes damage organic materials. On the other hand, X-ray beams are much softer than electron beams. In the past ten years, lithographic fabrication techniques have enabled us to use optical devices such as Fresnel zone plates (FZP) whose spatial resolution is 20-100 nm. Scanning transmission X-ray microscopy (STXM) is a very powerful method for chemical imaging around the interface of composite materials [1]. In this study, we performed C *K*-edge absorption imaging applied to the analysis of distributions of chemical states in CFRP.

2. Experimental
CFRP thin film was fabricated by focused-ion-beam (FIB) processing and finished by milling using an argon ion beam to remove the damage to surfaces by FIB. The thickness of the film was adjusted to about 80 nm. The CFRP analyzed in this study is a commercial product that consists of CF made from

polyacrylonitrile and epoxy resin. The estimated tensile strength and elastic modulus of this CFRP are 1760 MPa and 125 GPa, respectively. Fig. 1(a) shows a schematic illustration of a cross section (CS) and a longitudinal section (LS) of the CF. Fig. 1(b) presents a backscattered electron image of the thin film. Note that the angle between the incident direction of electron beams and the normal vector of the thin film is 35 degrees. We collected image stacks of C *K*-edge XANES in the red-dotted square regions ((I) the interface region between CF and resin, and (II) CS of CF) in Fig. 1(b). Fig. 1(c) shows a transmission electron microscope (TEM) image of CF (CS) and resin. An inner-ring texture was observed in CF (CS), and details will be discussed in the next section.
STXM measurements were performed at BL-13A of the Photon Factory, IMSS, KEK, Japan. The compact STXM instrument [2] realizes XANES imaging with a spatial resolution of 40 nm by using FZP, and is especially suited to measuring XANES spectra around C *K*-edges. Moreover, the undulator of BL-13 can generate a radiation light source switched to horizontal and vertical linear by APPLE-II-type undulators [3]. In this study, linear horizontal ρ/2 (LHR) and linear vertical σ/2 (LVS) modes of X-rays were used. The aXis2000 software was used for analysis of the image stacks [4].

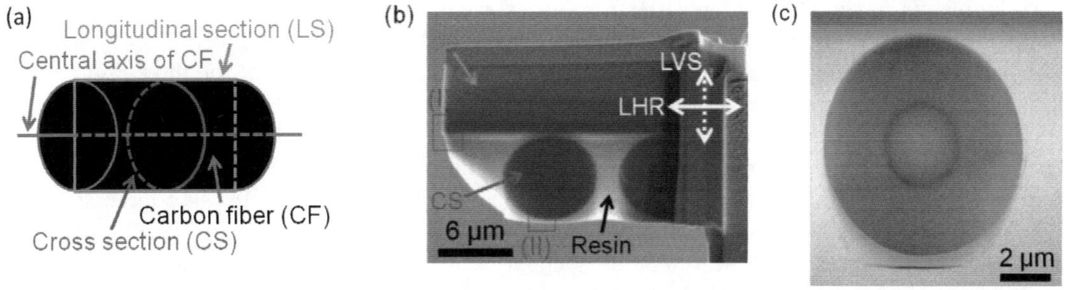

Figure 1. (a) Schematic illustration of a cross section (CS) and a longitudinal section (LS) of CF. (b) BSE image of the thin film. (c) Transmission electron microscope image of CS of CF and resin.

3. Results and discussion

Fig. 2(a) shows an X-ray absorption contrast image of (I) in Fig. 1(b) taken at *E*=286.5 eV of LHR mode of X-rays. The energy corresponds to the transitions from 1*s* to π* (phenolic (C*-OH)). This image consists of 70×70 pixels (pixel size: 40×40 nm^2). Around the interface between CF and resin, another absorption contrast accounted for by neither CF nor resin was observed. This suggests the presence of another phase around the interface. We consider that the fact that the absorbance of the interface region is higher than those of CF and resin can be explained by the difference in density among them because we adjusted the entire thickness uniformly by FIB processing and argon milling. Image stack measurements around the C *K*-edge were carried out in the view shown in Fig. 2(a). Fig. 2(b) shows the C *K*-edge XANES spectra of resin, CF (LS), and the interface. These are the averaged spectra of the black-dashed square regions of (i) resin, (ii) CF (LS), and (iii) the interface in Fig. 2(a), respectively. These spectra are normalized by the intensity of the peak top of the transitions from 1*s* to σ* of C-C at 292 eV. The split peak in the XANES spectra of the regions of (i) and (iii) around 284.8 eV corresponds to the transition from 1*s* to π* of C=C, while the other peaks at 286.5 and 288.3 eV correspond to the transitions from 1*s* to π* (phenolic (C*-OH)) and those from 1*s* to Rydberg/σ* (O(C*-H)R), respectively [5]. Moreover, the peak at 285.2 eV of the spectrum of (ii) corresponds to the transitions from 1*s* to π* of C=C. These are consistent with the typical chemical structure of epoxy resin [6]. In the case of the spectrum of (ii), only two peaks were observed at 285.2 eV and around 292 eV, which correspond to the transitions from 1*s* to π* of C=C and those from 1*s* to σ* of C-C, respectively. This result is consistent with the X-ray diffraction (XRD) study of CF [7], which shows XRD peaks originating from graphite. Therefore, this suggests that CF consists of stacks of graphene sheets. On the other hand, the peak positions of the spectrum of (iii) are the same as those of (i). However, the intensity ratio among the peaks is different from those of (i). These observations suggest that the organic material around the interface consists of the same functional groups with different compositional ratios from those of resin. This result is in accordance with the general manufacturing process of CFRP, in which CF is coated with organic material before being embedded in resin to improve the affinity between CF and resin.

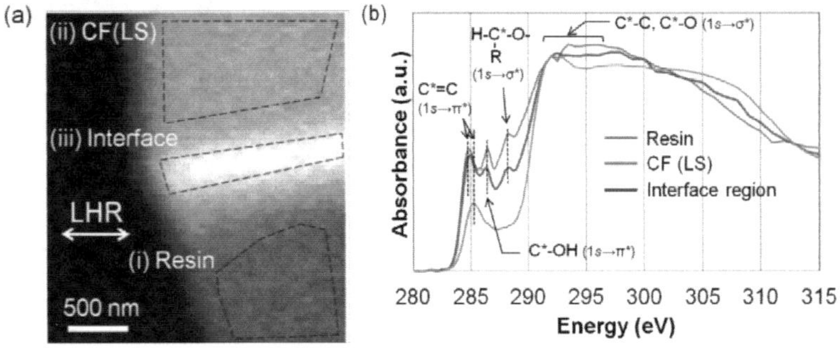

Figure 2. (a) The black-dashed squares show regions of averaged XANES spectra of (i) resin, (ii) CF (LS), and (iii) the interface used for RGB deconvolution. (b) Averaged XANES spectra of resin, CF (LS), and the interface in each region shown in Fig. 2(a).

Firstly, in order to deliberate whether a third phase is present between CF and resin, two types (A and B) of singular-value decomposition (SVD) analyses were carried out: (A) by using (i) and (ii) excluding (iii) shown in Fig. 2(a); and (B) by using the averaged spectra of (i), (ii), and (iii), respectively. These averaged spectra are shown in Fig. 2(b). Residual curves of the three regions after SVD analyses of (A) and (B) are shown in Fig. 3(a) and (b), respectively. Both regions of (i) and (ii) fit in very well in both cases. By contrast, the region of (iii) could fit in only in the case of (B). From the perspective of spectra analysis, it is concluded that a third phase is present in the interface, which is neither CF nor resin. An RGB color-deconvoluted image of the SVD of (B) using three spectra is shown in Fig. 3(c). The thickness of the third phase is about 100 nm. In addition, a transition region of about 300 nm in thickness was observed at the boundary of the resin and the third phase.

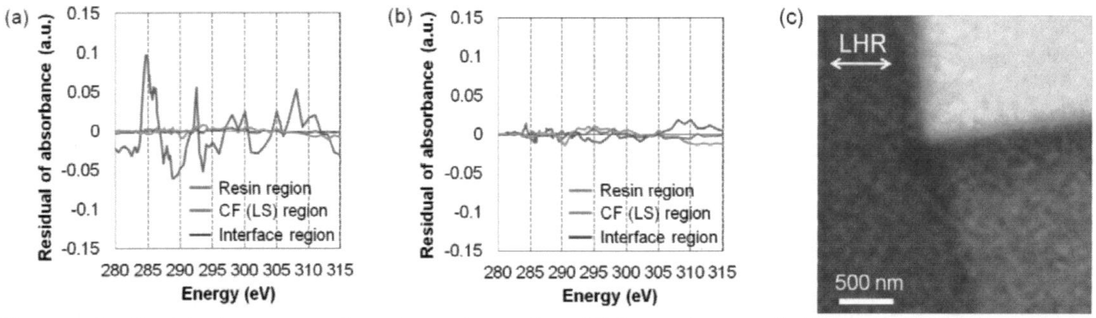

Figure 3. Residual curves of each region after SVD analyses of (A) resin/CF(LS) and (B) resin/CF(LS)/interface are shown in (a) and (b), respectively. (c) RGB (red: resin, green: CF (LS), blue: the third phase) color-deconvoluted image of the interface between resin and CF (LS). The two-way arrow is the polarization of LHR mode of X-ray beams.

Subsequently, we performed image stack measurements of (II) in Fig. 1(b) to investigate the chemical state of the inner ring observed in CF (CS). C *K*-edge spectra of resin, CF (CS), and inner ring in CF (CS) are shown in Fig. 4(a). These spectra are averages of each region and normalized at 292 eV as well as Fig. 2(b). The configuration of the inner ring is almost the same as that of CF (CS). This means that the chemical state of the inner ring was not different from that of CF. The contrast of the inner ring in CF (CS) observed in the TEM and X-ray absorption contrast images may originate from the difference in density between them.

Finally, the intensity ratio of the transitions from 1*s* to π^* to those from 1*s* to σ^* varied in the cases of LS (green spectrum in Fig. 2(b)) and CS (green spectrum in Fig. 4(a)) of CF. It is well known that the XANES peak of a particular electron orbital is sensitive to the relationship between the orientation of its electron orbital and the polarization of incident X-ray beams [8]. The XANES intensity originating from a particular orbital is enhanced when the orientation of its orbital is parallel to the polarization of the X-ray. Fig. 4(c) shows the averaged XANES spectra of CS and LS of CF using LHR and LVS. In the case of CS, there is no change in the XANES spectra of the C *K*-edge between two polarized X-rays. On the other hand, in the case of LS, the absorption peak intensity originating from the transitions from 1*s* to π^* using LVS was higher than that using LHR. This indicates that the π orbital in CF is oriented to the central axis of CF. From these results, it is suggested that the graphene sheets almost preferred to orient to the fiber axis direction as schematically illustrated in Fig. 4(d).

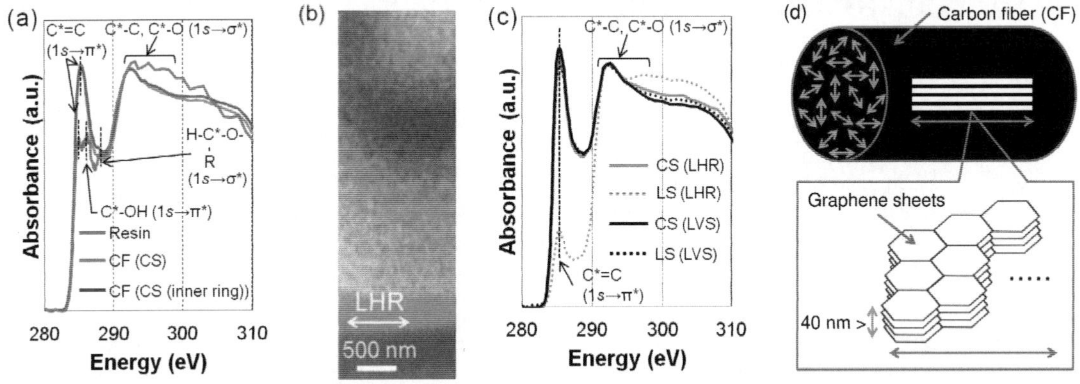

Figure 4. (a) Averaged XANES spectra of resin (red), CF (CS) (green), and inner ring in CF (CS) (blue) in (II) of Fig. 1(b). (b) RGB (red: resin, green: CF (CS), blue: inner ring in CF (CS)) color-deconvoluted image. The two-way arrow shows the polarization of LHR of X-ray beams. (c) Averaged XANES spectra of CS and LS of CF using two types of polarized X-ray. These spectra are normalized by the intensity at 292 eV. (d) Schematic illustration of the distribution of graphene sheets in CF. The blue and red two-way arrows are the thickness and plane directions of graphene sheets, respectively.

4. Conclusion

STXM studies of CFRP were performed to reveal the distributions of the chemical states in CFRP and to evaluate STXM as a characterization technique for chemical imaging of CFRP. From the image stacks of C *K*-edge absorption, the presence of a coating layer between CF and resin is suggested on the basis of the spectral deconvolution analysis of the spectra. The inner region of CF (CS) has the same chemical structure as the outer region of CF. Orientation of the π orbital of C=C in CF could be observed by linear polarized X-rays. STXM is a useful technique for analyzing chemical-state imaging of carbon-bearing multi-composite materials, providing additional and essential information regarding enhancement mechanisms of tensile strength and elastic modulus of CFRP.

Acknowledgements

We would like to thank Prof. K. Mase (KEK) for his technical support. The experiment was conducted with the approval of the Photon Factory Program Advisory Committee (proposal no. 2015C206).

References

[1] Ade H and Hsiao B 1993 *Science* **262** 1427
[2] Takeichi Y, Inami N, Suga H, Miyamoto C, Ueno T, Mase K, Takahashi Y, Ono K 2016 *Rev. Sci. Instrum.* **87** 013704
[3] Tsuchiya K, Shioya T, Aoto T, Harada K, Obina T, Sakamaki M, Amemiya K 2013 *J. Phys.: Conf. Ser.* **425** 132017
[4] Hitchcock A P 2009 http://unicorn.mcmaster.ca/aXis2000.html
[5] Moffet R C 2011 *LBNL Paper* LBNL-4245E
[6] Hedrick J L, Yilgor I, Jurek M, Hedrick J C, Wilkes G L and McGrath J E 1991 *Polymer* **32** 2020
[7] Qin X, Lu Y, Xiao H, Wen Y, Yu T 2012 *Carbon* **50** 4459
[8] Watts B,Schuettfort T, McNeill C R 2011 *Adv. Funct. Mater.* **21** 1122

Characterisation of speckle-based X-ray phase-contrast imaging

M Zdora[1,2], P Thibault[3], C Rau[1] and I Zanette[1]

[1] Diamond Light Source, Harwell Science and Innovation Campus, OX11 0DE, UK
[2] Department of Physics & Astronomy, University College London, WC1E 6BT, UK
[3] Department of Physics & Astronomy, University of Southampton, SO17 1BJ, UK

E-mail: marie-christine.zdora@diamond.ac.uk

Abstract. We present a study on the influence of different scan and reconstruction parameters on the image quality of the differential phase signals obtained from speckle-based X-ray phase-contrast imaging measurements in single-shot as well as 2D and 1D speckle-stepping modes. In particular, the effects of the analysis window size and the number of diffuser steps on the spatial resolution and signal sensitivity of images of a phantom sample are investigated and discussed. It is shown that the trade-off between spatial resolution, scan time and simplicity of the setup has to carefully be addressed for each specific experiment.

1. Introduction

In the last decades X-ray phase-contrast imaging has shown vast potential in the fields of biomedical imaging, material science and metrology. Among the different techniques speckle-based X-ray phase-contrast imaging [1, 2, 3], has recently seen increased interest due to its simple setup and its compatibility with polychromatic X-rays [4, 5].The sample-induced modulations of an X-ray near-field speckle pattern created by a random phase modulator are analysed to retrieve the differential phase signal, the transmission and the small-angle scattering signal. X-ray speckle imaging is typically performed in two different operational modes, speckle tracking [1, 2] and speckle scanning [3]. While single-shot speckle tracking is fast and dose efficient, it suffers from low spatial resolution. The speckle-scanning mode can achieve significantly higher resolution, but requires several hundreds of projections, an extremely stable setup and high-precision scanning motors. Recently, also a one-dimensional stepping approach has been proposed [6], and other operational modes are currently being explored aiming to further reduce the number of steps and simplify the stepping scheme [7]. Here, we characterise the different existing operational modes of X-ray speckle-based imaging on a phantom sample exploring the effects of different scan and reconstruction parameters, in particular the analysis window size and the number of diffuser steps.

2. Experimental setup and signal reconstruction

Measurements were performed at the I13-2 Diamond-Manchester Imaging Beamline at Diamond Light Source, UK using a monochromatic X-ray beam of energy 19 keV with the setup shown in Fig. 1(a). The sample - silicon spheres (diameter 0.48 mm) and quartz spheres (diameter 0.35 mm) in a polypropylene pipette tip filled with water - was mounted on a hanging rotation

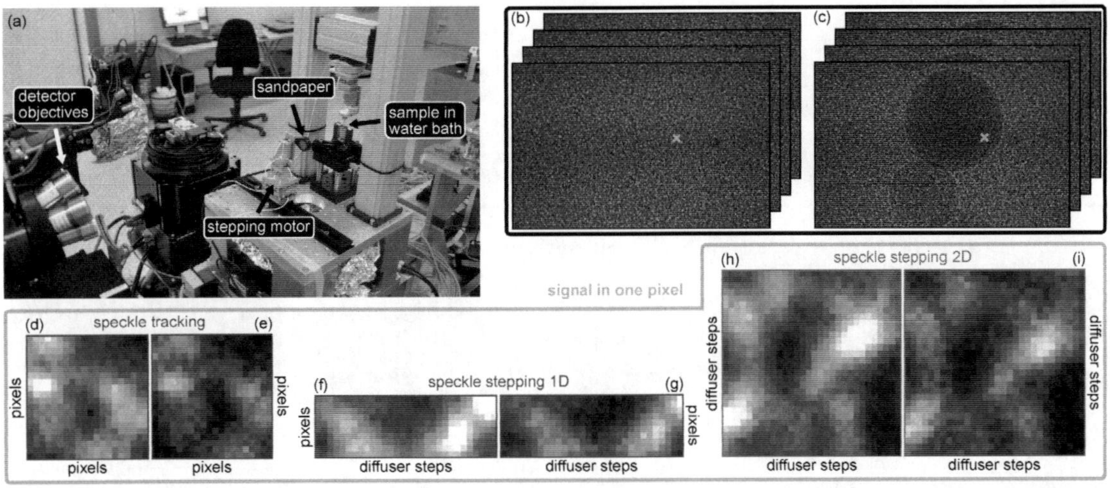

Figure 1. Experimental setup (a). Reference images (b) and sample images (c) are collected for different diffuser steps. The visibility[1] of the pattern averaged over the whole field of view is 42%. The signal in one pixel (cyan cross) is shown for single-shot tracking analysis (d, e), for 1D speckle-stepping mode (f, g) and for 2D speckle-stepping mode (h, i).

stage and immersed in a water tank. The diffuser, here a piece of P800 sandpaper, was placed on a piezo xy-translation stage at a distance of approximately 10 cm downstream of the specimen. The detector system (150 μm-thick CdWO4 scintillation screen, optical microscope with 20× magnification, pco.4000 CCD camera with 9 μm pixel size) giving an effective pixel size of $p_{\text{eff}} = 0.45\,\mu$m was positioned 20 cm downstream of the diffuser.
Reference and sample images were acquired for 30×30 translational steps of the diffuser (step size: $d_{\text{step}} = 0.45\,\mu$m) with an exposure time of 2.0 s per projection.

The reconstruction of the image signals in single-shot mode as well as in 2D and 1D stepping modes is based on cross-correlation in real space [4] delivering the displacement s of the speckle pattern caused by the refraction of X-rays in the sample, which can be converted into a refraction angle $\alpha = s\zeta/z$, where z is the diffuser-detector distance, $\zeta = p_{\text{eff}}$ for the single-shot mode and $\zeta = d_{\text{step}}$ for the speckle-stepping mode. In case of 1D stepping, $\zeta = d_{\text{step}}$ for the refraction in the scan direction and $\zeta = p_{\text{eff}}$ for the signal perpendicular to the scan direction. For the single-shot analysis a windowed cross-correlation is applied to a single reference and sample image, see Figs. 1(d)-(e), while the analysis of the 2D speckle-stepping scan is performed pixel-wise in the diffuser plane over the sample and reference signals at the different diffuser steps, see Figs. 1(h)-(i). A 1D stepping analysis was conducted considering only one row of 30 steps in the horizontal and a few pixels in the vertical direction for cross-correlation (see Figs. 1(f)-(g)). From the two differential phase signals the phase shift can be obtained, e.g. via 2D Fourier integration [8].

3. Effects of different scan and reconstruction parameters

The different parameters used for reconstruction in single-shot speckle-tracking, in conventional 2D speckle-stepping and in 1D horizontal speckle-stepping modes are summarised in Tab. 1.

For the speckle-tracking analysis the window size was varied from 5×5 pixels to 30×30 pixels. Figures 2(a)-(d) show the reconstructed refraction angle signals in the horizontal direction.

[1] We define the visibility of the speckle pattern as $(I_{\max} - I_{\min})/(I_{\max} + I_{\min})$, where $I_{\max}$ and $I_{\min}$ are the maximum and minimum intensities of the pattern in a 100 × 100 pixels region of interest in the background area.

tracking	**5×5 px**	**10×10 px**	**20×20 px**	**30×30 px**
$\sigma_x \mid \sigma_y$ [nrad]	1982 | 1824	1309 | 1157	352 | 270	153 | 144
2D stepping	**5×5 steps**	**10×10 steps**	**20×20 steps**	**30×30 steps**
$\sigma_x \mid \sigma_y$ [nrad]	1979 | 1875	1346 | 1219	399 | 312	180 | 150
1D stepping	**30 steps, 5 px**	**30 steps, 10 px**	**30 steps, 20 px**	**30 steps, 30 px**
$\sigma_x \mid \sigma_y$ [nrad]	764 | 675	435 | 331	241 | 184	188 | 139

Table 1. Different reconstruction and scan parameters and respective angular sensitivities.

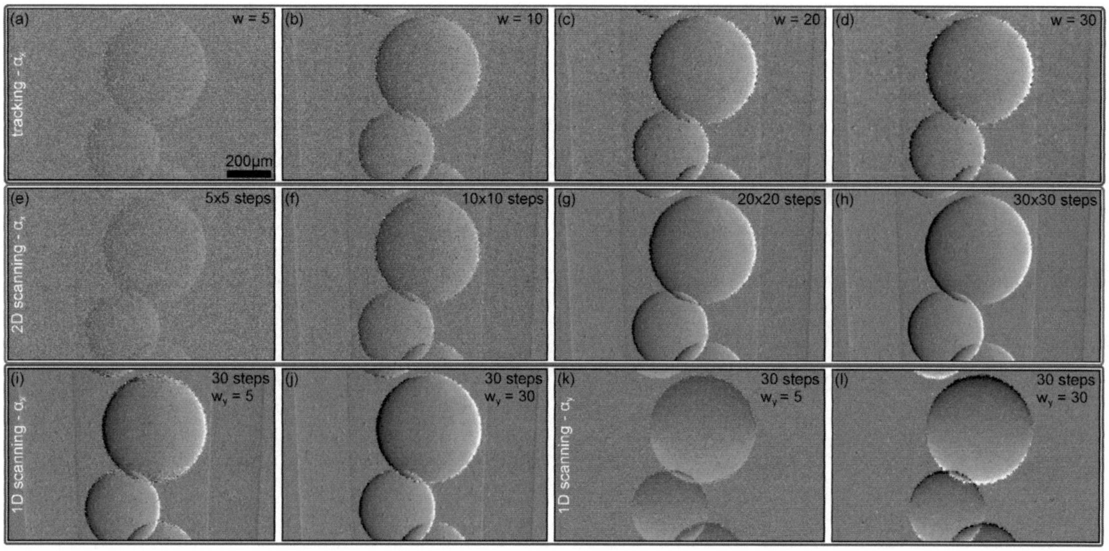

Figure 2. Refraction α_x in the horizontal reconstructed in single-shot mode with different window sizes (a-d), in 2D speckle-stepping mode with different numbers of steps (e-h) and in 1D stepping mode with 30 steps in the horizontal and different numbers of pixels in the vertical (i,j). Refraction α_y in the vertical for 1D scanning (k, l). Intensity windows: -6 μrad to 6 μrad.

A clear improvement in image quality with increasing window size can be observed. This is confirmed by the increase in angular sensitivity defined as the standard deviation of the refraction angle in a region of 100 × 100 pixels in the background area without sample, which can be found in Tab. 1. However, a larger window results in a loss in resolution as visible in Figs. 2(a)-(d). The trade-off between spatial resolution limited by the window size and image quality has to carefully be considered for the desired results.

A significant increase in spatial resolution can be achieved with the speckle-stepping mode by performing a 2D scan of the diffuser in steps smaller than the speckle size. However, this comes at the cost of a large number of acquisition frames and the need for very accurate stepping motors to ensure equidistant step sizes. The refraction angle signals in the two directions were reconstructed for different numbers of steps keeping the step size constant at 0.45 μm, see Figs. 2(e)-(h). The image quality significantly improves with more steps (see Tab. 1), however the immense increase in scan time and hence dose to the sample needs to be considered for practical implementation.

As a third operational mode providing some trade-off between the two previous methods, the diffuser can be scanned in only one direction, which significantly reduces the number of acquired projections. However, for a 2D analysis of the speckle shift, we have to take several pixels in the direction perpendicular to the scan direction. Commonly several tens of scan positions are

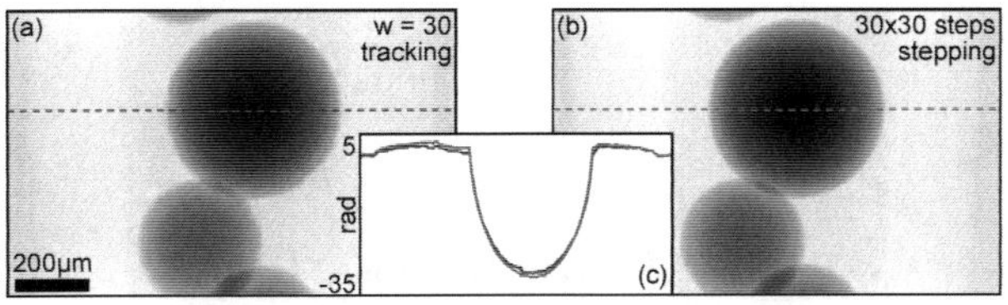

Figure 3. Integrated phase shift for the single-shot (a) and the 2D stepping case (b). Intensity window: -33 rad to 4 rad. Horizontal line profiles through the centre of the silicon sphere (c).

used for this mode to allow a small window size of only several pixels [6]. Here, a scan of 30 steps along the horizontal direction was considered and w_y pixels in the vertical direction were taken for analysis. The reconstructed refraction angle signals in the horizontal and the vertical directions in Figs. 2(i)-(j) and 2(k)-(l), respectively, demonstrate that taking a larger number of pixels in the vertical direction for analysis improves the image quality at the cost of a reduced resolution in the vertical direction.

The observations from the differential phase signals translate to the integrated phase shift (see Figs. 3(a) and (b) for single-shot imaging with a window size of 30 $\times$ 30 pixels and 2D speckle scanning with 30 $\times$ 30 steps, respectively). The resolution of the speckle-stepping scan is superior to the single-shot case. Moreover, artefacts at the edges of the spheres can be observed in single-shot mode, which are not present for the stepping case.

4. Conclusion and outlook

We have presented a study on the effects of scan and reconstruction parameters on the image quality of the differential phase signal obtained from speckle-based single-shot tracking, 2D scanning and 1D scanning measurements.

For single-shot speckle imaging the size of the analysis window needs to be in the order of 20 $\times$ 20 pixels for an acceptable angular sensitivity, which, however, leads to a loss in spatial resolution of 20$\times$ the pixel size. A comparable angular sensitivity at much higher resolution can be achieved with 2D stepping over at least 20 $\times$ 20 steps, which also yields accurate results at the edges of features where the single-shot analysis fails. 1D scanning requires less projections, but leads to a reduced resolution in the direction perpendicular to the scan direction.

When choosing the preferred operational mode for a speckle imaging experiment, there is an inevitable trade-off between spatial resolution, scan time and simplicity of the setup. To overcome this and combine the advantages of the techniques, investigations of a mixed approach are currently being undertaken and are expected to yield high-resolution, high-sensitivity results with a simple setup and only few required projections.

Acknowledgments

We thank J. Vila-Comamala (ETH Zurich), R. Gradl (TU Munich), K. S. Morgan (Monash University), K. Wanelik and S. Logan (Diamond Light Source) for support in the experiment. We acknowledge Diamond Light Source for time on I13-2 under proposal NT13211-1.

References

[1] Berujon S, Ziegler E, Cerbino R and Peverini L 2012 *Phys. Rev. Lett.* **108** 158102
[2] Morgan K S, Paganin D M and Siu K K W 2012 *Appl. Phys. Lett.* **100** 124102
[3] Berujon S, Wang H and Sawhney K 2012 *Phys. Rev. A* **86** 063813
[4] Zanette I, Zhou T, Burvall A, Lundström U, Larsson D H, Zdora M, Thibault P, Pfeiffer F and Hertz H M 2014 *Phys. Rev. Lett.* **112** 253903
[5] Zhou T, Zanette I, Zdora M, Lundström U, Larsson D H, Hertz H M, Pfeiffer F and Burvall A 2015 *Opt. Lett.* **40** 2822–5
[6] Wang H, Kashyap Y and Sawhney K 2016 *Sci. Rep.* **6** 20476
[7] Berujon S and Ziegler E 2015 *Phys. Rev. A* **92** 013837
[8] Kottler C, David C, Pfeiffer F and Bunk O 2007 *Opt. Express* **15** 1175–1181

IOP Conf. Series: Journal of Physics: Conf. Series **849** (2017) 012025 doi:10.1088/1742-6596/849/1/012025

Near-Field Diffraction Imaging from Multiple Detection Planes

L Loetgering*, M Golembusch, R Hammoud, T Wilhein

Institute for X-Optics, RheinAhrCampus Remagen - Joseph-Rovan-Allee 2, 53424 Remagen, Germany

*loetgering@hs-koblenz.de

Abstract. We present diffraction imaging results obtained from multiple near-field diffraction constraints. An iterative phase retrieval algorithm was implemented that uses data redundancy achieved by measuring near-field diffraction intensities at various sample-detector distances. The procedure allows for reconstructing the exit surface wave of a sample within a multiple constraint satisfaction framework neither making use of a priori knowledge as enforced in coherent diffraction imaging (CDI) nor exact scanning grid knowledge as required in ptychography. We also investigate the potential of the presented technique to deal with polychromatic radiation as important for potential application in diffraction imaging by means of tabletop EUV and X-ray sources.

1. Introduction

With the construction of high-coherent flux synchrotron sources and the development of fast and high-capacity computation facilities, the X-ray community witnessed a recent advent of coherent imaging techniques [1]. A variety of different approaches has emerged that spans from single- to multi-shot data acquisition [2, 3]. While single-shot techniques such has CDI and holography principally allow for imaging fast processes [4], scanning techniques such as ptychography, where a specimen is translated transversal to a beam, allow to use redundant information to infer additional experimental parameters, such as illumination [5] and coherence [6]. Another technique that utilizes redundant information is propagation-based phase retrieval, where diffraction patterns are recorded at multiple sample-detector distances that may be used to provide multiple constraints for iterative phase retrieval algorithms [7].

We recently presented a technique [8] to overcome limited detector size and detector translation alignment in propagation-based phase retrieval using a coregistration approach that is reviewed in section 2. In this paper, we present experimental results on an extension of this technique to polychromatic illumination.

2. Algorithm

2.1. Algorithm

Multiple constraint satisfaction problems are conveniently described in a divide and concur framework [9]. In the case of propagation-based phase retrieval, each recorded diffraction pattern provides a constraint that is imposed on the exit surface wave field by means of projection onto sets, i.e.

$$\pi_M(\psi) = \mathcal{D}_z^{-1}\left(\sqrt{I_z}\frac{\mathcal{D}_z(\psi)}{|\mathcal{D}_z(\psi)|}\right), \tag{1}$$

where ψ is the exit-surface wave field, $\mathcal{D}_z$ is an angular spectrum near-field propagator and I_z is the intensity recorded at distance z from the sample. Given multiple diffraction intensities, the exit-surface wave can be updated via a weighted average of multiple projections

$$\psi^{n+1} = \frac{\sum_k c_k\, \pi_k(\psi^n)}{\sum_k c_k}, \tag{2}$$

where π_k denotes projection onto the kth diffraction intensity constraint and c_k is a weighting factor (see subsection 2.2). The weighted average of projections in a three constraint satisfaction problem in two-dimensional Euclidean space is illustrated in Figure 1 on the left.

2.2. Coregistration procedure

The weighting factors in Equation 2 are the maxima of the normalized cross-correlation between the squared magnitude of the near-field propagated iterate and the measured intensity, i.e.

$$c_k(\Delta x, \Delta y) = max\left\{\frac{\iint I_k(x+\Delta x, y+\Delta y)\left|\mathcal{D}_z(\psi(x,y))\right|^2 dx\, dy}{\iint I_k(x,y) dx\, dy\ \iint\left|\mathcal{D}_z(\psi(x,y))\right|^2 dx\, dy}\right\}. \tag{3}$$

3. Experiment

3.1. Data acquisition

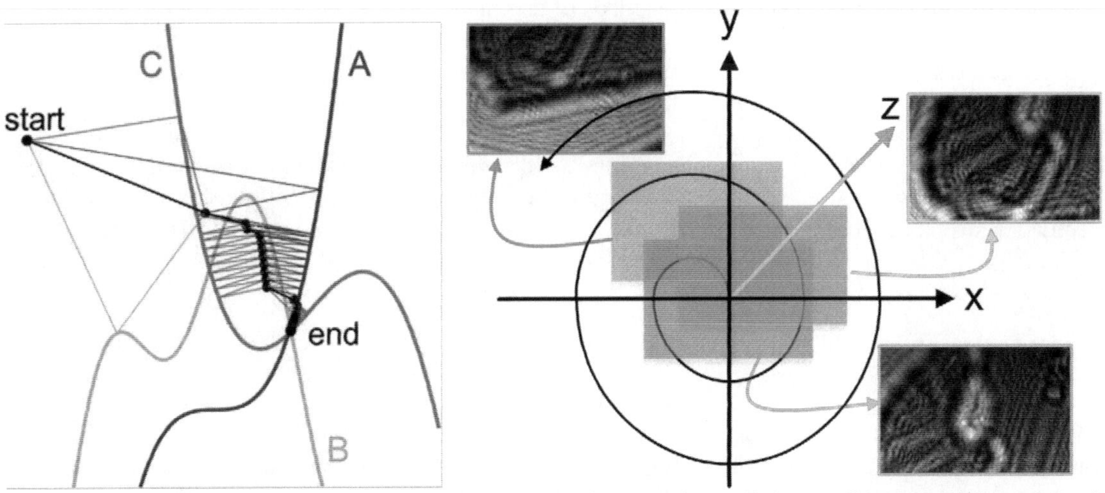

Figure 1 (left) illustration of the weighted average updating rule: Each iterate (black dots) is projected onto constraints A, B and C (colored curves) and then averaged to form the subsequent iterate, (right) the diffraction data is acquired along a screw spiralling around the optical axis.

To extend the numerical aperture of the detected data, the diffraction patterns where acquired in a spiralling motion along the optical axis, or in other words: the trajectory of each pixel on the detector prescribed a screw with extending diameter. The diffraction data was recorded on a UI-1226LE-M, 8 Bit CMOS detector with 480 x 752 pixels à 6 μm. The detection geometry is shown on the right of Figure 1.

3.2. Monochromatic illumination

The left half of Figure 2 shows a reconstruction of a 10 μm thick histological section of a mouse brain. For this reconstruction 105 diffraction patterns were recorded (not shown here) at distances ranging from 100 mm to 308 mm (Δz = 2 mm) with an exposure time of 2 ms. The total lateral displacement of the detector was 123 mm in each transversal direction (2048 x 2048 pixels à 6 μm). The light source used for plane wave illumination was an Argon-Ion-Laser with a wavelength of 488 nm. The reconstruction procedure was the same as described in [8].

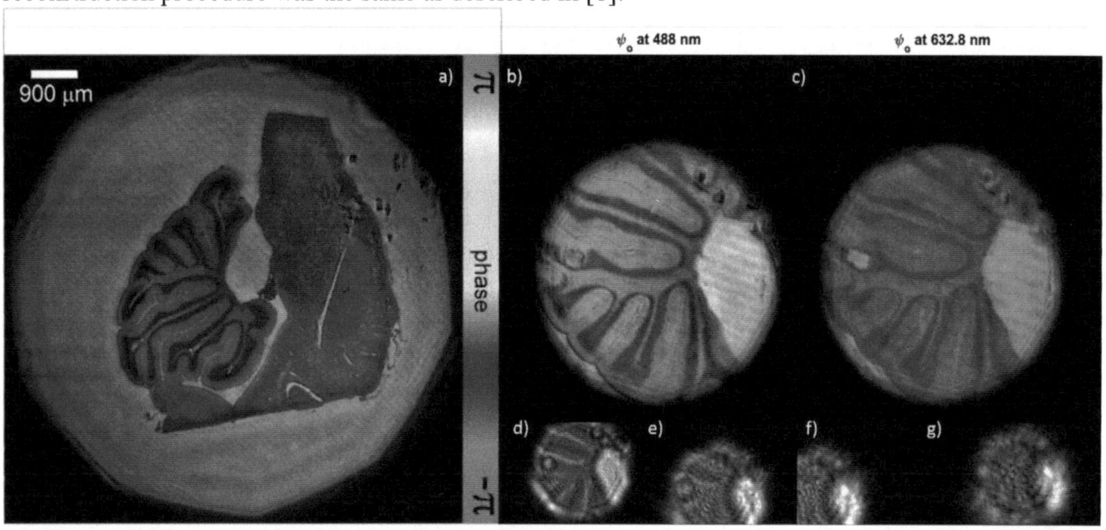

Figure 2 (a) propagation-based reconstruction of a mouse brain using monochromatic radiation at 488 nm, (b,c) polychromatic reconstructions of another slice of the mouse brain at 488 nm and 632 nm, respectively, (d-g) 4 out of 42 diffraction patterns used for the reconstruction of b) and c), transmissivity and phase shift of the specimen are encoded as brightness and hue, respectively.

3.3 Polychromatic illumination.

The right half of Figure 2 shows a polychromatic (488 nm and 632 nm) reconstruction of a similar slice of the same specimen as depicted in Figure 2a. Here a total of 42 diffraction patterns where recorded at distances 100 mm to 400 mm (Δz = 5 mm, exposure time 20 ms). The total lateral displacement of the detector was 24 mm in each transversal direction. The updating rule in Equation 1 had to be adapted to polychromatic radiation as proposed in [10], i.e.

$$\pi_M(\psi_l) = \mathcal{D}_z^{-1}\left(\frac{\sqrt{I_z} \cdot \mathcal{D}_z(\psi_l)}{\sqrt{\sum_l |\mathcal{D}_z(\psi_l)|^2}}\right), \tag{4}$$

where l is an index to label wavelength. For practical implementation it is noted that if the updating rule in Equation 2 converges slowly, initial updates can be implemented sequentially while for later iterations the averaged version in Equation 2 yields better photon statistics.

4. Discussion

4.1. Validity of the coregistration procedure

As pointed out in [8], the coregistration of adjacent diffraction patterns may fail if the detector is shifted too far along the optical axis. Shifting the detector within the longitudinal speckle size $l_z = 8\lambda(z/D)^2$, guarantees the wave field to exhibit a sufficient degree of temporal autocorrelation resulting in a reliable coregistration of diffraction patterns.

4.2. Requirements for polychromatic near-field reconstructions

For the polychromatic reconstruction depicted in Figure 2, an aperture was deliberately placed upstream the sample to limit the extent of the plane wave incident on the sample. This prevents the lowest spatial frequencies in the angular spectrum representation of the exit wave to overlap on a single pixel in the detector plane. Mathematically, this relation is expressed by $FOV \leq \Delta\lambda z/p$, where FOV is the field of view or the size of the aperture, $\Delta\lambda = \lambda_2 - \lambda_1$ is the wavelength separation of the two lasers and p is the pixel size of the detector. We believe that a similar relation holds for ptychographic information multiplexing [10].

5. Conclusion

Propagation-based phase retrieval does neither rely on a priori knowledge as required in CDI nor accurate mechanical scanning as required in ptychography. A disadvantage of propagation-based phase retrieval is that – without additional knowledge - it allows only to reconstruct the exit-surface wave emanating from the specimen as opposed to ptychography where both object transmission function and illumination can be retrieved.

Acknowledgment

L Loetgering would like to thank Prof. Dr. K Schilling from the Institute of Anatomy and Cell Biology at University of Bonn for the histological specimens.

References

[1] Thibault P, Elser V 2010 *Annu. Rev. Condens. Matter Phys.* **1** 237-55.
[2] Miao J, Charalambous P, Kirz J, Sayre D 1999 *Nature* **400** 342-4.
[3] Rodenburg JM, Faulkner HM 2004 *Applied Physics Letters* **85** 4795-7.
[4] Büttner F, Moutafis C, Schneider M, Krüger B, Günther CM, Geilhufe J, Schmising CV, Mohanty J, Pfau B, Schaffert S, Bisig A 2015 *Nature Physics* **11** 225-8.
[5] Thibault P, Dierolf M, Menzel A, Bunk O, David C, Pfeiffer F 2008 *Science* **321** 379-82.
[6] Thibault P, Menzel A 2013 *Nature* **494** 68-71.
[7] Cloetens P, Ludwig W, Baruchel J, Van Dyck D, Van Landuyt J, Guigay JP, Schlenker M 1999 *Applied physics letters* **75** 2912-4
[8] Loetgering L, Hammoud R, Juschkin L, Wilhein T 2015 *EPL (Europhysics Letters)* **111** 64002.
[9] Gravel S, Elser V. 2008 *Physical Review E.* **78** 036706.
[10] Batey DJ, Claus D, Rodenburg JM 2014 *Ultramicroscopy* **138** 13-21.

Multimodal imaging of the human knee down to the cellular level

G. Schulz[1], C Götz[1], M. Müller-Gerbl[2], I. Zanette[3], M.-C. Zdora[3,4], A. Khimchenko[1], H. Deyhle[1], P. Thalmann[1], and B. Müller[1]

[1] Biomaterials Science Center, Department of Biomedical Engineering, University of Basel, Allschwil, CH.
[2] Musculoskeletal Research, Department of Biomedicine, University of Basel, CH.
[3] Diamond Light Source, Harwell Science & Innovation Campus, Didcot, Oxfordshire OX11 0DE, UK.
[4] Department of Physics & Astronomy, University College London, London WC1E 6BT, UK.

E-mail: `georg.schulz@unibas.ch`

Abstract. Computed tomography reaches the best spatial resolution for the three-dimensional visualization of human tissues among the available nondestructive clinical imaging techniques. Nowadays, sub-millimeter voxel sizes are regularly obtained. Regarding investigations on true micrometer level, lab-based micro-CT (μCT) has become gold standard. The aim of the present study is firstly the hierarchical investigation of a human knee post mortem using hard X-ray μCT and secondly a multimodal imaging using absorption and phase contrast modes in order to investigate hard (bone) and soft (cartilage) tissues on the cellular level. After the visualization of the entire knee using a clinical CT, a hierarchical imaging study was performed using the lab-system nanotom® m. First, the entire knee was measured with a pixel length of 65 μm. The highest resolution with a pixel length of 3 μm could be achieved after extracting cylindrically shaped plugs from the femoral bones. For the visualization of the cartilage, grating-based phase contrast μCT (I13-2, Diamond Light Source) was performed. With an effective voxel size of 2.3 μm it was possible to visualize individual chondrocytes within the cartilage.

1. Introduction

The human musculoskeletal system consists of a variety of hard and soft tissue components. One of the big challanges in micro computed tomography (CT) is the simultaneous visualization of soft and hard tissue components, since the X-ray attenuation of soft and hard tissue significantly differs [1]. An optimized choice of the photon energy for bone, for example, would result in almost fully transparent cartilage. On the other hand an optimized energy for soft tissues would lead to streak artifacts due to the high absorbing hard tissues. For such specimens the use of phase tomography is of advantage [2]. During the present study at different length scale ranges the authors used a conventional CT, as currently available at radiology departments, an advanced conventional CT system, as available at the Biomaterials Science Center in Basel, and the synchrotron radiation facility Diamond in absorption and phase contrast modes. The goal is to demonstrate the performance of these CT systems in the morphological characterization of a human knee. The results allow for the proper choice of modality to characterize any other joint from mammalians on the desired length scale down to the cellular level.

2. Materials and methods

2.1. Specimen preparation and medical CT

The donated body of an 87 years old female was fixed in 10% formalin within 24 h after death following the standard protocol of the Institute of Anatomy at the University of Basel. The extraction of the knee was perfomred after fixation with transversal cuts about 10 cm above and below the femorotibial joint space. The extracted knee was scanned at this point using a Siemens Somatom Emotion® 16 medical CT scanner with an anisotropic voxel size of $0.3 \times 0.3 \times 0.6$ mm^3. For the scan the acceleration voltage of 130 kV and a beam current of 25 mA were chosen. A smooth reconstruction kernel (B31s) was chosen. For this scan, the knee was taken out of the formalin solution and measured in air. To prevent the tissue from drying out, the sample was then stored in a 4% formalin solution. During all further scans, the specimens were kept in the container with the formalin. After the first two scans the skin, fat and muscle tissues were removed. A femoral cylinder of 5 mm was extracted using a metal hollow punch from 2 cm sections for further higher resolved measurements with the laboratory CT-system. The location of the extracted cylinder was around the contact area of the femorotibial joint where a relatively thick cartilage layer was preserved.

2.2. X-ray microtomography

After medical CT scans the entire knee was measured using the nanotom® m (phoenix|x-ray, GE Sensing & Inspection Technologies GmbH, Wunstorf, Germany). Due to the size the resulting of 65 μm pixel size was chosen. The scan was perfomed using an acceleration voltage of 180 kV and a beam current of 30 μA. The mean photon energy of the X-ray spectrum was increased by adding a 0.3 mm Cu-filter between source and specimen. The scan had a duration of 17 h. This data set was registered to the medical CT data set using a three-dimensional rigid algorithm with six degrees of freedom, namely three translation and three rotation degrees [3]. After extraction of the femoral cylinder a high-resolution scan with a pixel length of 3 μm was performed. Accelration voltage of 40 kV and a beam current of 350 μA were adjusted. The reconstruction of the radiographs was performed using the software phoenix datos|x 2.0.1 - RTM (GE Sensing & Inspection Technologies GmbH, Wunstorf, Germany) where a cone beam filtered back-projection algorithm is used.

2.3. X-ray grating interferometry

The femoral cylinder was measured at I13-2 (Diamond Light Source, Harwell, UK) with a pixel lenght of 2.3 μm. The photon energy was set to 19 keV in order to be sensitive enough to visualize small electron density differences within the cartilage. The currently established interferometer there consisted of a beam-splitter grating g_1 with a periodicity of $p_1 = 4.79$ μm and a Si-structure height of 23 μm and an absorption grating having a periodicity of $p_2 = 2.40$ μ and a Au-structure height of 109 μm. The distance between the two gratings was set to 48 cm which corresponds to the 9th Talbot order. A total number of 5 phase steps were acquired with an exposure time of 5 s at each projection angle. 1600 projections were recorded over 360. During the scanning procedure, the sample was again kept in a small water tank with parallel polymethylmethacrylat plates. The data was phase-retrieved using pixel-wise Fourier analysis. The processed phase projections were then reconstructed using a modified filter kernel (Hilbert transform) in combination with standard filtered back-projection algorithm [4, 5].

3. Results and discussion

The nanotom® m scan (Figure 1, right) shows the trabecular structure of the bone in detail and shows sharp edges betwee tissue borders in comparison to the medical CT (Figure 1 left). The sagittal slices show all three composing bones of the knee joint (i.e. femur, tibia and patella).

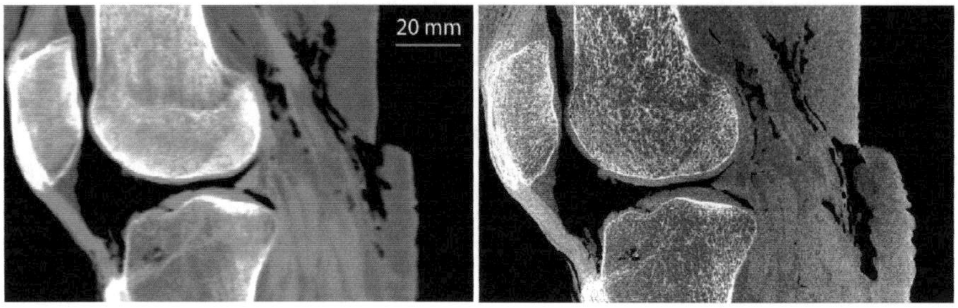

Figure 1. Saggital slices through the human knee acquired using a medical CT scaner (left) and the nanotom® m

The cartilage is visible in both medical CT and μCT data sets. However the higher resolution of the μCT data allows a much clearer discrimination of it with the background.

Figure 2 shows orthogonal slices through the femoral cylinder acquired using nanotom® m with a pixel length of 3 μm. The zoom-in image offers insights into the anatomy of the subchondral bone. The intermediate layer between bone and cartilage marked with the yellow arrows is associated with the so called mineralized cartilage zone [6]. The demarcation towards the cartilage marked with the right yellow arrows presumably represents the tide mark line. Due to their location, the bright circular structures are assumed to be calcified chondrocytes.

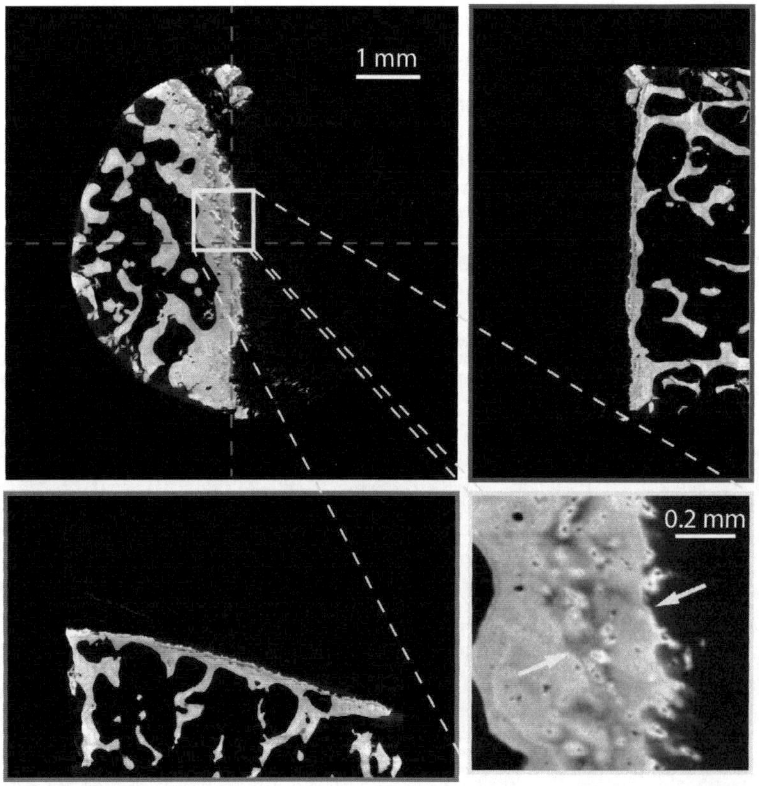

Figure 2. Three orthogonal slices through the femoral cylinder including a zoom-in obtained with the nanotom® m and a pixel length of 3 μm.

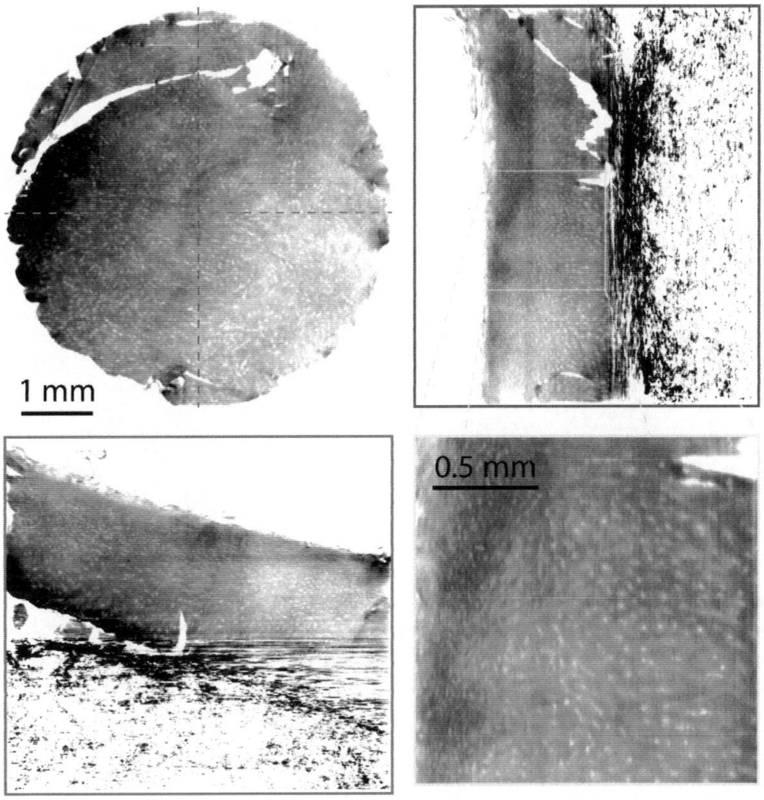

Figure 3. Three orthogonal slices through the femoral cylinder (also shown in Figure 2) including a zoom-in obtained using gratnig-based phase contrast at Diamond Light Source.

Althogh the higher resolution of the data set leads to sharp insights of the hard tissues, almost no contrast is present within the cartilage.

Grating-based phase contrast results are shown in Figure 3 by three orthogonal slices through the specimen. However, due to the low energy, which was used in order to achieve the highest possible contrast, huge artefacts are present within the bone. On the other hand the internal structures of the cartilage are visible in detail. Individual chondrocytes respectively show up inside the cartilage layer. Often more than one chondrocyte are located inside the same lagunae resulting in an ellipsoidal shape. A perpendicular orientation towards the surface in the vicinity of bone of them is present in order to gain more stiffness of the cartilage tissue In the outer region of the cartilage they are generally smaller and rather oriented parallel to the surface.

Acknowledgments
The authors gratefully acknowledge the financial support of the Swiss National Science Foundation (SNSF) projects 147172 and 133802.

References
[1] Drews, S. *et al* 2008 *Proc. SPIE* **7078** 70780C.
[2] Holme, M. N. *et al* 2014 *Nature Prot.* **9** 1401-15.
[3] Andronache, A. *et al* 2008 *Med. Image Anal.* **12** 3-15.
[4] Faris, G. W. and Byer, R. L. 1988 *Appl. Opt.* **27** 5202-12.
[5] Pfeiffer, F. *et al* 2007 *Nucl. Instrum. Methods Phys. Res. Sect. A* **580** 925-8.
[6] Buckwalter, J. *et al* 1997 *J. Bone Joint Surg.* **79** 600-11.

IOP Conf. Series: Journal of Physics: Conf. Series **849** (2017) 012027 doi:10.1088/1742-6596/849/1/012027

High-dynamic-range water window ptychography

Max Rose[1], Dmitry Dzhigaev[1], Tobias Senkbeil[2], Andreas R. von Gundlach[2], Susan Stuhr[2], Christoph Rumancev[2], Ilya Besedin[1], Petr Skopintsev[1], Jens Viefhaus[1], Axel Rosenhahn[2] and Ivan A. Vartanyants[1,3]

[1]Deutsches Elektronen-Synchrotron DESY, Hamburg, 22607, Germany
[2]Analytical Chemistry - Biointerfaces, Ruhr-University Bochum, Bochum, 44780, Germany
[3]MEPhI (Moscow Engineering Physics Institute), Moscow, 115409, Russia

E-mail: max.rose@desy.de

Abstract. Ptychographic imaging with soft X-rays, especially in the water window energy range, suffers from limited detector dynamic range that directly influences the maximum spatial resolution achievable. High-dynamic-range data can be obtained by multiple exposures. By this approach we have increased the dynamic range of a ptychographic data set by a factor of 76 and obtained diffraction signal till the corners of the detector. The real space half period resolution was improved from 50 nm for the single exposure data to 18 nm for the high-dynamic-range data.

1. Introduction

Ptychography is a scanning coherent diffractive imaging (CDI) method for extended samples [1, 2]. It provides a self consistent solution to the phase problem in CDI by overlapping illuminations on the measured object. In addition, it can also recover the complex valued illumination function. Especially in the field of X-ray microscopy (XRM), which lacks efficient and high quality imaging optics, ptychography is a powerful lens-less imaging technique [3]. Ptychography also decouples the illumination beam size from the final resolution of the picture, which is advantageous over scanning transmission X-ray microscopy (STXM).

The so-called water window energy range from 280 eV to 530 eV contains the absorption edges of oxygen, nitrogen and carbon and proteins and carbon rich material are visible with high contrast against the aqueous environment of biological samples [4]. This makes the water window unique and especially well suited for high resolution XRM [5] and subsequently also for X-ray ptychography [6] of unstained biological samples.

In the soft X-ray range silicon based integrating pixel detectors are common devices for measuring diffraction patterns. Unfortunately, they have limitations in the context of coherent diffractive imaging. Due to the integrating nature of the detector and electronic imperfections, noise and dark-current and the maximum photon number per pixel (898 photons/pixel) the effective dynamic range (DR) of the detector is limited. This is a severe limitation for CDI and water window ptychography of weakly scattering samples due to the large DR of diffraction patterns [7]. Different approaches exist to circumvent this problem. Among them are i) pixel capacity enlargement ii) active pixel gain switching [8] and iii) multiple exposure merging [9]. Pixel capacity enlargement requires pixel size enlargement. This is often not tolerable in diffractive imaging due to the required oversampling criterion [10]. Detectors with active pixel gain switching are being under development but they are currently not available to a broader community.

2. Experiment

Here we present a ptychographic experiment on a known test sample (NTT-AT, XRESO-50HC) in order to characterize the detailed beam properties on a high resolution scale. We measured multiple exposures in our experiment to obtain high-dynamic-range (HDR) diffraction patterns by the combination of multiple data sets. The experiment was performed at the XUV Beamline P04 at PETRA III storage ring [11].

The diffraction patterns from the Siemens star test sample were measured with an ANDOR 16 bit detector (DODX436-BN) at 500 eV with the detector at a distance of $z = 26.2$ cm. The incident beam on the sample was originating from a 3 µm pinhole at a distance of 1 mm. We scanned our sample on a equidistant rectangular grid (8×9 positions) with a step size of 1.2 µm. We used the following strategy for the measurement: The first scan was obtained without beam stop, the second scan with a semi-transparent beam stop (BS1) of 1.6 mm diameter and the third scan with an opaque beam stop (BS2) of 3.4 mm. The resulting diffraction patterns (background and dark current corrected) in analog-digital units (73 ADU/photon) are shown on a logarithmic scale in Figure 1. We clearly see the far-field diffraction signal that points outward from the center of the diffraction pattern (the center of the reciprocal space). The diffraction signal gets brighter in the longer exposed image of Figure 1b) and c).

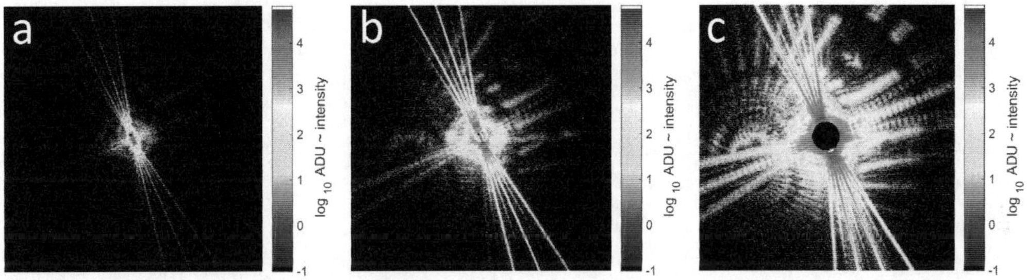

Figure 1. Siemens star far-field diffraction patterns of different exposure time. a) Short exposure $t_{\text{noBS}} = 0.15$ sec, b) long exposure $t_{\text{BS1}} = 15 \cdot t_{\text{noBS}}$ with semi-transparent beamstop and c) long exposure $t_{\text{BS2}} = 150 \cdot t_{\text{noBS}}$ with opaque beam stop.

We filled the data behind the beamstop in the center of each long exposures (Figure 1b) and c)) by the corresponding central part of the short exposure (Figure 1a)). The following definition was used to obtain a HDR diffraction pattern: $I_i^{\text{HDR}} = (I_i^{\text{noBS}} + I_i^{\text{BS}} \cdot \text{mask}_{\text{BS}}) \cdot t_{\text{BS}} / (t_{\text{noBS}} + t_{\text{BS}} \cdot \text{mask}_{\text{BS}})$. Here i denotes the index of two corresponding diffraction patterns and mask_{BS} is a binary mask that selects only the pixels outside the beam stop area. The HDR diffraction patterns were scaled to the long exposure time (t_{BS1}) with beam stop and the exposure time t_{noBS} without beam stop. Here only the relative exposure times are relevant. The combined data from Figure 1a) and b) was combined in a subsequent step with a second beam stop mask and the longest exposure (t_{BS2}) of Figure 1c).

A DR estimate of the measured diffraction signal can be obtained by evaluating its power spectral density (PSD) in photon units. This is the averaged signal over angle as a function of the distance to the center of the diffraction pattern. The PSD of all three data sets is shown in Figure 2. The PSD analysis was applied on the average of all diffraction patterns with the same exposure. The red curve shows the natural signal decay for increasing spatial frequencies measured in line-pairs per micro meter (lp/µm). The noisy background dominates the signal

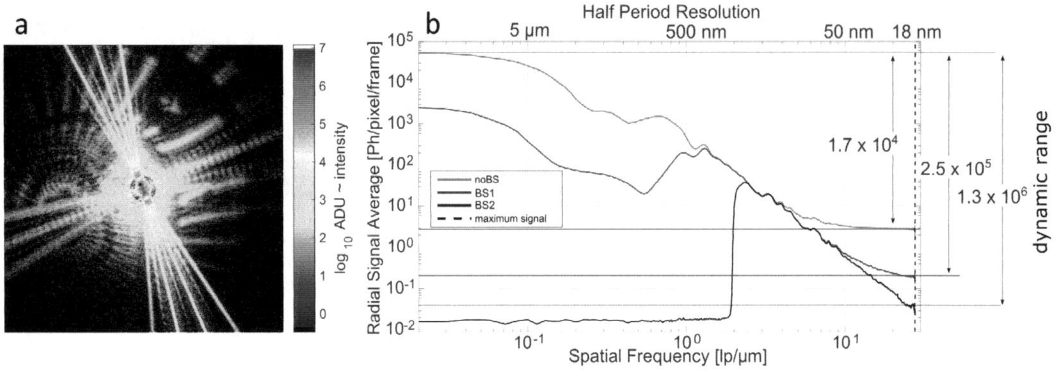

Figure 2. a) Merged Siemens star diffraction pattern of Figure 1. b) Power spectral density with (blue and black) and without beam stop (red) and indication of the dynamic range.

beyond 10 lp/µm (equal to 50 nm half period resolution). The blue curve is the diffraction signal weighted with the characteristic profile of the semi-transparent beam stop. The signal decays exponentially until the dark current noise dominates beyond 20 lp/µm. The black curve follows the form of the blue line except that the signal is dominating over noise till the maximum measured spatial frequency of 28 lp/µm. By scaling according to the relative exposures of the red (t_{noBS}), blue (t_{BS1}) and the black (t_{BS2}) PSD curve from different exposures we got an effective increase of the DR from $1.7{\cdot}10^4$ to $1.3{\cdot}10^6$. Our measurement strategy has led to a gain in DR by a factor of 76. As a consequence of the DR improvement we were able to measure diffraction signal till the corners (note factor $\sqrt{2}$ in the following equation) of the square detector. We cropped to $N_{\mathrm{px}} = 1920$ pixels in each dimension to center the diffraction patterns. The detector pixel size was $\Delta_{\mathrm{D}} = 13.5\,\mu\mathrm{m}$ and the corresponding maximum half period resolution of 18 nm was calculated with $\Delta_{\mathrm{res}} = \lambda z/\sqrt{2}N_{\mathrm{px}}\Delta_{\mathrm{D}}$.

3. Ptychographic reconstruction from HDR data

The reconstruction was done with the ePIE algorithm and 4000 iterations on the single exposure data without beam stop (see Figure 3a)). The relatively weak signal close to the detector edges shown in Figure 1a) led to a reduced reconstruction quality for the very fine 50 nm structures in the center of the test pattern. To increase resolution we have combined the data with and without beam stop. For the HDR ptychographic reconstruction we started with the merged data set. After 10 initial iterations we mask these values from the modulus constraint step in order to let the algorithm find the best estimate for the values inside the beam stop area. This modification to the original ePIE algorithm can correct for a limited amount of scaling and position imperfections between the data sets used for the HDR merging. Missing data in the beam stop region, due to low statistics (see Figure 2a)), often acts as a high pass filter. Reconstructed edges are enhanced and large uniform areas that are slightly non-uniform. However, the HDR data reconstruction resulted in an improved image of our test sample shown in Figure 3b). The 50 nm structures are clearly resolved with two pixels per line feature.

Together with the object reconstruction we have obtained an illumination function at the sample plane. The retrieved complex valued illumination was propagated back to the pinhole in order to determine a precise distance between pinhole and sample. In Figure 3 c) the illumination in the pinhole and sample plane and a horizontal slice along the propagation distance is shown. At the pinhole plane the illumination takes the shape of the pinhole with 3 µm in diameter. At the sample plane downstream of the pinhole the illumination has a full width of half maximum of 1.25 µm.

IOP Conf. Series: Journal of Physics: Conf. Series **849** (2017) 012027 doi:10.1088/1742-6596/849/1/012027

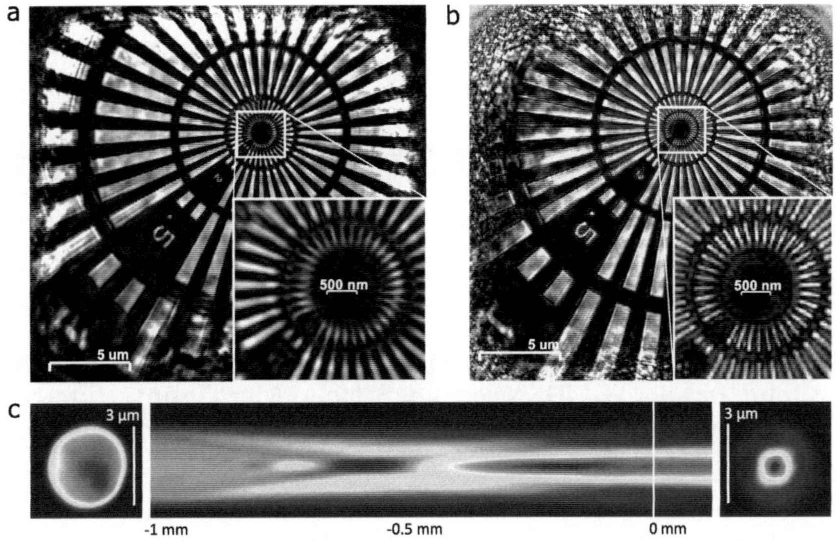

Figure 3. Results of ptychographic reconstruction. Siemens star amplitude reconstruction from a) the single exposure data set and b) the HDR data. c) Reconstructed illumination function in the pinhole plane −1 mm (left) and propagated to the object plane at 0 mm (right).

4. Conclusions

We have presented a HDR ptychographic experiment in the water window photon energy range. Our HDR data were obtained by merging multiple increasing exposures into a single data set. To protect the detector from the highly intense X-ray beam we used beam stops. The measured DR was improved by roughly two orders of magnitude. The DR improvement led directly to an improvement of the reconstructed test object. Together with the object we obtained a high quality complex valued illumination profile.

Due to the radiation hardness of our test sample we do not suffer from radiation damage and the multiple exposure scheme is a good tool to obtain very high resolution diffraction patterns from a comparably low-dynamic-range detector. For more radiation sensitive biological objects we may restrict the number of exposures to two instead of three. Our HDR method can be applied without complications to cryo-fixated and hydrated biological samples to obtain high resolution phase and absorption contrast images.

We acknowledge the financial support from BMBF 05K10VH4, the Virtual Institute VH-VI 403 and the Joachim Herz Foundation in cooperation with PIER Helmholtz Graduate School.

5. References

[1] Rodenburg J M *et al.* 2007 *Phys. Rev. Lett.* **98**(3) 034801
[2] Thibault P *et al.* 2008 *Science* **321** 379
[3] Vartanyants I A and Yefanov O M 2014 *X-ray Diffraction: Modern Experimental Techniques* (Pan Stanford Publishing) chap Coherent X-ray Diffraction Imaging of Nanostructures, p 269
[4] Schneider G 1998 *Ultramicroscopy* **75** 85
[5] Larabell C A and Nugent K A 2010 *Current Opinion in Structural Biology* **20** 623
[6] Giewekemeyer K *et al.* 2011 *Opt. Express* **19** 1037
[7] Rose M *et al.* 2015 *Journal of Synchrotron Radiation* **22** 819
[8] Wunderer C B *et al.* 2014 *Journal of Instrumentation* **9** C03056
[9] Takahashi Y *et al.* 2011 *Physical Review B* **83** 214109
[10] Miao J *et al.* 1999 *Nature* **400** 342
[11] Viefhaus J *et al.* 2013 *Nuclear Instruments and Methods in Physics Research Section A* **710** 151

IOP Conf. Series: Journal of Physics: Conf. Series **849** (2017) 012028 doi:10.1088/1742-6596/849/1/012028

Ptychographic Imaging of Nano-Materials at the Advanced Light Source with the Nanosurveyor Instrument

David A. Shapiro, Rich Celestre, Peter Denes, Maryam Farmand, John Joseph, A.L.D. Kilcoyne, Stefano Marchesini, Howard Padmore, Singanallur V. Venkatakrishnan, Tony Warwick and Young-Sang Yu

Lawrence Berkeley National Laboratory, Berkeley, CA 94720, USA

E-mail: `dashapiro@lbl.gov`

Abstract. We present a new ptychographic x-ray microscope dedicated to soft x-ray tomography and spectromicroscopy of nano-materials at the Advanced Light Source. The microscope utilizes an ultra-stable, high performance scanning mechanism with laser interferometer feedback for sample positioning and a fast frame rate charge-coupled device detector for soft x-ray diffraction measurements. The microscope can achieve point scan rates of 120 Hz with 1 nm RMS positioning accuracy. A high performance data pipeline has been developed which enables real time ptychographic reconstructions and user-friendly operation. This instrument, called The Nanosurveyor, can achieve a spatial resolution at least 10 times finer than the x-ray spot size in both two and three dimensions with sensitivity to electronic and magnetic states of nano-materials. In this paper we demonstrate the tomographic and spectromicroscopic capability of the Nanosurveyor instrument. At high brightness x-ray sources this instrument will enable spectromicroscopy and tomography of materials with diffraction limited spatial resolution.

1. The Nanosurveyor Instrument

The Advanced Light Source has developed a dedicated ptychographic microscope [1, 2, 3] which is optimized for use with soft x-rays in the energy range 250 to 2500 eV and has been commissioned at the bending magnet beamline 5.3.2.1. The core components of the instrument are a piezo rotation stage for tomographic measurements, a custom fast-frame rate CCD for coherent scattering measurements, a laser interferometric feedback system for nanometer scale positioning of the of focusing optic with respect to the sample stages [4] and a piezo flexure stage for scanning the zone plate relative to the sample.

1.1. Scanning System

The scanning mechanism in a ptychographic microscope should be fast, in order to maximize throughput and minimize the effects of any long term drift in sample position or illumination, and it should be precise so as to ensure adequate position overlap and minimize reconstruction error. Here, the scanning mechanism uses a combination of stepper motors for coarse positioning of the sample with respect to the focusing optics and piezo motors for fine positioning during ptychography scans. The sample and zone plate focusing optic are separated by a long

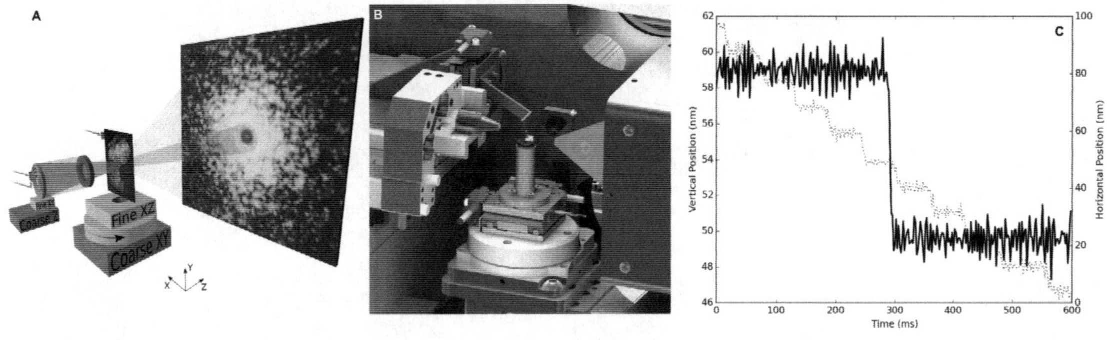

Figure 1. Schematic (A) and CAD rendering (B) of the sample scanning region of the Nanosurveyor Instrument. (C) Position data during a ptychography scan with 10 nm steps. Data for both the horizontal axis (thin line) and vertical axis (thick line) are shown. The RMS error for each is 1 nm and the piezo settling time (which dominates the system overhead per point) is 7 milliseconds.

mechanical path, so in order to achieve the highest degree of stability special care must be taken to account for low frequency drift and high frequency vibration. In order to minimize vibrations, the system has been designed to have a low center of gravity. In a standard scanning transmission x-ray microscope (STXM), the fine scanning is typically achieved by a two axis piezo flexure stage mounted on top of the coarse motion stepper motors. Here, we minimize the mass around the sample by moving the fine motion to the zone plate optic. The sample's region of interest is positioned on the x-ray beam using the coarse positioning of the stepper motors and then the x-ray beam is scanned across the sample with high precision using a two axis piezo flexure stage. In our current scheme, the order sorting aperture (OSA) is stationary and thus limits the effective fine scanning range to about 20 microns while the OSA diameter is 40 microns. It is possible that larger scan ranges would suffer from probe instability while scanning because of the finite illumination of the zone plate. The Long term stability and precise positioning is achieved through interferometric feedback control. A differential interferometer measures relative displacements between the sample stage and the zone plate and maintains closed loop positioning of the zone plate piezo stage [5].

1.2. Detector and Data Acquisition

Ptychographic data is several orders of magnitude larger than conventional scanning microscopy data because of the dramatic increase in angular resolution and range in reciprocal space. This increase in data volume demands a high bandwidth data acquisition (DAQ) system and high performance computing for near realtime analysis. Currently, a local machine running a Linux operating system collects raw data frames from the CCD interface node over a dedicated 10 Gigabit/s network and saves the data into separate TIFF files on a fileshare hosted by a small, remote cluster of graphical processing units (GPUs). The local machine collecting the data and the remote cluster also communicate via a dedicated 10 Gigabit interface. The local linux machine can initiate analysis jobs on the remote cluster which perform a pre-processing step on the data frames followed by the ptychographic reconstruction using the SHARP software across 16 GPUs [6, 7].

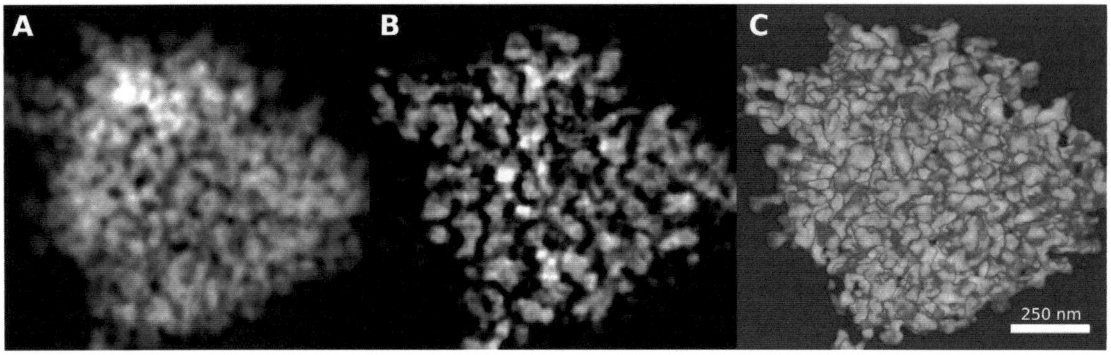

Figure 2. Ptychographic phase projection (A), tomographic ortho-slice (B) and volume rendering (C) of a ptycho-tomographic reconstruction of a particle of sintered Yttria-stabilized zirconia. The volume was reconstructed using the rMBIR algorithm from 160 projections spanning +/− 80 degrees and has a 6 nm voxel size. The ptychographic projections were obtained using 800 eV x-rays and have 10 nm resolution.

2. Tomography

X-rays have sufficient penetrating power and short wavelengths for imaging through microscopic materials with nanometre scale spatial resolution. Ptycho-tomography proceeds by first using x-ray ptychography to reconstruct two-dimensional projections of an object while rotating around a single axis [2]. We demonstrate ptycho-tomography of a nano-porous particle of Yttria-stabilized Zirconia using the Nanosurveyor instrument (Figure 2). The particle is approximately 1 μm in diameter with about 50% fill fraction. Projections were acquired using 800 eV x-rays, a 60 nm outer zone width Fresnel zone plate for illumination and were registered using an iterative alignment scheme that corrects for image translation and rotation. Scans proceed on a rectangular grid with 50 nm step size and 300 ms dwell time per scan point. 160 projections were collected spanning an angular range of +/− 80 degrees and the reconstructed volume, generated using the rMBIR algorithm, has a voxel size of 6 nm [8]. A rich internal pore network is clearly visualized.

3. Spectromicroscopy

The power of soft x-rays is sensitivity to electronic, magnetic and bond orientation contrast. Conventional scanning transmission x-ray microscopes measure a material's x-ray transmission as a function of position and energy. Thus, at each image pixel an x-ray absorption spectrum (XAS) can be measured and compared to reference spectra for quantitative composition analysis. Since x-ray ptychography is sensitive to both the absorption and the phase shift of x-rays it provides for a spectroscopy based on a material's full complex refraction index. Though the absorption and phase shift spectrum can be inferred from each other by the Kramer's-Kronig relation, the ability to directly measure both quantities should increase increase the overall signal-to-noise ratio and chemical sensitivity. Figure 3 shows images of a reference sample, an agglomerate of $FePO_4$ nanoparticles sized 100 × 80 × 20 nm, imaged at various energies across the Fe L_2 and L_3 absorption edges. The spectrum was obtained from a total of 65 images but only three are shown for brevity. At 707.5 eV, the phase shift of the $FePO_4$ relative to vacuum is negative thus the interior of the particles is black while the carbon coating of each particle appears with positive phase shift. The phase shift goes through zero at the absorption resonance and stays positive at higher energies though the achieved spatial resolution for the complex image, routinely below 10 nm as calculated by the Fourier Ring Correlation method [3],

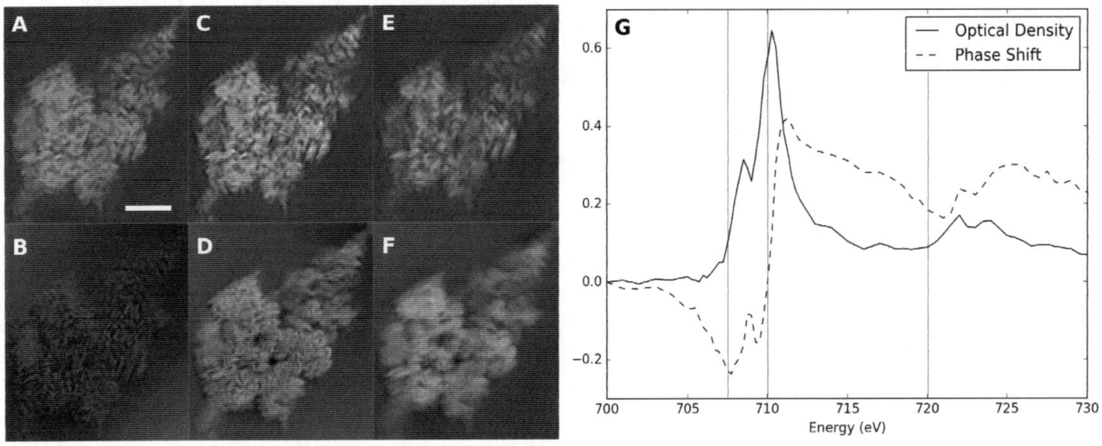

Figure 3. (A-F) Images of the $FePO_4$ reference sample at three different energies: 707.5 eV (A,B), 710 eV (C,D) and 720 eV (E,F). The top images are optical density $(2\pi\beta t/\lambda)$ and the bottom are phase shift $(2\pi\delta t/\lambda)$. (G) Spectra obtained from a sequence of images of the reference sample taken at 65 separate energies. The pixel size is 5 nm and the scale bar is 500 nm.

is highest on resonance where the total scattering cross-section is also highest. It is interesting to note that the contrast, and apparent spatial resolution, of the phase image changes drastically across the resonance as the real part of the refractive index goes through zero. The absorption spectrum agrees well with standard XAS measurements.

4. Conclusions

The ALS has commissioned an instrument for ptychographic imaging of materials combined with tomography and spectroscopy. The high performance scanning system and fast frame rate detector allow for ptychographic imaging that is limited only by the x-ray dwell time which is currently hundreds of milliseconds but which will be below 10 milliseconds at new beamlines optimized for coherent imaging [9]. We demonstrate the high resolution tomographic and spectroscopic imaging capabilities with a variety of nanomaterials relavent to energy research. The development of new diffraction limited x-ray sources will bring factors of 100 increased brightness and will both dramatically increase the time resolution of the measurements and allow for wavelength limited spatial resolution [1, 3, 10].

Acknowledgments

Soft X-ray ptychographic microscopy was carried out at beamline 5.3.2.1 at the Advanced Light Source. The Advanced Light Source is supported by the Director, Office of Science, Office of Basic Energy Sciences, of the U.S. Department of Energy under Contract No. DE-AC02-05CH11231. This work is partially supported by the Center for Applied Mathematics for Energy Research Applications (CAMERA), which is a partnership between Basic Energy Sciences (BES) and Advanced Scientific Computing Research (ASRC) at the U.S Department of Energy.

References

[1] Thibault P, Dierolf M, Menzel A, Bunk O, David C and Pfeiffer F 2008 *Science* **321** 379–382
[2] Dierolf M, Menzel A, Thibault P, Schneider P, Kewish C M, Wepf R, Bunk O and Pfeiffer F 2010 *Nature* **467** 436–439 ISSN 0028-0836 URL http://www.nature.com/nature/journal/v467/n7314/abs/nature09419.html

[3] Shapiro D A, Yu Y S, Tyliszczak T, Cabana J, Celestre R, Chao W, Kaznatcheev K, Kilcoyne A L D, Maia F, Marchesini S, Meng Y S, Warwick T, Yang L L and Padmore H A 2014 *Nature Photonics* **8** 765–769 ISSN 1749-4885 URL `http://www.nature.com/nphoton/journal/v8/n10/full/nphoton.2014.207.html`

[4] Denes P, Doering D, Padmore H A, Walder J P and Weizeorick J 2009 *Review of Scientific Instruments* **80** 083302 ISSN 0034-6748, 1089-7623 URL `http://scitation.aip.org/content/aip/journal/rsi/80/8/10.1063/1.3187222`

[5] Kilcoyne A L D, Tyliszczak T, Steele W F, Fakra S, Hitchcock P, Franck K, Anderson E, Harteneck B, Rightor E G, Mitchell G E, Hitchcock A P, Yang L, Warwick T and Ade H 2003 *Journal of Synchrotron Radiation* **10** 125–136 URL `http://dx.doi.org/10.1107/S0909049502017739`

[6] Marchesini S, Krishnan H, Daurer B J, Shapiro D A, Perciano T, Sethian J A and Maia F R N C 2016 *Journal of Applied Crystallography* **49** 1245–1252 ISSN 1600-5767 URL `http://scripts.iucr.org/cgi-bin/paper?S1600576716008074`

[7] Daurer B J, Krishnan H, Perciano T, Maia F R N C, Shapiro D A, Sethian J A and Marchesini S 2016 *arXiv:1609.02831 [physics]* ArXiv: 1609.02831 URL `http://arxiv.org/abs/1609.02831`

[8] Venkatakrishnan S V, Farmand M, Yu Y S, Majidi H, Benthem K v, Marchesini S, Shapiro D A and Hexemer A 2016 *IEEE Signal Processing Letters* **23** 944–948 ISSN 1070-9908

[9] Shapiro D, Roy S, Celestre R, Chao W, Doering D, Howells M, Kevan S, Kilcoyne D, Kirz J, Marchesini S, Seu K A, Schirotzek A, Spence J, Tyliszczak T, Warwick T, Voronov D and Padmore H A 2013 *Journal of Physics: Conference Series* **425** 192011 ISSN 1742-6596 URL `http://iopscience.iop.org/1742-6596/425/19/192011`

[10] Lim J, Li Y, Alsem D H, So H, Lee S C, Bai P, Cogswell D A, Liu X, Jin N, Yu Y s, Salmon N J, Shapiro D A, Bazant M Z, Tyliszczak T and Chueh W C 2016 *Science* **353** 566–571 ISSN 0036-8075, 1095-9203 URL `http://science.sciencemag.org/content/353/6299/566`

IOP Conf. Series: Journal of Physics: Conf. Series **849** (2017) 012029 doi:10.1088/1742-6596/849/1/012029

Deformation of a bismuth ferrite nanocrystal imaged by coherent X-ray diffraction

Marcus C Newton, Adam Pietraszewski and Anthony Kenny

E-mail: M.C.Newton@soton.ac.uk
Physics & Astronomy, University of Southampton, SO17 1BJ, UK

Ulrich Wagner and Christoph Rau

Diamond Light Source, Harwell Oxford Campus, Didcot OX11 0DE, UK

Abstract. Perovskite materials that contain transition metal-oxides often exhibit multifunctional properties with considerable utility in a device setting. $BiFeO_3$ is a multiferroic perovskite material that exhibits room temperature anti-ferromagnetic and ferroelectric ordering. Optical excitation of $BiFeO_3$ crystals results in an elastic structural deformation of the lattice with a fast response on the pico-second time scale. Here we report on dynamic optical excitation coupled with Bragg coherent X-ray diffraction measurements to investigate the structural properties of $BiFeO_3$ nanoscale crystals. A continuous distortion of the diffraction speckle pattern was observed with increasing illumination. This was attributed to strain resulting from photo-induced lattice deformation.

1. Introduction

Transition metal oxides with perovskite crystal structure are emerging as enabling materials for next generation photovoltaics, power electronics, energy harvesting, light emitting diodes and multiferroic devices.[1–3]. Bismuth Ferrite ($BiFeO_3$) is a perovskite metal-oxide semiconducting material with a band-gap of 2.8 eV that shows both antiferromagnetic and ferroelectric ordering at room temperature due to its two differing Fe-O bond lengths. Coexistence of antiferromagnetic and ferroelectric ordering is generally observed in perovskite structures where the ferroelectric displacement from centrosymmetry occurs in the non-magnetic ion. In the case of $BiFeO_3$, the ferroelectric displacement is primarily due to a distortion of the Bi ions from a centrosymmetry configuration. In $BiFeO_3$ each Fe ion is surrounded by six neighbouring Fe ions with antiparallel spin orientation leading to G-type antiferromagnetic ordering and a coupled long-range incommensurate spin-cycloid antiferromagnetic ordering along the [110] direction. Light-mediated photostriction is also observed in $BiFeO_3$ with magnitude dependent on the intensity and polarisation of the impinging light.[4]

Bragg coherent X-ray diffraction imaging (BCXDI) is a lens-less imaging technique suited to the study of nanoscale crystals. It is largely non-destructive and provides information on how atoms are displaced from a reference position at the surface and throughout the bulk of a material.[5] BCXDI experiments are performed by illuminating a sample with a spatially coherent X-rays that exceed the dimensions of the crystal. Coherent light scattered from the entire volume of the nanoscale crystal will interfere and produce a speckle pattern, the

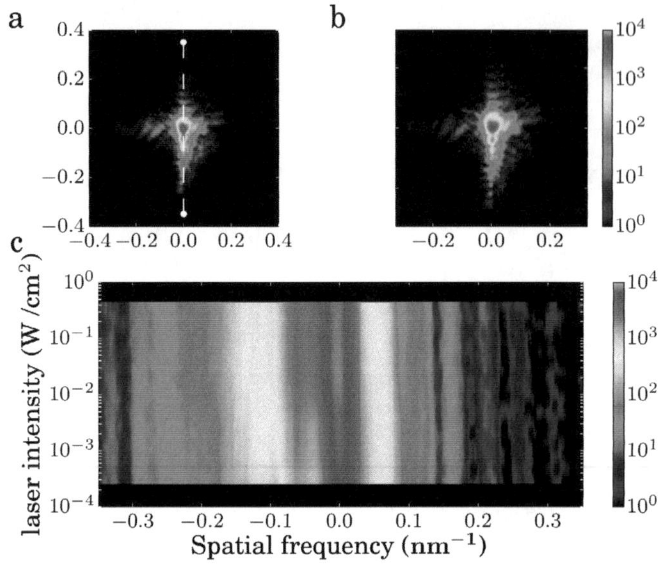

Figure 1. Two-dimensional slices of the speckle pattern of a single $BiFeO_3$ nanocrystal unilluminated (**a**) and illuminated with maximum intensity of 4.0 W/cm^2 of 632.8 nm wavelength laser light (**b**). (**c**), Collated line scans of each speckle pattern along the path indicated in (**a**)for increasing laser illumination intensity.

intensity of which is routinely acquired using a photon counting area detector.[6] If the detector is positioned far enough away from the sample so that the finest fringes are resolved and the speckle pattern is sufficiently oversampled, inversion of the speckle pattern by using iterative projection phase retrieval techniques permits recovery of the complex three-dimensional electron density of the object and its phase information.[7–10] We are then able to directly relate the recovered phase information to lattice displacements in the direction of the wave transfer vector $\mathbf{Q}$ as $\phi(\mathbf{r}) = \mathbf{Q} \cdot \mathbf{u}(\mathbf{r})$, where $\mathbf{u}(\mathbf{r})$ is the average unit cell displacement vector.[11–14]

2. Experiment

$BiFeO_3$ nanocrystals were synthesised on r-plane sapphire substrates using a pulsed laser deposition (PLD). The source material consisted of a solid target of $BiFeO_3$. The target was ablated using a 248 nm, 20 mJ KrF excimer laser system in an oxygen partial pressure of 0.03 hPa. X-ray diffraction measurements confirmed single phase $BiFeO_3$ growth constituents when PLD synthesis was performed at 550°C. The distribution of nanocrystal sizes was largely uniform and dependent on the duration of the synthesis process and the laser fluence. Synthesis performed for 30 minutes with a repetition rate of 200 Hz produced highly faceted nanocrystals of $BiFeO_3$ approximately 200 nm in size. Experiments were performed in air at the coherence branch of beamline I13 at the Diamond Light Source using 9.1 keV photons.[15] A sample of sparsely arranged $BiFeO_3$ nanocrystals was mounted onto the goniometer assembly, a single nanocrystal aligned to the eucentric point and the (110) specular reflection located. A polarised 632.8 nm Helium-Neon (HeNe) laser was focussed via a 0.28 NA microscope objective on to the eucentric point of the sample goniometer and aligned to the X-ray beam. The intensity was attenuated using a linearly graded neutral density filter. The X-ray beam was condensed to 10 µm using a defocussed zoneplate assembly (D = 400 µm, Δr = 150 nm, efficiency 35%). Data was acquired using the 3 mega-pixel photon counting Excalibur detector[16]. Experimental results were gathered by identifying the speckle pattern of a single nanoscale crystal of $BiFeO_3$

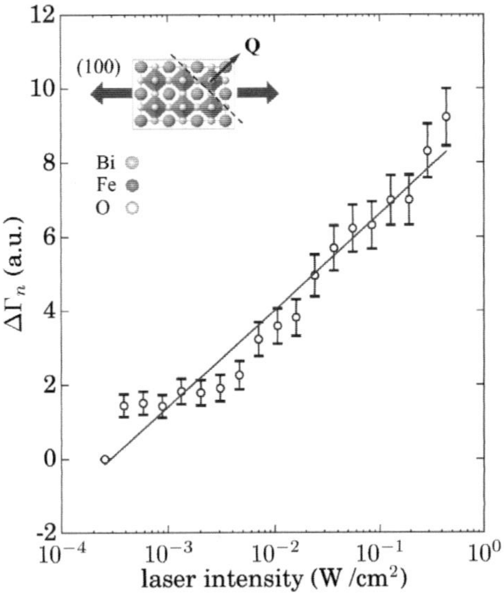

Figure 2. The change in intensity of the cross-correlation measure ($\Delta\Gamma_n$) for each speckle pattern.

on the area detector and then acquiring speckle patterns at varying laser illumination intensity.

3. Results and discussions

Two-dimensional speckle patterns in the absence of laser illumination and with maximum laser illumination are shown in figure 1(**a**) and (**b**). Figure 1(**c**) shows collated line scans of each speckle pattern along the path indicated in (**a**) for increasing laser illumination intensity. A progressively asymmetric distortion of the speckle pattern is clearly observed with increasing illumination intensity. This asymmetry is understood to result from lattice strain in the material. [14] Photo-induced structural deformation will therefore result in an atomic displacement field $\mathbf{u}(\mathbf{r})$.

We can quantify changes in each speckle pattern by contrasting the intensity sum of the cross-correlation for each successive speckle pattern with that of the unperturbed speckle pattern. If the cross-correlation of the nth successive speckle pattern $I_n(\mathbf{q})$ with that of the unperturbed is denoted as $\Gamma_n(\mathbf{q}) = I_n(\mathbf{q}) \otimes I_0(\mathbf{q})$, then the normalised change in sum $\Delta\Gamma_n$ is given by:

$$\Delta\Gamma_n = \sum_{\mathbf{q}} \frac{\Gamma_n(\mathbf{q}) - \Gamma_0(\mathbf{q})}{\Gamma_0(\mathbf{q})}. \quad (1)$$

Here, $\Gamma_0(\mathbf{q})$ is the auto-correlation for the unperturbed speckle pattern. This result is shown in figure 2 along with a linear fit to the change in correlation sum with the logarithm of the laser intensity. Deformation of the speckle pattern appears to rapidly increase with an increase in the intensity of the laser. Our observations suggest that structural distortion results from photo-excitation as a result of light-mediated photostriction. This effect is thought to result from photo excited electron-hole pairs that are separated by the ferroelectric polarisation field within the crystal. Their separation yields an internal voltage that further induces a reverse piezoelectric effect. The photostrictive effect is also known to occur with below band-gap excitation.[4, 17]

This can be understood if a magneto-electric coupling component that additionally couples to Fe^{3+} ions in the lattice is included. [18–21]

4. Conclusion

The possibility of harnessing multiferroic compounds such as $BiFeO_3$ that can operate at room temperature is desirable for the realisation of numerous technological applications. We have demonstrated that photostriction can occur in nanocrystals of $BiFeO_3$ and results in lattice strain that increases with illumination intensity.

Acknowledgments

This work was supported in part by the Royal Society RG/130498 and the Japan Society for the Promotion of Science (JSPS) 24681014. The Diamond Light Source is kindly acknowledged for the provision of beam time.

References

[1] Cheong S W and Mostovoy M 2007 *Nature Materials* **6** 13–20
[2] Bibes M and Barthelemy A 2008 *Nature Materials* **7** 425–426
[3] Choi T, Lee S, Choi Y J, Kiryukhin V and Cheong S W 2009 *Science* **324** 63–66
[4] Kundys B, Viret M, Colson D and Kundys D O 2010 *Nat Mater* **9** 803–805 ISSN 1476-1122
[5] Miao J, Ishikawa T, Robinson I K and Murnane M M 2015 *Science* **348** 530–535
[6] von Laue M 1936 *Annalen der Physik* **26** 55–85
[7] Fienup J 1982 *Applied Optics* **21** 2758–2769
[8] Bates R 1982 *Optik* **61** 247–262 ISSN 0030-4026
[9] Bauschke H H, Combettes P L and Luke D R 2002 *J. Opt. Soc. Am. A* **19** 1334–1345
[10] Newton M C 2012 *Phys. Rev. E* **85**(5) 056706
[11] Miao J, Kirz J and Sayre D 2000 *Acta Crystallographica Section D* **56** 1312–1315
[12] Robinson I and Miao J 2004 *MRS Bulletin* **29** 177–181
[13] Robinson I and Harder R 2009 *Nature Materials* **8** 291–298 ISSN 1476-1122
[14] Newton M C, Leake S J, Harder R and Robinson I K 2010 *Nature Materials* **9** 120–124 ISSN 1476-1122
[15] Rau C, Wagner U, Pei Z and De Fanis A 2011 *physica status solidi (a)* **208** 2522–2525
[16] Tartoni N, Dennis G, Gibbons P, Gimenez E, Horswell I, Marchal J, Pedersen U, Pesic Z, Plackett R, Rau C, Somayaji R, Spiers J, Thompson J, Willis B, Angelsen C, Booker P, Burge S, Lipp J, Nicholls T, Taghavi S and Thorpe M 2012 *Nuclear Science Symposium and Medical Imaging Conference (NSS/MIC), 2012 IEEE* pp 530–533 ISSN 1082-3654
[17] Wen H, Chen P, Cosgriff M P, Walko D A, Lee J H, Adamo C, Schaller R D, Ihlefeld J F, Dufresne E M, Schlom D G, Evans P G, Freeland J W and Li Y 2013 *Phys. Rev. Lett.* **110**(3) 037601
[18] Matsuda M, Fishman R S, Hong T, Lee C H, Ushiyama T, Yanagisawa Y, Tomioka Y and Ito T 2012 *Phys. Rev. Lett.* **109**(6) 067205
[19] Lebeugle D, Colson D, Forget A, Viret M, Bataille A M and Gukasov A 2008 *Phys. Rev. Lett.* **100**(22) 227602
[20] Xu Z, Wen J, Berlijn T, Gehring P M, Stock C, Stone M B, Ku W, Gu G, Shapiro S M, Birgeneau R J and Xu G 2012 *Phys. Rev. B* **86**(17) 174419
[21] Jeong J, Goremychkin E A, Guidi T, Nakajima K, Jeon G S, Kim S A, Furukawa S, Kim Y B, Lee S, Kiryukhin V, Cheong S W and Park J G 2012 *Phys. Rev. Lett.* **108**(7) 077202

IOP Conf. Series: Journal of Physics: Conf. Series **849** (2017) 012030 doi:10.1088/1742-6596/849/1/012030

Hard X-ray submicrometer tomography of human brain tissue at Diamond Light Source

A Khimchenko[1], C Bikis[1], G Schulz[1], M-C Zdora[2,3], I Zanette[2], J Vila-Comamala[2], G Schweighauser[4], J Hench[4], S E Hieber[1], H Deyhle[1], P Thalmann[1] and B Müller[1]

[1] Biomaterials Science Center, Department of Biomedical Engineering, University of Basel, Allschwil, CH.

[2] Diamond Light Source, Harwell Science & Innovation Campus, Didcot, Oxfordshire OX11 0DE, UK.

[3] Department of Physics & Astronomy, University College London, London WC1E 6BT, UK.

[4] Institute of Pathology, Department of Neuropathology, Basel University Hospital, Basel, CH.

E-mail: anna.khimchenko@unibas.ch

Abstract. There is a lack of the necessary methodology for three-dimensional (3D) investigation of soft tissues with cellular resolution without staining or tissue transformation. Synchrotron radiation based hard X-ray in-line phase contrast tomography using single-distance phase reconstruction (SDPR) provides high spatial resolution and density contrast for the visualization of individual cells using a standard specimen preparation and data reconstruction. In this study, we demonstrate the 3D characterization of a formalin-fixed paraffin-embedded (FFPE) human cerebellum specimen by SDPR at the Diamond-Manchester Imaging Branchline I13-2 (Diamond Light Source, UK) at pixel sizes down to 0.45 µm. The approach enables visualization of cerebellar layers (*Stratum moleculare* and *Stratum granulosum*), the 3D characterization of individual cells (Purkinje, stellate and granule cells) and can even resolve some subcellular structures (nucleus and nucleolus of Purkinje cells). The tomographic results are qualitatively compared to hematoxylin and eosin (H&E) stained histological sections. We demonstrate the potential benefits of hard X-ray microtomography for the investigations of biological tissues in comparison to conventional histology.

1. Introduction

The most common cell imaging technique is microscopy of stained tissue sections. It can be a simpler set up like a hematoxylin and eosin (H&E) stained sections for classical optical microscopy, or a more complex one like labelling with fluorescent dyes or antibody staining for epifluorescence or confocal microscopy. Advanced microscopy approaches have been successfully applied to study the three-dimensional (3D) cerebellar microstructure [1, 2]. The main disadvantage of these methods is the requirement for a chemical agent to obtain information on the cellular level, whose use may have effects on the tissue morphology.

X-ray tomography can overcome some disadvantages of the histology, providing high-resolution 3D data of a comparable quality, without staining and non-destructively. Synchrotron radiation based grating interferometry enabled the identification of non-stained Purkinje cells [3, 4]. As it was recently demonstrated, the synchrotron radiation based in-line tomography allows for the 3D characterization down to subcellular structures [5]. Thus, X-ray microtomography is emerging as a promising solution for the visualization of cellular structures in 3D, offering a simpler specimen preparation, shorter experiment time and complexity, compared to histology, albeit at the cost of inferior lateral spatial resolution.

In-line phase contrast X-ray tomography [6–8] with single distance phase recovery (SDPR) using Paganin's methods [9] is one of the simplest phase contrast methods [10] in terms of data acquisition, as it is based only on free-space propagation of X rays between specimen and detector [11]. In this study, we demonstrate the 3D characterization of a formalin-fixed paraffin-embedded (FFPE) human cerebellum specimen by synchrotron radiation based SDPR at the Diamond-Manchester Imaging Branchline I13-2 (Diamond Light Source, UK) [12] at pixel sizes down to 0.45 µm. The approach enables the visualization of cerebellar layers (*Stratum moleculare* and *Stratum granulosum*), individual cells (Purkinje, stellate and granule cells) and subcellular structures (nucleus and nucleolus of Purkinje cells). The tomographic results are qualitatively compared to H&E-stained histological sections.

2. Materials and methods

2.1. Tissue preparation

The *post mortem* human cerebellum specimen was extracted from a 73-year-old male. The specimen preparation was performed according to the standard histological protocols [5], but without the staining procedure. All procedures were conducted according to the Declaration of Helsinki and approved by the Ethikkommission Nordwestschweiz. After paraffin embedding the tomography specimen 4.0 mm in height and 2.6 mm in diameter was extracted from a histological cassette by a metal punch.

2.2. Synchrotron radiation based in-line single distance tomography

Tomography experiments were performed at the Diamond-Manchester Imaging Branchline I13-2. Two scans of the same specimen were acquired. A monochromatic X-ray beam with a photon energy of 19 keV was extracted from a silicon <111> double crystal monochromator. After being transmitted through the sample, the X-ray beam was collected by a scintillator-based X-ray detector pco.4000 (PCO AG, Kelheim, Germany) located at distances of 5 cm and 8.5 cm. The effective pixel size was set by the optical magnification of the detector to 0.45 µm for the propagation distance of 5 cm and to 1.1 µm for the propagation distance of 8.5 cm. Propagation distances were selected based on the contour plot for used pixel sizes and photon energy [13]. The measurement with 0.45 µm pixel size was performed in a local tomography configuration as the sample diameter was bigger than the detector field of view for the selected magnification. Due to the detector size (4008 × 2672 pixels), 2400 equiangular projections [14] over 180 degrees with exposure times of 8.00 s and 1.58 s, respectively, were acquired. The exposure time for 0.45 µm measurement was increased in order to preserve comparable photon counts between scans. Data preprocessing and phase recovery were performed using the software tool ANKAphase [13] with an input parameter $\delta/\beta = 2406$ [5] for 1.1 µm effective pixel size and $\delta/\beta = 414$ for 0.45 µm effective pixel size. The second δ/β was decreased in comparison to theoretical value for paraffin in order to reduce gradient artefact in the data. The tomographic reconstruction was done in Matlab R2014b (Simulink, The MathWorks, Inc., USA).

2.3. Histology

A comparable cerebellum specimen was investigated by means of histology. A section of 4 µm thickness was cut using a microtome, mounted on a glass slide, and stained with H&E. Image of the resulting slide was taken on a light microscope with optical magnification 60×. The histological section was cropped around the region of interest and registered to the selected region of the tomography slice with the effective pixel size of 1.1 µm using the Demon registration tool [15] with affine transformation constraint (rotation and scaling).

2.4. Data registration
The tomographic datasets were registered using an automatic 3D/3D registration tool [16, 17] with rigid transformation constraints. The SDPR with effective pixel size 0.45 µm was set as the reference dataset.

3. Results and discussion
Figure 1 shows selected slices of non-registered and registered FFPE human cerebellum measured with effective pixel sizes of 1.1 µm (A, B) and 0.45 µm (C). Magnifications (E, F) visualize subcellular structures, such as nuclei of Purkinje cell. The microstructure of the human brain is comparably recognized in histology and tomography. Analysing the magnified images in Figure 1 (D-E) one can conclude that already with an effective pixel size of 1.1 µm tomography is comparable to histological sectioning in terms of cellular structure visualization. The image with a pixel size of 0.45 µm demonstrates the potential to surpass H&E-stained histology in terms of subcellular structures.

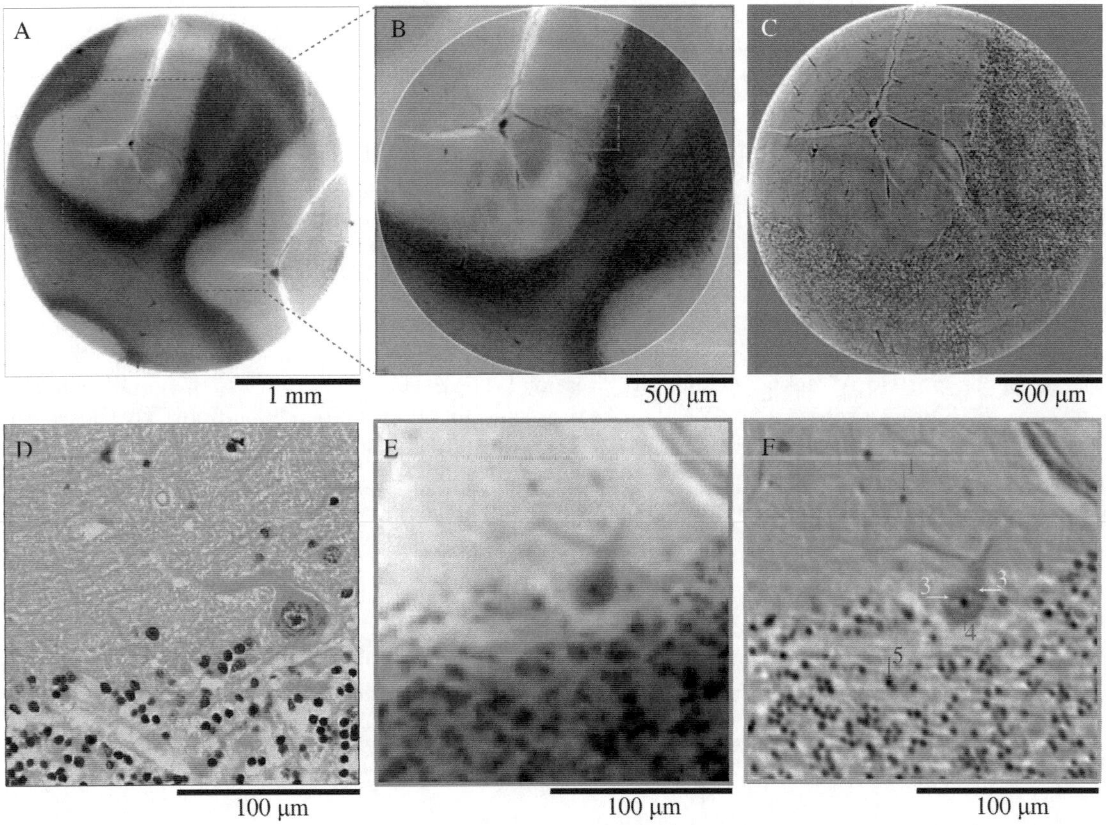

Figure 1. FFPE human cerebellum slices with magnifications highlighting an individual Purkinje cell. Selected tomography slices measured with effective pixel sizes of 1.1 µm (non-registered: A and registered: B) and 0.45 µm (C) with magnifications (E, F) and cropped histological section (D). Green (1): stelatte cell; Blue (2): dendrite of Purkinje cell; Yellow (3): dense structure within nucleus of Purkinje cell; Red (4): nucleolus of Purkinje cell; Violet (5): granular cell.

Figure 1 (D-F) show a large Purkinje cell with an average diameter of 30 µm situated between the *Stratum granulosum* and *Stratum moleculare*. The Purkinje cell axon takes off after axon hillock and branches further in the *Stratum moleculare*.

While the spatial resolution of the data with the effective pixel size of 0.45 µm is superior with respect to the data with the effective pixel size of 1.1 µm, the density contrast is superior in the data with the effective pixel size of 1.1 µm. The decrease of the density contrast of the higher resolution data can be a result of the use of local tomography, in which the regions of the sample outside of the field of view generate a background that may decrease the quality of the reconstruction. The strategy to increase the contrast can lay in data reconstruction with a higher δ/β, although it may cause a radial gradient and image blurring.

Tomography scanning time only accounts for a fraction of the time required for serial sectioning histology. In addition, to produce histological slices, see Figure 1 (D), the sample was sectioned into 4 µm thick sections, stained with H&E and visualized by light microscopy. Preparation steps induce local deformations within the inhomogeneous tissue, which can be corrected during post-processing [18]. To produce 3D data from individual sections a complex volume reconstruction is required [19].

4. Conclusion

We conclude that SDPR can reach the subcellular level, providing a compromise between spatial resolution, density contrast, and penetration depth. We present an approach for the visualization of cells within tissue of relatively large volumes without staining. For the specific case of the cerebellum, the clear visualization of various cell types, including Purkinje cells, stellate cells in *Stratum moleculare* and granule cells in *Stratum granulosum* was demonstrated. While histology still yields superior two-dimensional (2D) image quality, synchrotron radiation based tomography provides a valuable 3D compliment.

Acknowledgments

The authors gratefully acknowledge the financial support of the Swiss National Science Foundation (SNSF) project 147172. The authors highly appreciate the support of the team of Diamond Light Source, Didcot, UK, particularly C Rau and P Thibault.

References

[1] Benard M *et al* 2015 *J. Vis. Exp.* **99** 52810
[2] Wang H *et al* 2014 *NeuroImage* **84** 1007-17
[3] Schulz G *et al* 2010 *J. R. Soc. Interface* **7** 1665-76
[4] Schulz G *et al* 2012 *Sci. Rep.* **2** 826
[5] Hieber S E *et al* 2016 *Sci. Rep.* **6** 32156
[6] Snigirev A *et al* 1995 *Rev. Sci. Instrum.* **66** 548692
[7] Cloetens P *et al* 1996 *J. Phys.* D: *Appl. Phys.* **29** 13346
[8] Zanette I *et al* 2013 *Appl. Phys. Lett.* **103** 244105
[9] Paganin D *et al* 2002 *NeuroImage* **206** 3340
[10] Lang S *et al* 2014 *J. Appl. Phys.* **116** 154903
[11] Krenkel M *et al* 2016 *AIP Advances* **6** 035007
[12] Rau C *et al* 2011 *Phys. Stat. Sol.* **A208** 2522-25
[13] Weitkamp T *et al* 2011 *J. Synchrotron Rad.* **18** 617-29
[14] Holme M, *et al* 2014 *Nat. Protoc.* **9** 140115
[15] Dirk D-J *et al* 2009 *Proc. ISBI* **1** 963-6
[16] Andronache A *et al* 2008 *Med. Image Anal.* **12** 315
[17] Müller B *et al* 2012 *Int. J. Mater. Res.* **103** 242-249
[18] Germann M *et al* 2008 *J. Neurosci. Methods* **170** 14955
[19] Krauth A *et al* 2010 *NeuroImage* **49** 2053-62

PTYCHOTOMOGRAPHY AT DLS COHERENCE BEAMLINE I13

V.S.C. Kuppili[1,2], S. Sala[1,2], S. Chalkidis[1,2], A.M. Wise[6], A.D. Parsons[3], I. Zanette[3], C. Rau[3,4,5], P. Thibault[2]

[1]Department of Physics and Astronomy, University College London, WC1E 6BT London, UK
[2]Department of Physics and Astronomy, University of Southampton, Southampton, SO17 1BJ, UK
[3]Diamond Light Source, OX11 0DE Didcot, UK
[4]School of Materials, University of Manchester, Grosvenor St., Manchester M1 7HS, UK
[5]Department of Otolaryngology, Northwestern University Feinberg School of Medicine, Chicago, Illinois 60611, USA
[6]SSRL, SLAC National Accelerator Laboratory, Menlo Park, CA 94025, USA

E-mail: ucapvsc@ucl.ac.uk

Abstract. We describe the implementation and execution of ptychotomography at I13-1, the coherence branchline at Diamond Light Source. The data collection and image reconstruction protocol is demonstrated with the three dimensional reconstruction of a nanoporous gold sample.

1. Introduction

Ptychography is a scanning coherent diffraction imaging technique in which the sample is scanned perpendicular to the path of X-ray beam, collecting diffraction patterns at each point. The points are chosen such that there is enough overlap of probe between adjacent points. The resulting far-field diffraction patterns are then processed through iterative phase retrieval algorithms to obtain high resolution, phase contrast transmission maps of an extended sample [1,2]. Usually phase contrast is significantly higher than absorption contrast in the hard X-ray regime [8]. A natural extension of two dimensional ptychographic imaging is the three dimensional analog, ptychotomography [3,4,7]. We exploit the high resolution 2D ptychographic projections obtained at various angles (figure 1) as a starting point which are then taken through the conventional tomographic process of alignment and 3D reconstruction thus obtaining a three-dimensional complex transmission map of the sample under study.

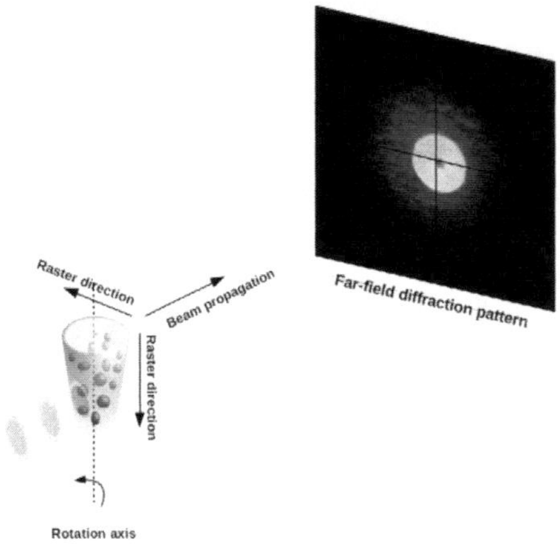

Figure 1. Schematic representation of the ptychotomographic experimental setup. The sample is rastered perpendicular to the X-ray beam collecting far-field diffraction patterns at each point. These patterns are taken through phase retrieval algorithms obtaining a ptychographic projection. Such projections collected at various angles constitutes a typical ptychotomographic dataset.

2. Experimental Details

3D nanoporous gold structures were prepared by selective dealloying of 33% gold-silver alloy micro whiskers (Palmer metals, UK). The micro whiskers were exposed to nitric acid (69% Sigma-aldrich) for 48 hours after which they were annealed for 18 hours at 375°C [5].

Figure 2. Scanning electron microscopy image of 33% gold-silver alloy micro whiskers dealloyed for 48 hours and then subjected to annealing for 18 hours at 375°C.

Ptychographic imaging was carried out at the I13-1, the coherence branchline of the I13 beamline at Diamond light source, UK [6]. The samples were imaged with X-rays at a photon energy 8.4keV and 0.1s exposure time per frame. The detector to sample distance was 6.7m. The photon counting Merlin detector with a pixel size of 55μm and 515x515 pixels wide was used. A zone plate, beam-stop, order sorting aperture ensemble was employed to obtain a confined (smaller than the field of view) X-ray probe in order to illuminate the sample. The reconstruction pixel size in this geometry was 35nm at the 512x512 cropping of the detector. 543 such equispaced angular projections were collected in interlaced format. The datasets were processed through

IOP Conf. Series: Journal of Physics: Conf. Series **849** (2017) 012031 doi:10.1088/1742-6596/849/1/012031

150 difference map iterations and 300 maximum likelihood iterations employing 3 probe modes for reconstruction (figure 3).

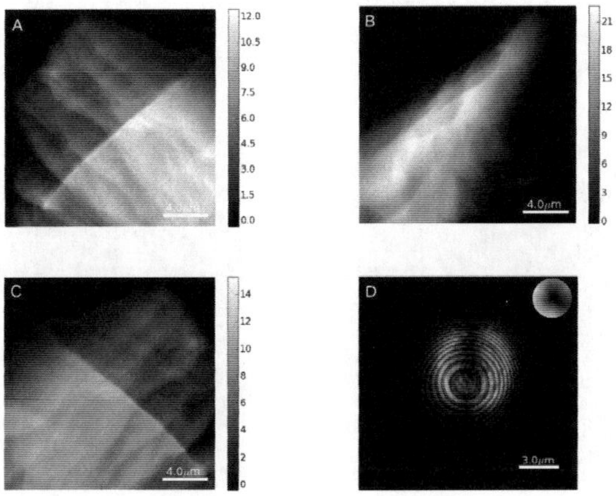

Figure 3. A,B,C. Phase unwrapped projections of nanoporous gold sample as seen in the figure 2 at 90, 0, 90 degrees respectively. The gray scale shows the phase shift in radians. D, is the primary mode of the orthogonalised probe. A colourwheel is used here to represent the complex-valued probe. The brightness of the pixel color corresponds to the magnitude of the probe while hue gives the relative phase of the probe.

In principle the raw ptychographic angular phase projections are corrected for phaseramp and phase unwrapped before taking them through the tomographic reconstruction routine. However presence of "residues", points in phase that produce inconsistencies in phase unwrapping lead to imperfect unwrapping of phase in the angular projection. Imperfect phase unwrapping over the entire range of angular projections would result in overall blurring of the tomographic reconstruction. While phase unwrapping is necessary for alignment purposes, the 3D reconstruction itself can be obtained from phase wrapped angular projections. In order to get around this limitation, we have followed the approach suggested in [4] where a thinner portion of the projection which can be phase unwrapped easily is chosen as a representative element of the projection (figure 4). All further alignment routines are carried out with the representative elements and applied to the complete projection. The vertical alignment was done using edge detection method [3] and horizontal alignment was done using tomographic consistency technique [7].

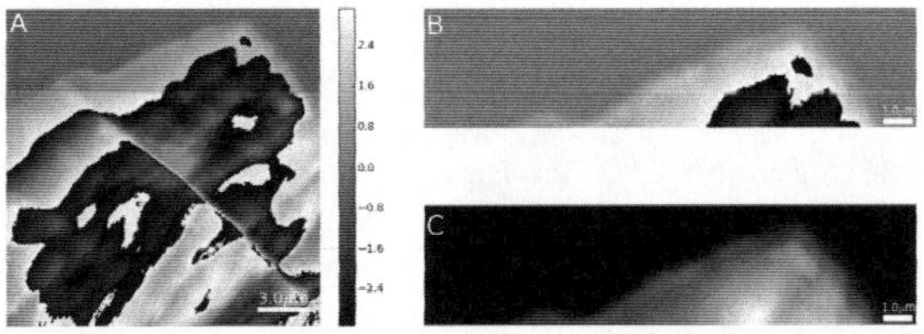

Figure 4. A, ptychographic projection after phase ramp removal. B, part of A used for alignment. C, same as B after phase unwrapping.

IOP Conf. Series: Journal of Physics: Conf. Series **849** (2017) 012031 doi:10.1088/1742-6596/849/1/012031

3. Results and Discussion

3D slices (figure 5) were then tomographically reconstructed by taking the derivatives of the aligned phase wrapped projections through the filtered back-projection technique [4]. We observe artefacts due to sharp edges (streak artefacts) and the fact that the sample is not completely in the field of view. Optimisation and standardisation of the ptychotomographic reconstruction routine is currently in progress.

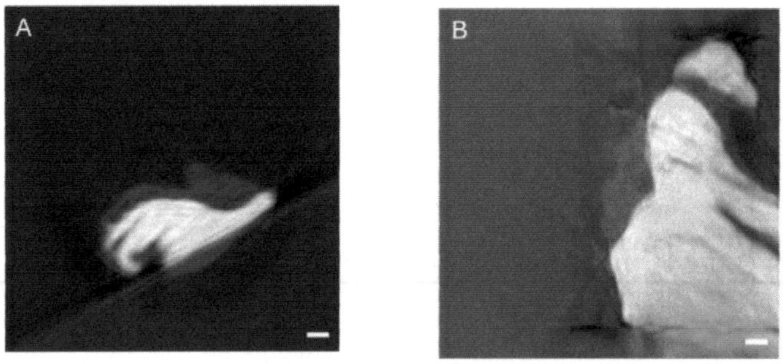

Figure 5. A, axial slice corresponding to 77th row of the phase projection. B, longitudinal slice. The scale bar in A and B corresponds to 1 micron.

4. Conclusions

We were successful in collecting and reconstructing three dimensional ptychotomographic data from the nanoporous gold sample, which is presently in the optimisation stage. The spatial resolution achievable with the current experimental setup is limited by the beam stability and the performance of the sample stages. Overheads mainly from the motor movements increase the data acquisition times and therefore (without any X-ray shutter) the risk of radiation damage is increased. We are currently in the process of achieving faster scan rates and making further improvements to the existing 3D data acquisition and reconstruction routines. We are heading towards execution and optimisation of multi-wavelength ptychographic imaging and on-the-fly scanning ptychotomography.

5. Acknowledgments

We acknowledge the financial support of ERC starting grant, OptimaX. We would like to thank Dr. Ulrich Wagner for facilitating the beamtimes at I13 coherence branch. We would like to thank Dr. Kaz Wanelik, Simon Logan, Andrew Wilson for their technical support. We thank Diamond Light Source for access to beamline I13-1 (MT13228) that contributed to the results presented here.

6. References

[1] H. M. L. Faulkner and J. M. Rodenburg 2004 *Physical Review Letters* **93** 23901–23903
[2] P. Thibault, M. Dierolf, A. Menzel, O. Bunk, C. David and F. Pfeiffer 2008 *Science* **379** 379–383
[3] M. Dierolf, A. Menzel, P. Thibault, P. Schneider, C. M. Kewish, R. Wepf, O. Bunk and F. Pfeiffer 2010 *Nature* **467** 436–439
[4] M. Guizar-Sicairos, A. Diaz, M. Holler, M. S Lucas, A. Menzel, R. A. Wepf and O. Bunk 2011 *Optics Express* **19** 21345
[5] Y. Ding, Y. J. Kim and J. Erlebacher 2004 *Advanced Materials* **16** 1897–1900
[6] Z. D. Pešić, A. D. Fanis, U. Wagner and C. Rau 2013 *Journal of Physics: Conference Series* **425** 182003
[7] M. Guizar-Sicairos, J. Boon, K. Mader, A. Diaz, A. Menzel and O. Bunk 2015 *Optica* **2** 259–266
[8] F. van der Veen and F. Pfeiffer 2004 *Journal of Physics: Condensed Matter* **16** 5003–5030

IOP Conf. Series: Journal of Physics: Conf. Series **849** (2017) 012032 doi:10.1088/1742-6596/849/1/012032

Ptychographic imaging for the characterization of X-ray free-electron laser beams

S Sala[1,2,3], BJ Daurer[4], MF Hantke[4], T Ekeberg[4], ND Loh[5], FRNC Maia[4] and P Thibault[3]

[1] Department of Physics & Astronomy, University College London, London WC1E 6BT, UK
[2] Diamond Light Source, Harwell Science & Innovation Campus, Didcot OX11 0DE, UK
[3] Department of Physics & Astronomy, University of Southampton, Southampton SO17 1BJ, UK
[4] Department of Cell and Molecular Biology, Uppsala University, 75124 Uppsala, Sweden
[5] Department of Physics, National University of Singapore, Singapore 117551

E-mail: simone.sala.14@ucl.ac.uk

Abstract. We present some preliminary results from a study aimed at the characterization of the wavefront of X-ray free electron laser (XFEL) beams in the same operation conditions as for single particle imaging (or flash X-ray imaging) experiments. The varying illumination produced by wavefront fluctuations between several pulses leads to a partially coherent average beam which can be decomposed into several coherent modes using ptychographic reconstruction algorithms. Such a decomposition can give insight into pulse-to-pulse variations of the wavefront. We discuss data collected at the Linac Coherent Light Source (LCLS) and FERMI.

1. Introduction

Many X-ray experiments require precise knowledge of the beam. One example is the characterization of the focusing optics in operation conditions aimed at selecting a plane where optics-induced aberrations are minimized. Such aberrations may have a strongly detrimental effect on any experiment performed at high power density and with a small focal spot, which is often the case at X-ray free-electron lasers (XFELs). Time-resolved characterization of the beam is also beneficial in the case of pulsed sources in general, as these might show shot-to-shot fluctuations potentially affecting any experiment extended in time. Furthermore, wavefront sensing is also instrumental in producing an accurate initial estimate of the complex-valued illumination function, useful to some phase retrieval algorithms for imaging experiments as it is the case of flash X-ray imaging [1] which has been the target application for this study.

XFELs often combine all these requirements. These extremely bright sources most frequently exploit self-amplified spontaneous emission (SASE). The stochastic nature of the spontaneous microbunching of the electron beam translates into a small but measurable stochastic variation of both temporal and spectral distribution between X-ray pulses.

Many approaches are available for beam diagnostics at XFELs such as Shack-Hartmann sensors, ablative imprints [2], Young's double-slit experiments [3], grating interferometry [4], aerosol spheres diffraction [5], coherent scattering speckles analysis [6] and ptychography [7]. This latter technique has been chosen to carry out this study as it has been already successfully applied to recover the full wavefront at high-brilliance X-ray sources [8, 9].

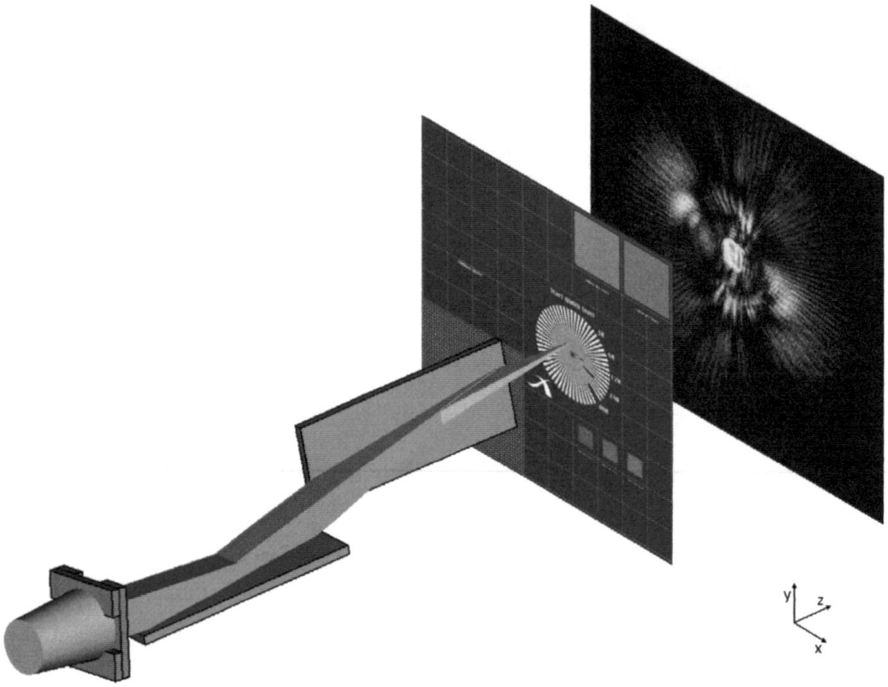

Figure 1. Sketch of the ptychographic setup used for the experiments performed at LCLS and FERMI. The beam is shaped by an aperture and focussed by KB-mirrors. The sample is scanned perpendicularly to the beam and diffraction patterns are collected by a detector downstream.

2. Methods

Ptychography is a lensless scanning diffraction technique [10]. It requires the beam to be scanned through a specimen non-destructively and its image reconstruction algorithm relies on the overlap among adjacent illuminated areas of the specimen. Such an overlap ensures data redundancy to a degree that makes phase retrieval algorithms robust enough to relax several constraints. Ptychography allows to achieve the simultaneous reconstruction of both the transmission function of the specimen (object) and the illumination function (probe) [11].
Several sources of decoherence can be accounted for by exploiting multimodal decomposition of the probe [12] which improves the quality of the reconstructed images in most experimental conditions [13]. This approach is applied here to XFEL data in order to retrieve the wavefront of the average beam and the individual pulses.
A first set of data has been collected at the AMO end station at the Linac Coherent Light Source (LCLS), SLAC's SASE XFEL, at 1.25 keV.
A second set of data has been collected at the DiProI beamline [14, 15] at FERMI, Elettra's seeded EUV / soft X-rays FEL, at 83 eV.
In both cases a setup based on the one shown in Fig. 1 has been used. Homogeneous attenuation of the beam's intensity by several orders of magnitude was needed to avoid damaging the sample. Except for this attenuation, all other conditions have been the same as those used in flash X-ray imaging experiments.
The sample used is a 100×100 μm^2 gold resolution and calibration test pattern including a 30×30 μm^2 Siemens star 150 nm thick and with a 30 nm smallest feature mounted on top of a 100 nm thick silicon nitride chip. A SEM image of the sample is shown within the sketch in Fig. 1.

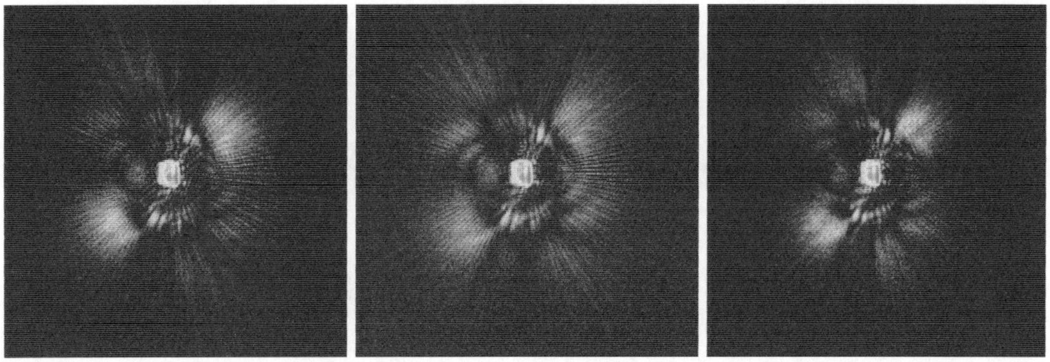

Figure 2. Three diffraction patterns collected at FERMI at the same sample position show that at least intensity and pointing vary between pulses. Each frame is plotted in a logarithmic scale and reaches a resolution of 83 nm on the edges of the frame.

3. Results

Several diffraction patterns have been collected at every scanning position in order to gather enough statistical significance and to make possible the analysis of the average beam before moving on to the single shot analysis.

FERMI data have been complicated by a significant jitter of the beam. Fig. 2 for example shows how different diffraction patterns have been generated by different pulses collected at the same sample position. The observed differences are the result of beam pointing variations as well as flux and intensity distribution fluctuations. The pulses produced by a seeded XFEL are expected to be more stable and reproducible than those produced by SASE XFELs, but due to the observed jitter a more elaborate position refinement analysis has been necessary. This position refinement can be included in the reconstruction algorithm as an orthogonal probe relaxation [16] which allows to correct for the difference between the recorded motor positions and the actual positions of the scanned beam onto the sample as well as for other pulse-to-pulse fluctuations.

At LCLS, the average beam has been investigated exploiting diffraction patterns obtained by averaging over 300 pulses collected at each sample position. Three of them are shown in Fig. 3. Using multimodal ptychographic reconstruction algorithms a first estimate of the average beam has been produced. The first three coherent modes of the average probe shown in Fig. 4 reveal an average beam profile a few micrometers wide which is consistent with the expected focal spot size. The relative intensity of each mode suggests that pulse-to-pulse fluctuations resulting in decoherence effects on the average probe are present though limited.

4. Conclusions

The data analysis is still in progress. The implementation of flexible position refinement methods within existing ptychographic algorithms will widen the experimental conditions in which these can be applied. Furthermore the outcome of these experiments is expected to shed new light on pulse-to-pulse fluctuations of the wavefront at XFELs and advance the development of fast and reliable beam diagnostic techniques for high-brilliance sources.

Acknowledgments

The research leading to these results has received funding from the European Research Council, the Swedish Research Council, the Knut and Alice Wallenberg Foundation and the Swedish Foundation for Strategic Research.

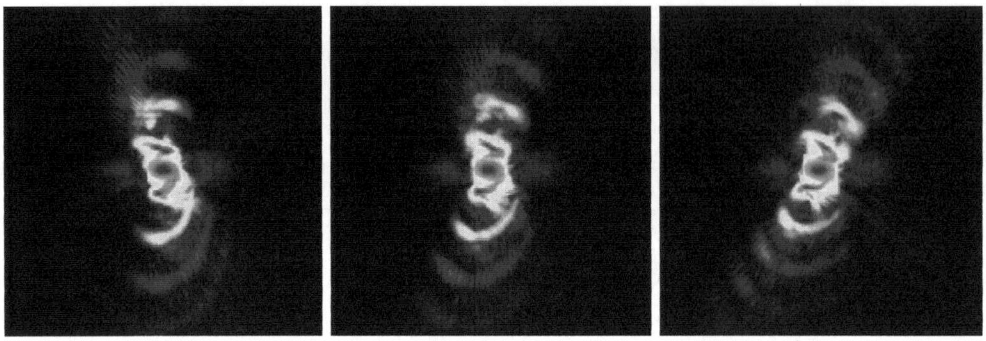

Figure 3. Preliminary analysis of LCLS average data: three diffraction patterns obtained by averaging over 300 pulses each. Each frame is plotted in a logarithmic scale and reaches a resolution of 100 nm on the edges of the frame.

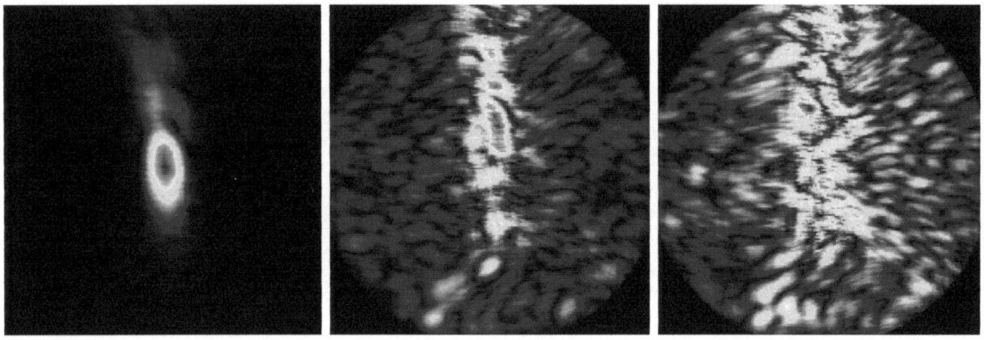

Figure 4. Preliminary analysis of LCLS average data: first three coherent modes of the probe obtained from a 10-mode ptychographic reconstruction. They contribute to the overall intensity of the probe by respectively 89.00%, 2.01% and 1.42%. Each frame is $9.7\times9.7\ \mu m^2$ and is plotted in a linear scale.

References

[1] Ekeberg T *et al.* 2015 *Physical Review Letters* **114** 1–6
[2] Chalupsky J *et al.* 2011 *Nucl. Instrum. Meth. A* **631** 130–133
[3] Vartanyants I A *et al.* 2011 *Physical Review Letters* **107** 1–5
[4] Rutishauser S, Samoylova L, Krzywinski J, Bunk O, Grünert J, Sinn H, Cammarata M, Fritz D M and David C 2012 *Nature communications* **3** 947
[5] Loh N D *et al.* 2013 *Optics express* **21** 12385–94
[6] Sikorski M *et al.* 2015 *Journal of Synchrotron Radiation* **22** 599–605
[7] Schropp A *et al.* 2013 *Scientific reports* **3** 1633
[8] Kewish C M, Thibault P, Dierolf M, Bunk O, Menzel A, Vila-Comamala J, Jefimovs K and Pfeiffer F 2010 *Ultramicroscopy* **110** 325–329
[9] Vila-Comamala J, Diaz A, Guizar-Sicairos M, Mantion A, Kewish C M, Menzel A, Bunk O and David C 2011 *Optics Express* **19** 21333
[10] Rodenburg J M and Faulkner H M L 2004 *Applied Physics Letters* **85** 4795–4797
[11] Thibault P, Dierolf M, Bunk O, Menzel A and Pfeiffer F 2009 *Ultramicroscopy* **109** 338–343
[12] Thibault P and Menzel A 2013 *Nature* **494** 68–71
[13] Enders B, Dierolf M, Cloetens P, Stockmar M, Pfeiffer F and Thibault P 2014 *Appl. Phys. Lett.* **104** 171104
[14] Pedersoli E *et al.* 2011 *Review of Scientific Instruments* **82**
[15] Capotondi F *et al.* 2015 *Journal of Synchrotron Radiation* **22** 544–552
[16] Odstrcil M, Baksh P, Boden S A, Card R, Chad J E, Frey J G and Brocklesby W S 2016 *Optics Express* **24** 8360

IOP Conf. Series: Journal of Physics: Conf. Series **849** (2017) 012033 doi:10.1088/1742-6596/849/1/012033

Low-dose, high-resolution and high-efficiency ptychography at STXM beamline of SSRF

Zijian Xu, Chunpeng Wang, Haigang Liu, Xulei Tao, Renzhong Tai

Shanghai Synchrotron Radiation Facility, Shanghai Institute of Applied Physics, CAS, 239 Zhangheng Road, Shanghai 201204, China

E-mail: xuzijian@sinap.ac.cn

Abstract. Ptychography is a diffraction-based X-ray microscopy method that can image extended samples quantitatively while remove the resolution limit imposed by image-forming optical elements. As a natural extension of scanning transmission X-ray microscopy (STXM) imaging method, we developed soft X-ray ptychographic coherent diffraction imaging (PCDI) method at the STXM endstation of BL08U beamline of Shanghai Synchrotron Radiation Facility. Compared to the traditional STXM imaging, the new PCDI method has resulted in significantly lower dose, higher resolution and higher efficiency imaging in our platform. In the demonstration experiments shown here, a spatial resolution of sub-10 nm was obtained for a gold nanowires sample, which is much better than the limit resolution 30 nm of the STXM method, while the radiation dose is only 1/12 of STXM.

1. Introduction

Since its first demonstration in 1999 [1], the coherent diffraction imaging (CDI) method has drawn extensive research interest within X-ray community [2, 3]. As a scanning version of CDI, Ptychographic CDI (PCDI, also called "ptychography") has the great potential to deliver diffraction-limited resolution imaging for extended samples[4, 5]. Similar to conventional CDI methods, the spatial resolution of ptychography is only limited by X-ray wavelength and the maximal scattering angle detectable. In this approach, a localized coherent illumination (also called "probe") scans across an extended sample by a small step size, ensuring that each scan position sufficiently overlaps with the adjacent ones. Therefore, structures in each illuminated area of the sample are expressed in multiple diffraction patterns. This redundancy makes it possible to retrieve both the probe and sample information simultaneously from the same dataset [5, 6]. Up to now, PCDI has been successfully applied in the study of a variety of samples in material science and biology with visible light [7], soft [8] and hard X-rays [5, 9], and electron beams [10].

The radiation damage is the key factor limiting the achievable resolution for biological samples in CDI. A PCDI platform with relatively low radiation dose, high resolution and high efficiency is developed at the scanning transmission X-ray microscopy (STXM) endstation of Shanghai Synchrotron Radiation Facility (SSRF). Compared to other FZP-based PCDI setups, we used a much larger illumination spot with 3-5 μm diameter by moving samples downstream from ZP focus, and a larger step size of 500-1000 nm in our platform, which results in fewer scanning positions and lower radiation dose when imaging the same sample area, though the larger light spot divergent from ZP focus will lead to a more difficult reconstruction due to the highly curved wavefront of the probe. By combining the multimode reconstruction method [11] and up-sampling method [12] with the position-correction algorithm [13] in our home-developed ptychography reconstruction software, we can obtain high-quality images from our PCDI platform with much lower radiation dose and less data acquisition time. Several demonstration experiment results are shown below, in which a spatial resolution as high as 8.5 nm has been obtained for a gold nanowires sample, which is much better than that of the STXM imaging (30 nm at best), while the radiation dose is only 1/12 of STXM, and the data collection time is only about 1/3 of STXM. These results indicate a great application potential of our PCDI platform especially in life science.

2. Experimental setup

The soft X-ray PCDI platform was developed based on the STXM endstation of BL08U1A beamline at SSRF. This beamline has a photon energy ranging from 200 to 2000 eV produced by an undulator. The experimental setup of soft X-ray ptychography [8, 14] shares a number of similarities with STXM [15]. Therefore, by replacing the photomultiplier tube detector with an X-ray CCD detector and keeping the rest unchanged, the STXM setup can be easily adapted for ptychography.

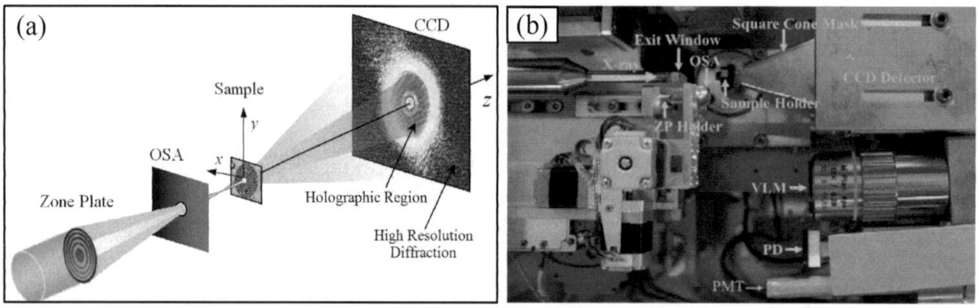

Figure 1. (a) A schematic diagram of the PCDI experiment setup based on STXM. The sample is moved downstream from the ZP focus, which increases the illumination spot to 3-5 um in diameter. (b) A photograph of the PCDI setup in the STXM chamber of SSRF. A square cone mask covers the CCD camera to reduce the noise on the diffraction signals.

As Figure 1 shows, A 200 μm diameter Fresnel zone plate (FZP) with 80 μm central stop and 30 nm outermost zone width is used to focus X-ray, and a 70 μm order sorting aperture is placed downstream of FZP to allow only the first order focused X-ray through. The sample (mounted on a 2D piezo stage) is placed 51 μm downstream of the ZP focus, resulting in an illumination spot of about 3 μm diameter on the sample for 700-710 eV energy (used for the experiments described below). The diffraction patterns are recorded by a 16-bit PI-MTE:2048B high vacuum compatible x-ray CCD detector, placed 70 mm downstream of the sample. A square cone mask covers the CCD front to prevent stray light from the laser interferometer. The CCD chip has a 2048×2048 imaging array with 13.5×13.5 μm^2 pixel size. A read-out time of 4.5 s/frame is needed in data acquisition.

3. Experimental results

We have imaged various samples using the PCDI method developed at the STXM beamline. To demonstrate the performance of the PCDI platform, we show two representative examples in this section: gold nanowires, and a bamboo tissue slice.

3.1. Gold nanowires

Gold nanowires with about 20 nm width placed on a Si_3N_4 window were imaged by a 3um illuminating spot with 704 eV photon energy. The sample was scanned with a step size of about 400 nm in a concentric circle grid of total 94 grid points. Each position was exposed for 200 ms. The dataset was reconstructed using 3 probe modes mixed-state reconstruction method, and the results are shown in figure 2. An STXM image of the same sample with 20 nm scanning steps is also shown for comparison. According to a power spectral density (PSD) analysis, the image resolution reaches 8.5 nm, much higher than the STXM resolution of 32 nm.

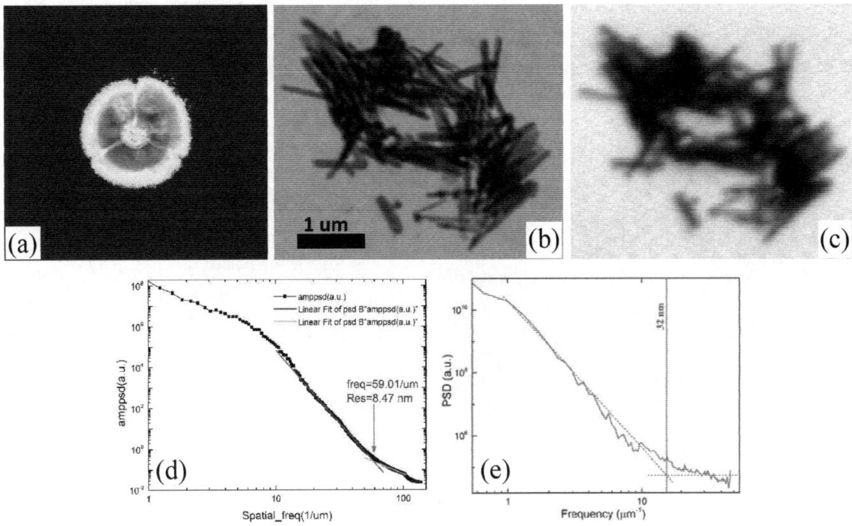

Figure 2. Ptychography imaging of Gold nanowires and its comparison with the STXM result. (a) is a typical diffraction pattern, the background noise of which was removed by subtraction of a dark CCD image and a thresholding process. (b) is the reconstructed amplitude image of the sample, (c) is the STXM image with 20 nm scanning steps. (d, e) are the PSD analyses of images (b, c), respectively, indicating a resolution of 8.5 nm for PCDI image and 32 nm for STXM image. The spatial resolution is determined by the critical frequency where the PSD curve significantly deviates from a power law decay into a random fluctuation regime.

Figure 2 shows a significant improvement of PCDI image over STXM image. This can be seen from both the image comparison of (b, c) and the PSD resolution analyses of (d, e). An overall performance comparison between PCDI and STXM is shown in table 1 based on this experiment. Table 1 shows that PCDI has great advantages over STXM not only in spatial resolution, but also in radiation dose and data collection efficiency. The radiation dose is proportional to the exposure time of the sample when the incident beam is stable. Therefore, we compared the radiation doses of PCDI and STXM by comparing the total exposure times in the two imaging processes, and found that PCDI only needs 1/12 exposure time (dose) of STXM to reach a resolution of about 10 nm which is 3-3.5 times better than that of STXM. In addition, due to much fewer scanning grid points in PCDI than in STXM, the experiment efficiency of PCDI is also higher than that of STXM. In the current configuration of our PCDI platform, the speed of PCDI data acquisition could be about 3 times that of STXM.

Table 1. The performance comparison between STXM and PCDI at STXM beamline of SSRF.

Performance	STXM	PCDI	STXM : PCDI
Exposure time (dose)	226 s	18.8 s	12:1
Data acquisition time	30 min	11 min	~3:1
Spatial resolution	>30 nm	8.5 nm	~3.5:1

3.2. Bamboo tissue slice

A frozen-dried bamboo tissue slice attached on a Si_3N_4 window was imaged using a 710 eV ZP-focused X-ray. The illuminating spot size on the sample was 3um in diameter, which scanned across the sample with a step size of 500 nm in a concentric circle grid of total 121 points. The sample-CCD distance was about 70 mm. The exposure time was 800 ms for each position. The dataset was reconstructed using 3 modes mixed-state algorithm, and the results are shown in figure 3. According to a PSD analysis, the image resolution reaches 28.6 nm, which is significantly higher the usual STXM resolution for biological samples (no better than 50 nm in our beamline). In addition, the phase image shows a better contrast than the amplitude image, which is usually the case for biological samples with weak X-ray absorption.

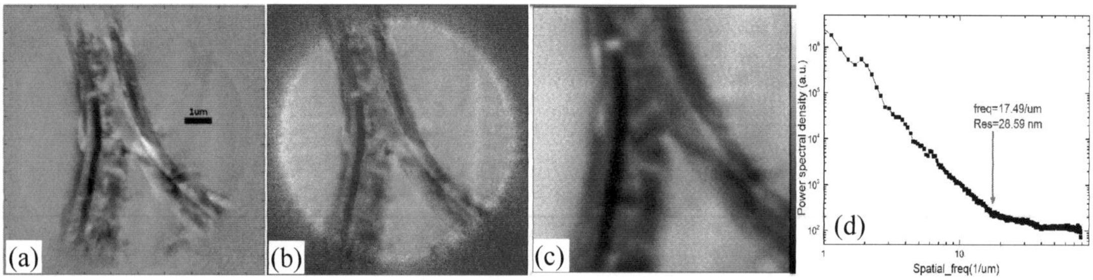

Figure 3. Ptychography imaging of Bamboo slice. (a) is the reconstructed phase image of the sample, (b) is the reconstructed amplitude, (c) is the STXM image with 30 nm scan steps. (d) is the PSD analysis of image (b), which indicates a resolution of 28.6 nm.

4. Summary and conclusion

We implemented a low-dose, high resolution and high efficiency ptychography platform at the STXM beamline of SSRF. A sub-10 nm resolution was obtained for material specimen by using relatively large illumination spot and large scanning steps and employing mixed-state reconstruction method combined with up-sampling and position-correction algorithms. Significant quality improvement was also observed for biological samples, which are especially sensitive to the radiation dose. A data acquisition rate of 3 times better than that of STXM was achieved, which can be further speeded up by using a faster CCD detector because at present over 90 percent of time is consumed by data read-out of CCD.

Acknowledgments

This work was supported by the National Natural Science Foundation of China (grant Nos. 11225527, 11575283, 11505277).

References

[1] Miao J, Charalambous P, Kirz J *et al* 1999 *Nature* **400**, 342-344
[2] Chapman H N and Nugent K A 2010 *Nature Photonics* **4**, 833-839
[3] Miao J, Ishikawa T *et al* 2015 *Science* **348**, 530-535
[4] Faulkner H M L and Rodenburg J M 2004 *Phys. Rev. Lett.* **93**, 023903
[5] Thibault P *et al* 2008 *Science* **321**, 379-382
[6] Maiden A M and Rodenburg J M 2009 *Ultramicroscopy* **109**, 1256-1262
[7] Claus D *et al* 2012 *Opt. Express* **20**, 9911-9918
[8] Shapiro D A *et al* 2014 *Nat. Photonics* **8**, 765-769
[9] Dierolf M *et al* 2010 *Nature* **467**, 436-439
[10] Putkunz C T *et al* 2012 *Phys. Rev. Lett.* **108**, 073901
[11] Thibault P and Menzel A 2013 *Nature* **494**, 68-71
[12] Batey D *et al* 2014 *Phys. Rev. A* **89**, 043812
[13] Zhang F *et al* 2013 *Opt. Express* **21**,13592-13606
[14] Maiden A M 2013 *Nat. Commun.* **4**, 1669
[15] Horowitz P and Howell J A 1972 *Science* **178**, 608-611

IOP Conf. Series: Journal of Physics: Conf. Series **849** (2017) 012034 doi:10.1088/1742-6596/849/1/012034

Quantitative Evaluation of Hard X-ray Damage to Biological Samples using EUV Ptychography

Peter Baksh[1],Michal Odstrcil[1],Aaron Parsons[2], Jo Bailey[3], Katrin Deinhardt[3], John E. Chad[3], William S. Brocklesby[1] and Jeremy G. Frey[4]

[1] Optoelectronics Research Centre, University of Southampton, UK
[2] Diamond Light Source, Didcot, OX11 0DE
[3] Biological Sciences, Faculty of Natural and Environmental Sciences, University of Southampton, UK
[4] Chemistry, Faculty of Natural and Environmental Sciences, University of Southampton, UK

E-mail: p.baksh@soton.ac.uk

Abstract.
Coherent diffractive imaging (CDI) has become a standard method on a variety of synchrotron beam lines. The high brilliance short wavelength radiation from these sources can be used to reconstruct attenuation and relative phase of a sample with nanometre resolution via CDI methods. However, the interaction between the sample and high energy ionising radiation can cause degradation to sample structure. We demonstrate, using a laboratory based high harmonic generation (HHG) based extreme ultraviolet (EUV) source, imaging a sample of hippocampal neurons using the ptychography method. The significant increase in contrast of the sample in the EUV light allows identification of damage induced from exposure to 7.3 keV photons, without causing any damage to the sample itself.

1. Introduction

The study of hippocampal neurons is of great interest to the scientific community to understand mechanisms that can lead to neuro-degenerative disease. To aid further understanding of biological processes, imaging at high spatial resolution and resolving internal structures beneath the sample surface is required.

Coherent diffractive imaging (CDI) is a rapidly evolving field capable of producing diffraction limited resolution of a sample's amplitude and phase, without the need for any imaging optics. A scanning variant of CDI known as ptychography is a proven robust imaging method to solve the phase retrieval problem. Ptychography records scatter patterns at multiple partially overlapping regions of the sample illuminated by spatially confined structured illumination [1]. The overlap provides redundancy in the collected dataset. The ptychography algorithms assume that there is identical sample transmission in the overlapping regions in combination with precise knowledge of relative sample shift and measured oversampled diffracted intensities to reconstruct both the complex exit wave field of the sample, and the complex probe illumination function.

Laboratory scale EUV sources such as plasma based sources and HHG sources have demonstrated the required coherence to successfully image via the ptychography technique [2,3]. Whilst the synchrotrons provide a much broader range of imaging wavelengths that spread from

IOP Conf. Series: Journal of Physics: Conf. Series **849** (2017) 012034 doi:10.1088/1742-6596/849/1/012034

IR to hard X-rays, the laboratory EUV sources are readily available, inexpensive, and in the case of HHG sources the coherent photon flux is approaching / comparable to synchrotron sources as shown in [4] where the flux was 1×10^{12} photons/sec .

The drawback of all short wavelength sources is that the sample is illuminated by ionising radiation that causes both primary and secondary damage effects in the sample. This radiation induced damage can significantly alter the sample structure, and in particular unfrozen biological samples are particularly sensitive to radiation damage. Below, we investigate the effect of radiation damage caused from an illumination of 7.3 keV synchrotron source using a 42 eV HHG laboratory source via the ptychography method.

2. Experimental method

The sample under investigation was a culture of mouse hippocampal neurons prepared on a 50 nm Si_3N_4 membrane coated in a thin layer of poly-D-lysine. The neurons extracted from the hippocampus of an E17 mouse embryo were cultured for 14 days before fixation. The cells were fixed using 4% paraformaldehyde fixing solution and 20% sucrose in phosphate buffered saline solution. Finally, they were fixed in 100% methanol for 20 minutes before air drying.

The sample was imaged at the I13 beam line at the Diamond Light Source. Photons generated from the undulator are monochromated using using a Si (1,1,1) crystal. 7.3 keV photons were focused down using a Fresnel zone plate of focal length 34 cm. The DC frequency is blocked using a central stop, and higher diffraction orders are removed using an order sorting aperture. The sample is placed after the focus, where the hard X-ray spot size was chosen to be $\approx 2\mu m$ in diameter. The scatter pattern from the sample was measured a distance of 7.1 cm from the sample using a Merlin MEDIPIX photon counting detector. The ptychography method was used to image the sample using a step size of 0.5 μm to ensure the ptychographic overlap constraint was fulfilled. The sample was positioned using piezoelectric actuators and the exposure times were set at 10 s. The photon flux on the sample was $1.4 \cdot 10^8$ photons/s.

After inducing radiation damage in a localised region, the sample was imaged using 42 eV photons from a HHG source at the University of Southampton. The coherent EUV radiation is generated from a highly non-linear interaction between ultra-fast pulses and an inert gas. The driving laser beam was generated from a 2 mJ, 40 fs, 800 nm, 1 kHz repetition rate CPA Ti:sapphire laser focused into a argon filled gas cell. The fundamental wavelength was removed by a single 200 nm Aluminium foil and partial spectral filtering and focusing of the EUV radiation was obtained using a curved MoSi multilayer mirror. After filtering, the spectrum illuminating the sample consists of three harmonics centered around 29 nm (27-th harmonic) with a 12% bandwidth [5]. The sample is placed at the focus of the EUV radiation, where the spot size is measured to be $\approx 12\,\mu m$. 80 μm upstream of the sample, a 10 μm aperture is placed to spatially confine the beam. The resulting flux incident upon the sample is $2 \cdot 10^8$ photons/s. Further detail on this experiment can be found in Ref. [3]. The scatter pattern of the sample was collected in the far field using an EUV sensitive cooled CCD, where the maximal collected NA was 0.26. The dynamic range of the CCD was numerically extended by 3 orders of magnitude, as described in Ref [3], from 10^4 to 10^7 counts, however despite this, the poor dynamic range of the CCD detectors is still a limiting factor, when imaging weakly scattering real biological samples.

The synchrotron datasets were reconstructed by the difference maps method and then refined by the ePIE algorithm [6]. However, for the case of the diffraction patterns collected using the HHG source, where both pointing stability and intensity stability are significantly poorer than the synchrotron source, the orthogonal probe relaxation ptychography (OPRP) method [7] was used for the reconstruction. This method relaxes the overlap constraint, and allows for beam fluctuation, reconstructing a different probe for each scanning position.

3. Experimental results

Fig. 1 shows the comparison of the reconstruction quality between the EUV ptychography reconstruction (Fig. 1(a)) and the X-ray ptychography (Fig. 1(b)). The detail of the EUV reconstruction is showing artefact-less reconstruction quality of the fine neural structures. On the other hand, any structure in the thick cell soma of the neurons are lost due to low penetration depth in biological material (≈50 nm). The X-rays easily penetrate through the thick cell bodies however, due to very low contrast, no details of the thinner neural connections can be seen.

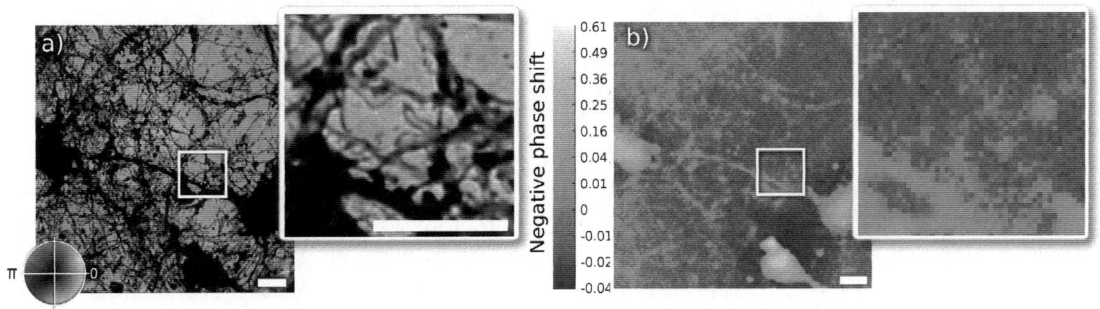

Figure 1. A reconstruction of neural structure in transitivity.(a) Complex colorscale reconstruction using 42 eV, (b) shows reconstructed phase from a ptychography scan at 7.3 keV where the sample dose was $1.4 \cdot 10^3$ Gy. Scalebar is 10 μm.

It was assumed that because the inelastic interaction of the hard X-ray photons with the sample is very weak, the dose deposited in the sample would be below the radiation damage threshold. Using the formula derived by Howells et al [8] the sample dose D in Gray (J/kg) is calculated

$$D = \frac{N_0 h\nu}{L_a n S} \quad , \tag{1}$$

where L_a is the attenuation length, $h\nu$ is the photon energy, N_0 is the number of incident photons per area S, and n is the average sample mass density.

The sample received a spatially dependent dose distribution. The white dashed box in Fig. 2(a) shows the region where the sample received a dose of $3.6 \cdot 10^5$ Gy, whereas outside this region the sample received a dose of only $1.4 \cdot 10^3$ Gy. The 42 eV ptychographic reconstructions shown in Fig. 2(a) show the region of radiation induced sample damage that can be clearly identified. A detailed image of the transition region is shown in Fig. 2(2). The damage can be approximately quantified when looking at the exponential intensity decay constant, α, of each EUV diffraction pattern extracted from a fit of $y = x^\alpha$, and plotting this as a function of spatial position, shown in Fig. 2(c). The region of higher radiation does has a more negative value as a function of spatial frequency, which indicates a steep exponential reduction in the high frequency signal.

Due to the strong photon/sample interaction, the EUV dose given to the sample is approximated to be $4.4 \cdot 10^7$ Gy during 15 s exposure, which is 2 orders of magnitude more than the X-ray dose. Despite the high dose of EUV light that is capable of free radical generation [9], no radiation damage was observed. We think the dose given from the 42 eV illumination is not causing any changes to the sample that can be measured using our microscope. It is also possible that other damage mechanisms occurred in the X-ray reconstruction such as thermal damage. Both the heat damage and secondary radiation damage from diffusion of free radicals within the sample can be mitigated through cryo sample preparation as discussed in Ref. [8].

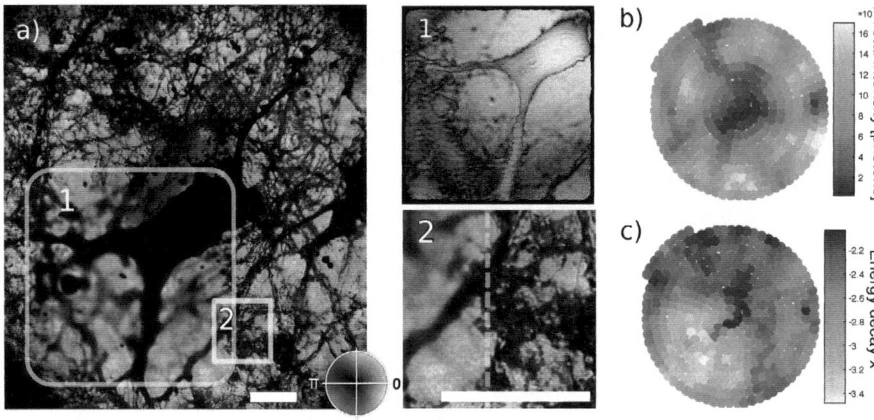

Figure 2. Radiation damage caused by X-ray ptychography at 7.3 keV. (a) EUV ptychography reconstruction after the X-ray ptychography scan. Subregion (1) was scanned by a 2 μm X-ray beam with 10 s exposure. The X-ray ptychography reconstruction of this subregion is shown in a contrast enhanced colorscale. The transition region around the damaged region from EUV ptychography image is shown in (2). The image (b) shows a scanning transmission (STXM) image of the sample, plotted at each real space scanning position, while (c) shows the measured exponential decay of the scattering energy in the Fourier (detector) plane against the spatial frequency. Scale bar denotes 10 μm.

4. Conclusion

We have shown that the damage threshold for hippocampal neurons prepared without any measures to mitigate radiation damage is between $1.4 \cdot 10^3$ and $3.6 \cdot 10^5$ Gy when measured at the resolution of our EUV microscope (lateral resolution =80 nm axial resolution 10 nm). We have confirmed radiation damage using highly sensitive EUV radiation from a coherent laboratory based EUV source, using the OPRP algorithm to overcome the shortcomings of HHG EUV stability.

Acknowledgments

This work builds on the investment from the EPSRC Basic Technology grant GR/R87307/01. M.O. acknowledges financial support from the EU FP7 Erasmus Mundus Joint Doctorate Programme EXTATIC under framework partnership agreement FPA-2012-0033 and P.B. acknowledges the EPSRC studentship. J.B. and K.D. acknowledge the BBSRC grant [BB/L007576/1]. The K5200 graphics card used for this research was donated by the NVIDIA Corporation.

References

[1] Rodenburg J 2008 *Adv. Imag. Elect. Phys.* **150** 87–184
[2] Odstrcil M, Bussmann J, Rudolf D, Bresenitz R, Miao J, Brocklesby W and Juschkin L 2015 *Opt. Lett.* **40** 5574–5577
[3] Baksh P D, Odstrčil M, Kim H S, Boden S A, Frey J G and Brocklesby W S 2016 *Opt. Lett.* **41** 1317–1320
[4] Butcher T, Anderson P, Chapman R, Horak P, Frey J and Brocklesby W 2013 *Physical Review A* **87** 043822
[5] Odstrčil M, Baksh P, Kim H, Boden S, Brocklesby W and Frey J 2015 *SPIE Opt. Eng.+ App.* p 958912
[6] Maiden A M and Rodenburg J M 2009 *Ultramicroscopy* **109** 1256–62 ISSN 1879-2723
[7] Odstrcil M, Baksh P, Boden S, Card R, Chad J, Frey J and Brocklesby W 2016 *Opt. Express* **24** 8360–8369
[8] Howells M R, Beetz T, Chapman H N, Cui C, Holton J, Jacobsen C, Kirz J, Lima E, Marchesini S, Miao H *et al.* 2009 *Journal of electron spectroscopy and related phenomena* **170** 4–12
[9] Beetz T and Jacobsen C 2003 *Journal of Synchrotron Radiation* **10** 280–283

Comparison of laboratory grating-based and speckle-tracking x-ray phase-contrast imaging

J. Romell[1], T. Zhou[1], M. Zdora[2,3], S. Sala[2,3], F.J. Koch[4], H.M. Hertz[1], A. Burvall[1]

[1] Biomedical and X-ray Physics, KTH Royal Intsitute of Technology, AlbaNova University Center, SE-106 91, Stockholm, Sweden
[2] Diamond Light Source, Harwell Science & Innovation Campus, Didcot, Oxfordshire OX11 0DE, UK
[3] Department of Physics & Astronomy, University College London, London WC1E 6BT, UK
[4] Institute of Microstructure Technology, Karlsruhe Institute of Technology, 76344 Eggenstein-Leopoldshafen, Germany

E-mail: jenny.romell@biox.kth.se

Abstract. Phase-contrast imaging with x-rays is a developing field for imaging weakly absorbing materials. In this work, two phase-contrast imaging methods, grating- and speckle-based imaging, that measure the derivative of the phase shift, have been implemented with a laboratory source and compared experimentally. It was found that for the same dose conditions, the speckle-tracking differential phase-contrast images have considerably higher contrast-to-noise ratio than the grating-based images, but at the cost of lower resolution. Grating-based imaging performs better in terms of resolution, but would require longer exposure times, mainly due to absorption in the grating interferometer.

1. Introduction

To overcome the limitations of low resolution and poor contrast when imaging weakly absorbing objects with x-rays, different phase-sensitive imaging techniques are being developed. While conventional absorption is readily measured, phase shift must be detected by finding the intensity variations caused by the shift. The well-established propagation-based imaging method (PBI) measures the second-order derivative of the phase, while grating-based imaging (GBI) [1] and the relatively new speckle-based imaging (SBI) [2, 3] measure the refraction angle. By tracking the shift of a known intensity pattern, a map of the refraction angle, which is proportional to the derivative of the phase shift, can be constructed. GBI and SBI use a Talbot pattern [4] and a near-field speckle pattern, respectively. These two methods have been implemented with a laboratory micro-focus x-ray source, and have been used to image different samples under the same dose conditions, with the goal of comparing the quality of the obtained images.

2. Method

The experimental arrangements of grating- and speckle-based imaging are shown in Fig. 1. The liquid-metal-jet source, similar to the source described in [5], was run at 40 kVp with an emission current of 0.6 mA, generating an x-ray beam with spot size $5.3 \times 7.7\ \mu\text{m}^2$. To keep the dose controlled, the source-to-object distance was kept constant at 65.6 cm at all times. The grating

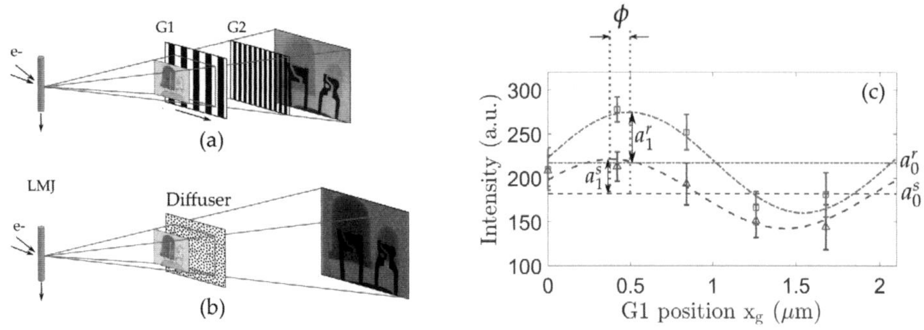

Figure 1. Experimental arrangements. (a) GBI. (b) SBI. (c) The intensity of a pixel during phase stepping in the GBI method, where deviations from the fit indicate photon noise.

interferometer used in GBI was designed with grating periods $p_1 = 4.08$ μm of the π-shifting phase grating G1 and $p_2 = 2.4$ μm of the absorption grating G2. Si with a thickness of 200 μm was used as substrate for both gratings, with 9.2 μm of Ni on G1 and 40 μm of gold on G2. Both G1 and G2 had a duty cycle of 50%. The distance between the gratings was $d = 12.4$ cm, and the detector was placed as close as possible after G2. For SBI, the speckle pattern was generated by a diffuser, here a piece of P800 sandpaper, placed after the sample. To reach sufficient speckle visibility [6], the detector was placed 1.5 m away from the source.

In GBI, G1 was scanned laterally over one period of the Talbot intensity pattern [1]. For each step, two images were recorded: one with a sample in the beam path, and one reference without. Processing of the stacks of images was carried out pixel-wise. The intensity variation of one pixel during the phase stepping can be seen in Fig. 1c, and can be described in terms of a cosine series. The superscripts s and r denote *sample* and *reference*, respectively. Differential phase contrast (DPC), which has been the main interest in this work, is proportional to $\phi = \phi_1^s - \phi_1^r$. For the grating interferometer, the angle of refraction α is related to the phase shift $\Phi(x, y)$ caused by the object, and to ϕ as [7]

$$\alpha_x = \frac{\partial \Phi}{\partial x}\frac{\lambda}{2\pi} = \frac{\phi p_2}{2\pi d}. \tag{1}$$

The parameters a_0, a_1 and ϕ_1, for both reference and sample, were extracted from the raw images by fast Fourier transform.

In x-ray speckle-tracking imaging, the type of SBI considered here, two images were recorded: one with only the diffuser in the beam path, and one with the diffuser and the sample. The arrangement is shown in Fig. 1b. Since no scanning process was carried out, the images were compared region-wise to find the shift of the speckle pattern, the absorption, and the loss of speckle visibility. Least-square minimisation of a cost function [8] gave the sample-induced local shift (δ_x, δ_y) around each pixel, in both horizontal and vertical directions. For comparison with GBI, only the horizontal shift has been considered. From the shift, the angles of refraction could be calculated as $\alpha_i = \delta_i/L$ for $i = x, y$, where L is the diffuser-to-detector distance. The resolution in the final images was mainly affected by the window size chosen for the reconstruction.

3. Results

For a plastic wedge with a small angle, specifically a LEGO® sword (Fig. 2), the contrast-to-noise ratios (CNR) of the DPC images were evaluated as $\text{CNR} = |I_A - I_B|/\sigma$. Here, σ is the standard deviation of the background and I the mean intensity. The figure was imaged with GBI and SBI for an exposure time of 540 s each, and for the SBI reconstruction a 24×24 pixel window

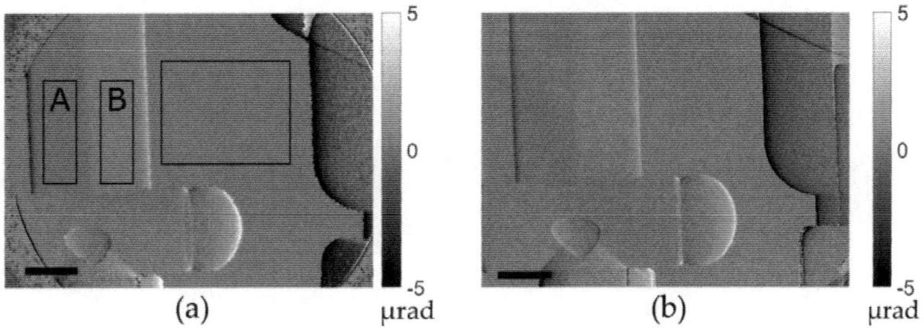

Figure 2. DPC images, expressed as refraction angle $\alpha_x \propto \partial\Phi/\partial x$. (a) SBI, with regions used in the CNR calculation. (b) GBI. The scale bar represents 2 mm in both figures.

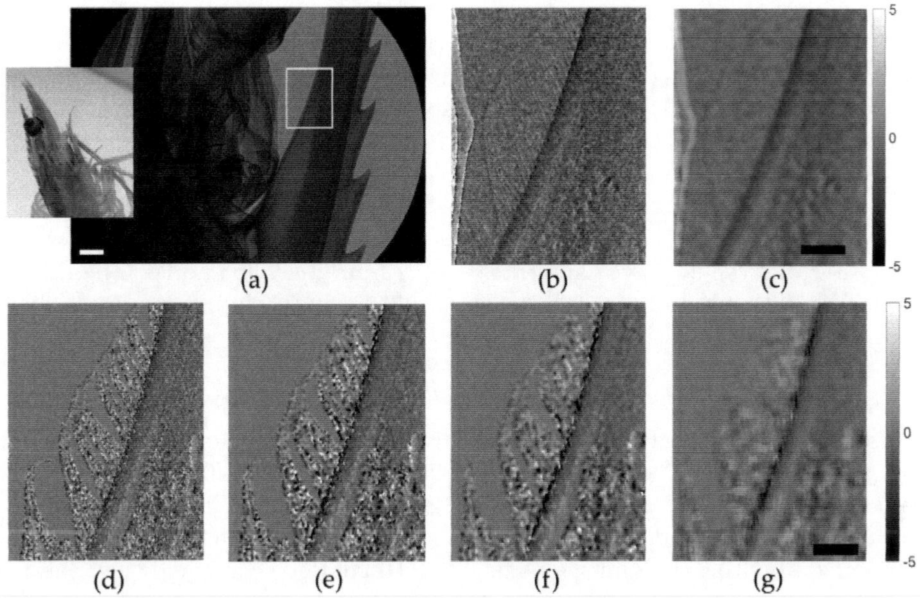

Figure 3. (a) Shrimp with marked region used for the DPC images (b)−(g). (b) GBI differential phase contrast. (c) Low-pass filtered GBI. (d)−(g) Speckle-based images, reconstructed with a window of $N \times N$ pixel^2, where N is (d) 4, (e) 8, (f) 12 and (g) 24. The white scale bar corresponds to 1 mm and the black 500 μm. The grey scale represents refraction angle in μrad.

was chosen for optimal noise [8]. Partly due to the inherent filtering in the speckle-tracking reconstruction method, the CNR was here 9.6, compared to 0.83 of GBI.

To evaluate the resolution, some small structures on a shrimp (see Fig. 3a) were imaged. The exposure time was 780 s. Since the window size of the SBI reconstruction greatly impacts not only noise but also resolution, different window sizes $N \times N$ pixel^2 were used, N ranging from 4 to 24. Shown in Fig. 3d-g are reconstructions for $N = 4$, 8, 12 and 24. It is clear from considering Fig. 3g that choosing a large window will filter out small features, while the hairs could be clearly seen, however noisy, when a window of 8×8 or 4×4 pixel^2 was used. The GBI DPC image, acquired during the same exposure time (Fig. 3b), showed strong background noise which made it difficult to distinguish the hairs at all. Comparing the methods from a noise perspective, by low-pass filtering the GBI image until the same background noise as in the 4×4 speckle image was reached (Fig. 3c), showed that the hairs were then blurred out entirely.

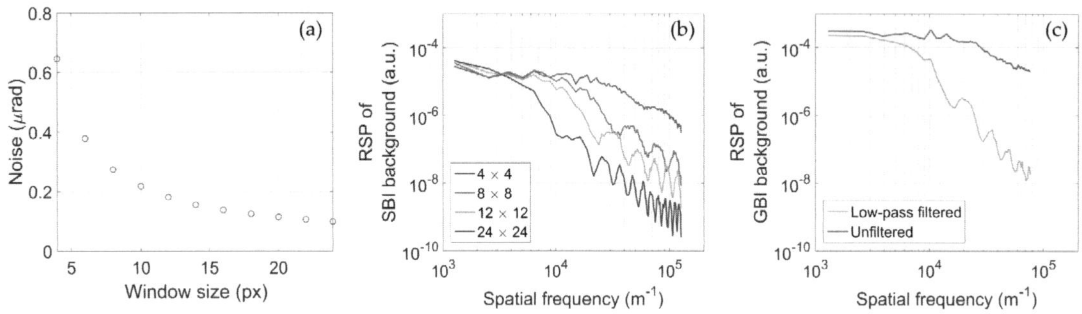

Figure 4. (a) Background noise of SBI DPC reconstructions of the shrimp (Fig. 3a) as a function of the window size. (b) Background RSP for SBI. (c) Background RSP for GBI.

Noise is clearly limiting the detectability of small features in the grating-produced image, while the reconstruction window, and hence resolution, is the main limiting factor for the speckle-tracking method. The noise dependence on the reconstruction window was evaluated by determining the background noise for different window sizes (Fig. 4a).

Finally, to appreciate the impact of the reconstruction method on the images, the average radial spectral power (RSP) of the background was evaluated for the four different window sizes used in Figs. 3d-g, and the unfiltered and filtered GBI from Figs. 3b and 3c. The RSP is shown in Figs. 4b and 4c. The effect of the rectangular window can clearly be seen in the fluctuations in Fig. 4b and the filtered GBI in Fig. 4c.

4. Conclusion and outlook

To summarise, the quality of the DPC images largely depends on both experimental arrangement and reconstruction method. The speckle-tracking method gives considerably higher CNR than GBI, but at the cost of lower resolution. With GBI, smaller details can be resolved, but it requires a larger dose, mainly due to the absorption in the interferometer.

As a continuation of this work, image quality in GBI could be compared to variations of SBI which can give higher resolution, such as speckle-scanning phase-contrast imaging.

Acknowledgments

We thank William Vågberg and Jakob Larsson for assistance in the lab, and Karlsruhe Institute of Technology (KIT) for providing the gratings for the interferometer.

References

[1] Pfeiffer F, Weitkamp T, Bunk O and David C 2006 Phase retrieval and differential phase-contrast imaging with low-brilliance X-ray sources *Nat. Phys.* **2** pp 258–61
[2] Morgan K S, Paganin D M and Siu K K W 2012 X-ray phase imaging with a paper analyzer *App. Phys. Lett.* **100** 124102
[3] Bérujon S, Ziegler E, Cerbino R, Peverini L 2012 Two-Dimensional X-Ray Beam Phase Sensing *Phys. Rev. Lett.* **108** 158102
[4] Talbot H F 1836 Facts related to optical science *Philos. Mag.* **9** pp 401–7
[5] Larsson D H *et al.* 2011 A 24 keV liquid-metal-jet x-ray source for biomedical applications *Rev. Sci. Instrum.* **82** 123701
[6] Zhou T *et al.* 2015 Speckle-based x-ray phase-contrast imaging with a laboratory source and the scanning technique *Opt. Lett.* **40** pp 2822–5
[7] Bech M 2009 *X-ray imaging with a grating interferometer*, PhD thesis, University of Copenhagen
[8] Zanette I *et al.* 2014 Speckle-based x-ray phase-contrast and dark-field imaging with a laboratory source *Phys. Rev. Lett.* **112** 253903

An improved acquisition scheme for the reverse projection method

C Wei[1], Z Wu[1*], F Wali[1], R Luo[1], and Y Tian[1*]

[1] National Synchrotron Radiation Laboratory, University of Science and Technology of China, Hefei, Anhui 230029, People's Republic of China
*Correspondence e-mail: wuzhao@ustc.edu.cn ychtian@ustc.edu.cn

Abstract. In recent years, efforts have been made on the information extraction method in x-ray grating-based phase contrast imaging. Among them, the reverse projection method proposed by P.P.Zhu in 2010 features as a fast, low-dose method to extract information without stepping the gratings. It enables to extend the phase contrast imaging system to the vivo studies. However, it has strict requirement on the system stability and the source stability. In this manuscript, an improved acquisition scheme has been proposed using a staggered grating making the reverse projection method more efficient with a lower requirement of the system and source stability.

1. Introduction

Thanks to the high contrast to the soft tissues, x-ray phase contrast imaging method has been studied for many years [1-7]. It has been developed four main methods named crystal interferometer imaging [5], propagation-based imaging [6], diffraction-enhanced imaging [7] and grating-based imaging [1-4]. Among them, grating-based imaging method is the one which can be carried out on the conventional x-ray tubes along with synchrotron sources. So it is the most promising method and has potential to be implemented in clinical applications.

Meanwhile, many information extraction methods have also been developed [8-15] and the most influential method is called phase stepping procedure [2, 15]. This method can extract absorption, differential phase and scattering signals simultaneously and accurately by stepping the analyzer grating two or more steps in its one period evenly. The main limitations of the stepping procedure is the requirement of additional scanning of the gratings thus high mechanical stability and calibration, compared with the continuous scanning mode used in attenuation based x-ray computed tomography. Therefore, information extraction method without stepping gratings is necessary in order to translate x-ray grating interferometry from laboratories to hospitals [8-12].

In the last decade, researchers have proposed many information extraction methods. Among these methods, reverse projection method [8] is of great significance because it utilizes the conjugate characteristic of the mutual reverse projections to extract absorption and differential phase signals by once circular scanning. The advantages of the reverse projection method are fast and efficient. However, it imposes a very strict requirement on source and system stability because of asynchronous reverse projections. In another word, time interval between two collections is so long that most of the conventional x-ray tubes can't reach the stability requirement. Meanwhile, a slight vibration or drift of the system in acquiring images, and the slight misalignment may cause inaccuracy.

In this manuscript, an improved acquisition scheme collecting the conjugate images directly in single shot using a staggered grating is proposed. This method thus reduces the strict restriction on the source and system stability.

2. Methods

To acquire the upslope image and downslope image in single shot, a staggered analyzer grating [12] with half period of the analyzer grating shift between each layer has been shown in figure 1. The height of the layer equaled to the size of the detector pixel. It is worth mentioning that if the upper layer can be precisely positioned at the upslope when collecting data, the down layer should be naturally positioned at the position of the downslope. Therefore, using this analyzer grating, the upslope image and downslope image can be acquired simultaneously and accurately in single shot. While the cost is that the spatial resolution in the direction parallel to the linear grating structures should decrease to half of that with the conventional analyzer grating. However in some fields, the effect caused by spatial resolution loss in the direction parallel to the linear grating structures can be compromised.

Figure 1. Structure of the staggered grating

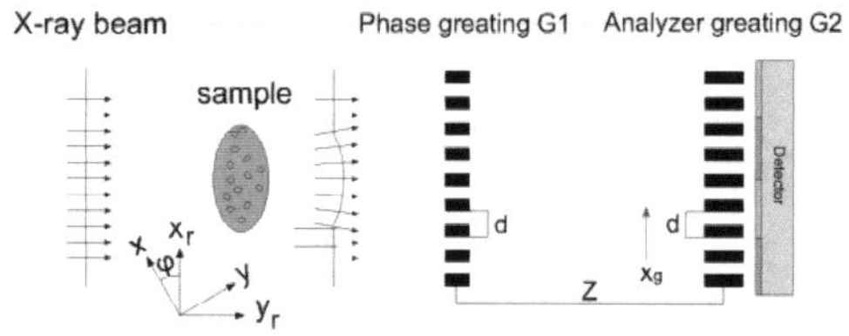

Figure 2. Illustration of grating interferometer

Using reverse projection method and our proposed staggered grating structures, the improved information extraction formula can be expressed as

$$M(x_r,2z-1,\varphi)=M(x_r,2z,\varphi)=\ln\left\{\frac{2I_0 S(\frac{x_g}{D})}{I(x_r,2z-1,\varphi)+I(x_r,2z,\varphi)}\right\} \quad (1)$$

$$\theta(x_r,2z-1,\varphi)=\theta(x_r,2z,\varphi)=\frac{1}{C}\frac{I(x_r,2z-1,\varphi)-I(x_r,2z,\varphi)}{I(x_r,2z-1,\varphi)+I(x_r,2z,\varphi)} \quad (2)$$

Where z-axis is in the direction parallel to the linear grating structures, and x_r-φ plane is shown in figure 2. M means the attenuation term of the sample while θ is defined as the refraction angle caused by the sample. I_0 is the initial x-ray intensity from the x-ray tube and I is the x-ray intensity recorded by the detector. $S(x_g/D)$ stands for the shifting curve, which can be expressed by a first-order Taylor expansion at its half slopes with $C=\frac{1}{S(x_g/D)}\frac{dS(x_g/D)}{d\theta}$.

3. Simulations

In this work, Talbot Interferometer imaging system, shown in figure 2, was simulated. It consisted of an x-ray source, a $\frac{\pi}{2}$ phase shift grating G1, an absorption staggered analyzer grating G2 and an image detector. G2 was positioned at the first fraction Talbot distance of G1.

In simulation, the mean energy of x-ray source was designed at 20keV. The periods of the phase grating and the analyzer grating were 4μm and the detector pixel equaled to 10μm. The sample consisted of four different materials which were Polypropylene, Polycarbonate, PMMA-formvar, Mylar respectively. The slice structure of the sample is shown in figure 3 (a). Four colours are used to distinguish different materials mentioned above, whose parameters have been listed in Table 1. The height of the sample, in the direction parallel to the linear grating structures, equaled to 1mm. The slice had been successfully reconstructed by using sinogram of the extracted data from equation (1) and equation (2) via FBP method. The comparisons of the theoretical and the extracted information using the improved method had also been performed.

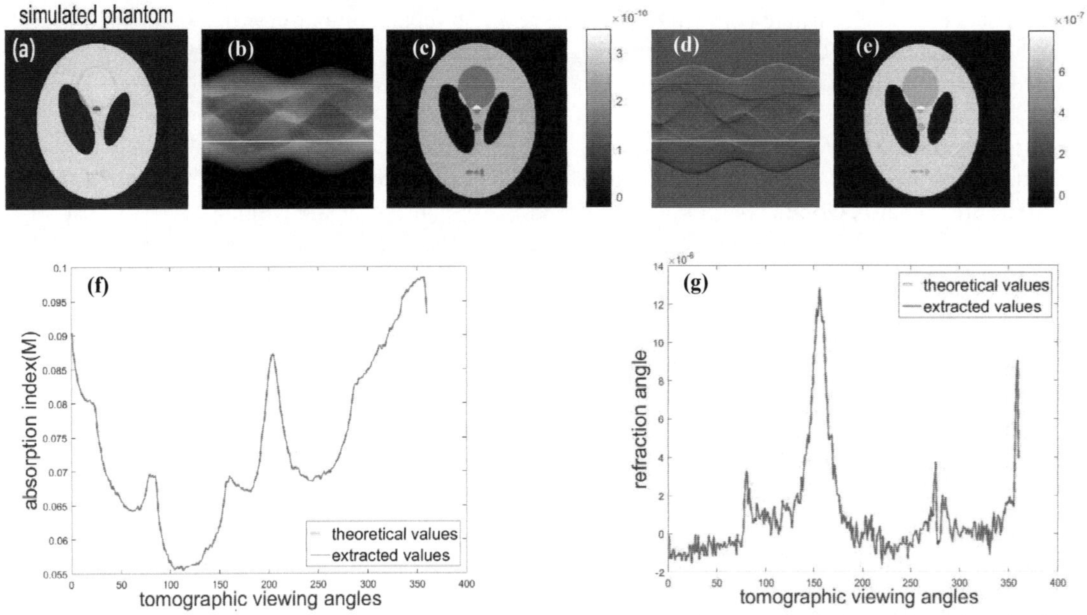

Figure 3. The slice structure of the sample (a), the sinogram of the extracted absorption signals (b) and refraction signals (d) and their slice reconstruction results (c, e). The comparisons of the profiles (f, g) at the positions indicated in (b, d) by the solid lines.

Table 1. Parameters of materials that make up the sample

Materials	Chemical formula	$\delta(10^{-7})$	$\beta(10^{-10})$	colour match
Polypropylene	C_3H_6	5.32936497	1.61908931	acid blue
Polycarbonate	$C_{16}H_{14}O_3$	6.56665406	2.58399607	yellow
PMMA-formvar	$C_5H_8O_2$	6.66553376	2.85367757	red
Mylar	$C_{10}H_8O_4$	7.56593067	3.39300005	orange

4. Results and Discussion

Figure 3(b) shows the sinogram of the extracted absorption and figure 3(c) displays the slice reconstruction of the imaginary part of the refractive index. While, figure 3(d) shows the sinogram of the extracted refraction angles and figure 3(e) displays the slice reconstruction of the real part decrement of the refractive index. Figure 3(f) and figure 3(g) present the comparisons of the profiles at the positions indicated in figure 3(b) and figure 3(d) by solid lines.

As it can be seen from figures 3(a, c, e), the reconstructed images show a high coincidence with the original sample structure. Meanwhile, the comparisons of the profiles shown in figure 3(f) and 3(g) also point out that the retrieved information of both absorption and refraction are in good agreement with their theoretical values, which proves the feasibility of the improved acquisition scheme. Although this new acquisition scheme reduces the requirements of source and system stability, however the reverse projection method still only extracts absorption and refraction signals. In the forthcoming works, a new fast information retrieval method may be proposed to exact scattering signal.

5. Conclusion

In this manuscript, an improved data acquisition scheme using a staggered grating has been proposed and simulated. In the reverse projection method, the conjugate reverse projections can be acquired in single shot and the strict requirement of the source and system stability can be reduced to some extent. Simulation has been performed to confirm the feasibility of the scheme. Finally, works in the near future may be done to get the scattering signal extracted by a novel method with the advantages of the proposed method.

Acknowledgements
This work was supported by grants from the Major State Basic Research Development Program of China (973 Program) (No.2012CB825804), the PhD Programs Foundation of the Ministry of Education of China (No.20123402130002) and the National Science Foundation of China (No.11275204, No. 11475175, No. 11405175).

References
[1] Pfeiffer F, Weitkamp T, Bunk O and David C, 2006, *Nature Physics* **2 (4)**, 258-61.
[2] Momose A, 2003, *Optics Express* **11 (19)**, 2303-14.
[3] Pfeiffer F, Kottler C, Bunk O and David C, 2007, *Phys. Rev. Lett* **98 (10)**, 108105.
[4] Momose A, Kawamoto S, Koyama I and Koyama Y, 2004, *SPIE proceedings Vol.***5535**, 352-60.
[5] Takeda T, Momose A, Hirano K, Haraoka S, Watanabe T and Itai Y, 2000, *Radiology* **214 (1)**, 298-301.
[6] Snigirev A, Snigireva I, Kohn V, Kuznetsov S and Schelokov I, 1995, *Rev. Sci. Instrum* **66 (12)**, 5486-92.
[7] Chapman D, Thomlinson W, Johnston R E, Washburn D, Pisano E, Gmur N, Zhong Z, Menk R, Arfelli F and Sayers D, 1997, *Phys. Med. Biol* **42 (11)**, 2015-25.
[8] Zhu P, Zhang K, Wang Z, Liu Y, Liu X, Wu Z, McDonald S A, Marone F and Stampanoni M, 2010, *Proc. Natl. Acad. Sci. U S A* **107 (31)**, 13576-81.
[9] Zanette I, Bech M, Pfeiffer F and Weitkamp T, 2011, *Applied Physics Letters* **98 (9)**, 094101.
[10] Bevins N, Zambelli J, Li K, Qi Z H and Chen G H, 2012, *Medical Physics* **39 (1)**, 424-8.
[11] Huang Z F, Chen Z Q, Zhang L, Kang K J, Ding F, Wang Z T and Ma H Z, 2010, *Optics Express* **18 (10)**, 10222-9.
[12] Ge Y, Li K, Garrett J and Chen G H, 2014, *Opt. Express* **22 (12)**, 14246-52.
[13] Wu Z *et al*, 2015, *Medical Physics* **42 (2)**, 741-9.
[14] Wu Z *et al* , 2013, *Medical Physics* **40 (3)**.
[15] Weitkamp T, Diaz A, David C, Pfeiffer F, Stampanoni M, Cloetens P and Ziegler E, 2005, *Optics Express* **13 (16)**, 6296-304.

IOP Conf. Series: Journal of Physics: Conf. Series **849** (2017) 012037 doi:10.1088/1742-6596/849/1/012037

The tomography beamline *ANATOMIX* at Synchrotron SOLEIL

T Weitkamp, M Scheel, JL Giorgetta, V Joyet, V Le Roux, G Cauchon, T Moreno, F Polack, A Thompson and JP Samama

Synchrotron SOLEIL, L'Orme des Merisiers, BP 48, Saint-Aubin, 91192 Gif-sur-Yvette, France

E-mail: weitkamp@synchrotron-soleil.fr

Abstract. ANATOMIX is a 200-m-long undulator beamline for full-field tomography techniques at photon energies from 5 to 25 keV. It is currently under construction at Synchrotron SOLEIL, the French national light source near Paris. ANATOMIX will feature experimental stations both for parallel-beam microtomography (with a beam of up to 40 mm width) and for zone-plate transmission X-ray microscopy (down to pixel sizes of 30 nm) in absorption and phase contrast. The location of ANATOMIX on a canted straight section of the SOLEIL storage ring implies specific challenges for the design and operation conditions of the beamline. In this paper we present general design aspects and the status of construction.

1. Introduction

Since the end of the 1990s, synchrotron microtomography methods have become widely-used tools in biomedical research, materials science and other fields [1–3]. Today, several beamlines at synchrotron radiation facilities in Europe respond to the needs of tomography users [4–12].

To complement these stations, three beamlines at Synchrotron SOLEIL—*PSICHÉ* [13], *PUMA* [14] and *ANATOMIX*—offer full-field radiography and tomography in the hard X-ray range, or will do so in the near future. Table 1 shows their key parameters.

In this paper we give an overview of the *ANATOMIX* beamline, currently under construction. It is the longest beamline at SOLEIL (200 m) and is fully dedicated to radiographic and tomographic methods in

Table 1. Beamlines at Synchrotron SOLEIL for X-ray full-field radiography and tomography.

Beamline	*PSICHÉ*	*PUMA*	*ANATOMIX*
X-ray source	Wiggler	Wiggler	Undulator
Energy range (keV)	20 to 100	15 to 60	5 to 25
Max. beam size (mm^2, h$\times$v)	15$\times$5	20$\times$10	40$\times$10
Min. pixel size (nm)	$\approx$ 500	$\approx$ 500	$\approx$ 30 (TXM); $\approx$ 500 (other)
Spectral beam modes	Filtered white, crystal mono	Filtered white, crystal mono	Filtered white, crystal mono, multilayer
Status	Operational	Expected 2018	Expected 2018

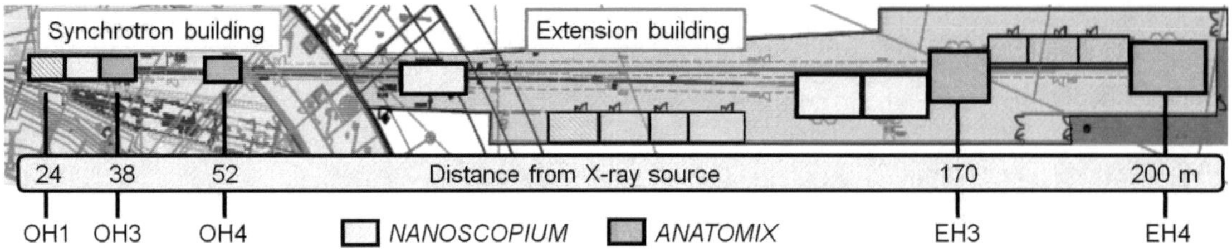

Figure 1. Floor plan of beamlines *NANOSCOPIUM* and *ANATOMIX*. Hutches are numbered jointly with *NANOSCOPIUM*. OH1, OH3, OH4, EH3, EH4 (blue) contain relevant instrumentation of *ANATOMIX*.

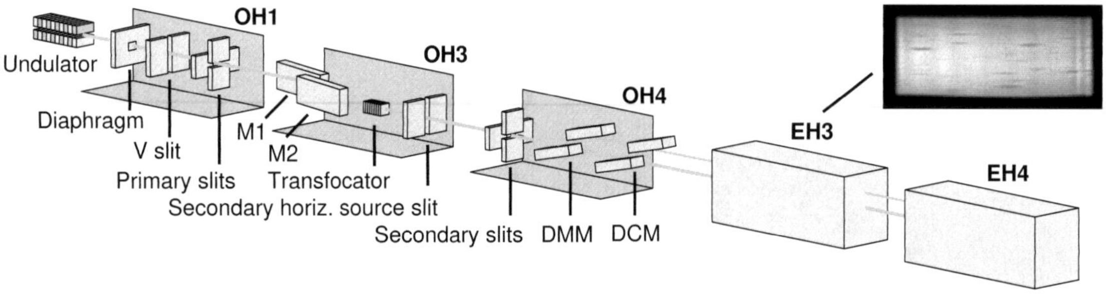

Figure 2. Schematic of *ANATOMIX* beamline optics. All elements shown except the diaphragm can be moved out of the beam path. (Attenuators and diagnostic devices not shown.) *Inset:* first beam in EH3, September 2016.

the hard X-ray range. It will provide stations for two methodologically different classes of experiments, to cover different ranges of length scales: (1) parallel-beam X-ray shadowgraphy, for spatial resolution down to the sub-micron range and (2) full-field transmission X-ray microscopy (TXM) down to a spatial resolution of 100 nm or less.

Both types of methods will be used both in absorption contrast and in phase contrast. For shadowgraphy, phase contrast will be accessible through propagation-based imaging [15, 16] and X-ray grating interferometry [17]. For the TXM station, Zernike-type phase contrast [18] is foreseen.

2. Technical details

ANATOMIX is located on a long straight section of the SOLEIL storage ring. It shares this section with the scanning nanoprobe beamline *NANOSCOPIUM* [19]. Each of the two beamlines takes photons from a canted undulator; the centers of both insertion devices are separated by 7 m, *ANATOMIX* being located on the upstream source. A specifically-developed electron-beam angle interlock [20] ensures that the stringent requirements on the stability of the electron-beam trajectory are always met. Together with a protective mask between the two insertion devices it prevents damage to the downstream undulator.

The photon emission axes of both insertion devices are separated by an angle of 6.5 mrad [21]. Two mirrors permanently inserted into the *NANOSCOPIUM* X-ray beam increase this separation to 26 mrad. The experimental hutches of both beamlines (and the monochromator hutch of *NANOSCOPIUM*) are located in a dedicated satellite building (Figure 1).

The optics concept is based on the principle that the direct white X-ray beam can be led all the way to the experimental stations—if desired, without any beam-steering optics or monochromatization. Optical elements to condition the beam spectrally and/or geometrically can be inserted as needed (Figure 2).

Table 2 lists the major beamline optics. In the first optics hutch (OH1), shared between the two

Table 2. List of important beamline elements. Distances are given from the center of the undulator.

Element	Distance (m)	Description	Hutch
Diaphragm	22.7	Rectangular aperture 2.5 × 2 mm^2 (h×v).	OH1
Vertical slit	23.2	Horizontal aperture 100...300 μm, removable. Increases horizontal transverse coherence where needed.	
Primary slits	25.9	Define footprint on mirror and other optics.	
Attenuators	34.6	Protect transfocator and other downstream elements.	OH3
Mirror M1	35.3	Horiz. deflection. Useful length 400 mm. Bender for horiz. focusing (focal length 3.5 m). Coatings: B_4C, Rh, Pt. Grazing angle ≈ 3 mrad.	
Mirror M2	36.3	Horiz. deflection. Plane mirror, steers beam back into direction parallel to direct beam, offset 5 mm. Same length, angle and coatings as M1.	
Transfocator	37.7	Beryllium lenses.	
Vertical slit	38.8	Defines horizontal secondary source when mirror is used.	
Secondary slits	48.4	Define footprint on monochromators.	OH4
Attenuators	48.8	Filter out low energies, especially when using DMM.	
DMM	49.8	Under design. Vertical deflection, beam offset 20 mm. Tentative specs: two strips Ru/B_4C, Ir/B_4C, $d = 2.4$ nm. 400-mm substrates.	
DCM	52.7	Vertical deflection, beam offset 20 mm. Si(111), LN_2-cooled.	
EH3	from 165.7	Guard slits; TXM sample stage and zone-plate table;	EH3
	to 174.1	microtomography sample stage; detector table	
EH4	from 197.2	Guard slits; microtomography sample stage;	EH4
	to 208.5	detector table; X-ray grating interferometer	

long beamlines, *ANATOMIX* has its diaphragm defining the opening aperture of the beamline, a pair of primary slits and a *DiagOn* beam imager [22] using a silicon Bragg crystal.

A total-reflection double mirror M1–M2 (mechanics: IRELEC, Saint-Martin-d'Hères, France; optics: WinlightX, Pertuis, France) with a bender on the first substrate can be used to create a secondary horizontal source. The divergent beam downstream of the focus is up to 40 mm wide at the experiment position. On the other hand, for experiments requiring high intensity but not a large beam, an assembly of refractive X-ray lenses (a *transfocator,* following ESRF terminology for similar devices [23]) can be used to collimate the beam. Both mirror and transfocator are located in OH3, roughly 35 m from source.

A double-crystal monochromator (DCM) with Si(111) Bragg crystals (AXILON AG, Cologne, Germany) and a double-multilayer monochromator (DMM) will be installed in the following optics hutch, OH4. The monochromators have vertical deflection geometry for an optimum match of the usable aperture to the beam dimensions and for minimum perturbation of the horizontal wavefront profile, as this is the in-plane direction for axial slices in tomography with a vertical rotation axis.

The X-ray flux density at the sample position using the DMM, without mirror or transfocator, is estimated to vary from more than 10^{12} ph/($mm^2 \cdot s$) below 18 keV to roughly 10^{11} ph/($mm^2 \cdot s$) at 25 keV. The DMM-beam flux at 18 keV will be around 10^{14} ph/s. Use of the mirror to obtain a wide beam reduces the flux density by about an order of magnitude; collimation with the transfocator, on the other hand, can increase it by a similar factor. The DCM will yield about 100× less flux than the DMM; the correspondingly narrower energy bandwidth of the beam is necessary for the TXM.

The first of the two experimental hutches of *ANATOMIX*, EH3, will host the TXM sample stage and zone plates, a sample stage for parallel-beam microtomography and a detector table. The second experiment hutch (EH4) will hold another microtomography sample stage and detector stage and a grating interferometer. It may also be used for a direct-detection pixel detector used with the TXM.

3. Current status and conclusion

ANATOMIX is currently under construction and expected to begin user operation in 2018. Since the final cryogenic U18 in-vacuum undulator [24] is still under construction, an in-vacuum U20 undulator is used for commissioning. The optical elements from the source to (and including) the mirror are operational or under commissioning; the other beam conditioning optics are ordered or under construction. The experimental stations are under construction and/or design. The first beam in experimental hutch EH3 was observed on 26 September 2016.

Acknowledgments

All support groups, the accelerator and source groups and the SOLEIL administration are gratefully acknowledged for their efforts for the *ANATOMIX* project. We thank A. King, J.-P. Itié and S. Schöder from the *PSICHÉ* and *PUMA* beamlines of SOLEIL for supplying data for Table 1.

The construction of *ANATOMIX* is largely funded by the French National Research Agency (ANR) through the EQUIPEX investment program, project *NanoimagesX*, grant no. ANR-11-EQPX-0031. The *NanoimagesX* consortium comprises 17 partner units from public research and industry. They are acknowledged for their contributions to the functional specifications of the experimental stations.

References

[1] Momose A 2005 *Jpn. J. Appl. Phys.* **44** 6355–6367
[2] Stock S R 2008 *Int. Mater. Rev.* **53** 129–181
[3] Maire E and Withers P J 2014 *Int. Mater. Rev.* **59** 1–43
[4] Rack A, Zabler S, Müller B R, Riesemeier H, Weidemann G, Lange A, Goebbels J, Hentschel M and Görner W 2008 *Nucl. Instrum. Meth. A* **586** 327–344
[5] Rack A *et al* 2009 *Nucl. Instrum. Meth. B* **267** 1978–1988
[6] Tromba G *et al* 2010 *AIP Conf. Proc.* **1266** 18–23
[7] Marone F, Mokso R, Modregger P, Fife J, Pinzer B, Thüring T, Mader K, Mikuljan G, Isenegger A and Stampanoni M 2011 *AIP Conf. Proc.* **1365** 116–119
[8] Schell N, King A, Beckmann F, Ruhnau H U, Kirchhof R, Kiehn R, Müller M and Schreyer A 2010 *AIP Conf. Proc.* **1234** 391–394
[9] Weitkamp T, Tafforeau P, Boller E, Cloetens P, Valade J P, Bernard P, Peyrin F, Ludwig W, Helfen L and Baruchel J 2010 *AIP Conf. Proc.* **1221** 33–38
[10] Rau C, Wagner U, Pešić Z and De Fanis A 2011 *Physica stat. sol. A* **208** 2522–2525
[11] Greving I *et al* 2014 *Proc. SPIE* **9212** 92120O
[12] Drakopoulos M *et al* 2015 *J. Synchrot. Radiat.* **22** 828–838
[13] King A *et al* 2016 *Rev. Sci. Instrum.* **87** 093704
[14] Bertrand L *et al* 2011 *J. Synchrot. Radiat.* **18** 765–772
[15] Spanne P, Raven C, Snigireva I and Snigirev A 1999 *Phys. Med. Biol.* **44** 741–749
[16] Weitkamp T, Haas D, Wegrzynek D and Rack A 2011 *J. Synchrot. Radiat.* **18** 617–629
[17] Weitkamp T, Bunk O, Pfeiffer F, David C, Bruder J and Cloetens P 2008 *Eur. J. Radiol.* **68** S13–S17
[18] Stampanoni M, Mokso R, Marone F, Vila-Comamala J, Gorelick S, Trtik P, Jefimovs K and David C 2010 *Phys. Rev. B* **81** 140105(R)
[19] Somogyi A, Kewish C M, Ribbens M, Moreno T, Polack F, Baranton G, Desjardins K and Samama J P 2013 *J. Phys. Conf. Series* **463** 012027
[20] Hubert N *et al* 2016 *Proceedings, 4th International Beam Instrumentation Conference (IBIC 2015)* TUPB009 URL `http://inspirehep.net/record/1480676/files/tupb009.pdf`
[21] Brunelle P, Briquez F, Loulergue A, Marcouillé O, Nadji A, Nadolski L S, Tordeux M A and Zhang J 2011 *Proceedings of the Second International Particle Accelerator Conference (IPAC 2011)* pp 2124–2126 URL `https://accelconf.web.cern.ch/accelconf/IPAC2011/papers/wepc050.pdf`
[22] Desjardins K *et al* 2007 *AIP Conf. Proc.* **879** 1101–1104
[23] Snigirev A, Snigireva I, Vaughan G, Wright J, Rossat M, Bytchkov A and Curfs C 2009 *J. Phys. Conf. Series* **186** 012073
[24] Valléau M *et al* 2016 *AIP Conf. Proc.* **1741** 020024

Launch of the I13-2 data beamline at the Diamond Light Source synchrotron

A J Bodey[1] and C Rau[1,2,3]

[1] Diamond Light Source, Oxfordshire, OX11 0DE, UK
[2] School of Materials, University of Manchester, Manchester, M13 9PL, UK
[3] Feinberg School of Medicine, Northwestern University, Chicago, IL 60611, USA

andrew.bodey@diamond.ac.uk

Abstract. Users of the Diamond-Manchester Imaging Branchline I13-2 commonly spend many months analysing the large volumes of tomographic data generated in a single beamtime. This is due to the difficulties inherent in performing complicated, computationally-expensive analyses on large datasets with workstations of limited computing power. To improve productivity, a 'data beamline' was launched in January 2016. Users are scheduled for visits to the data beamline in the same way as for regular beamlines, with bookings made via the User Administration System and provision of financial support for travel and subsistence. Two high-performance graphics workstations were acquired, with sufficient RAM to enable simultaneous analysis of several tomographic volumes. Users are given high priority on Diamond's central computing cluster for the duration of their visit, and if necessary, archived data are restored to a high-performance disk array. Within the first six months of operation, thirteen user visits were made, lasting an average of 4.5 days each. The I13-2 data beamline was the first to be launched at Diamond Light Source and, to the authors' knowledge, the first to be formalised in this way at any synchrotron.

1. Motivation

The Diamond-Manchester Imaging Branchline I13-2 at Diamond Light Source (DLS) is dedicated to imaging and tomography experiments that span numerous disciplines including biology, materials science, geology and palaeontology. The branchline provides partially-coherent 5-35keV X-rays and variable in-line phase contrast. The high flux of the beam enables datasets to be recorded rapidly, and the branchline is commonly used for high-throughput tomography and time-resolved radiography (eg. [1]) and tomography (eg. [2]), leading to very large data volumes.

Tomographic reconstructions can have very large file sizes. 32-bit reconstructions computed from images collected at I13-2 by a pco.edge 5.5 (PCO AG, Germany) detector operating in full-frame mode (2560 x 2160 pixels) are 57GB in size, while a pco.4000 detector can generate 172GB tomographic volumes. While these can be compressed by a factor of 2 or 3 by conversion to 16- or 8-bit respectively, loading such volumes into RAM (while leaving enough for image processing) still requires considerable computing power.

A similar problem exists with raw data. Inspecting raw datasets (sample, flat-field and dark-field images) allows users to check for experimental problems and is most efficiently achieved by loading complete datasets into RAM. A single-file raw dataset of 4,000 sample images from a pco.edge 5.5

IOP Conf. Series: Journal of Physics: Conf. Series **849** (2017) 012038 doi:10.1088/1742-6596/849/1/012038

detector is 42GB in size. While only 4,022 images are necessary for complete rotational sampling with this detector's full array, 12,000 were collected during a recent experiment at I13-2 to optimise the signal:noise ratio, yielding a raw dataset of 127GB.

Beamtimes at I13-2 often produce hundreds of tomographic scans and generate tens of terabytes of data. Beamtimes involving time-resolved studies can produce considerably more than 100TB of data. I13-2 currently produces approximately one third of all data generated at DLS.

In order to extract useful information from tomograms, visualisation and analysis are required. The latter commonly involves filtering, segmentation (partitioning a tomographic volume into multiple segments) and measurements of particle sizes and morphologies, porosity and connectedness. Some image processing operations are very computationally expensive. The non-local means filter for example, can be very effective in denoising tomograms (figure 1), but commonly requires several hours to process a complete tomographic volume – even with a powerful workstation.

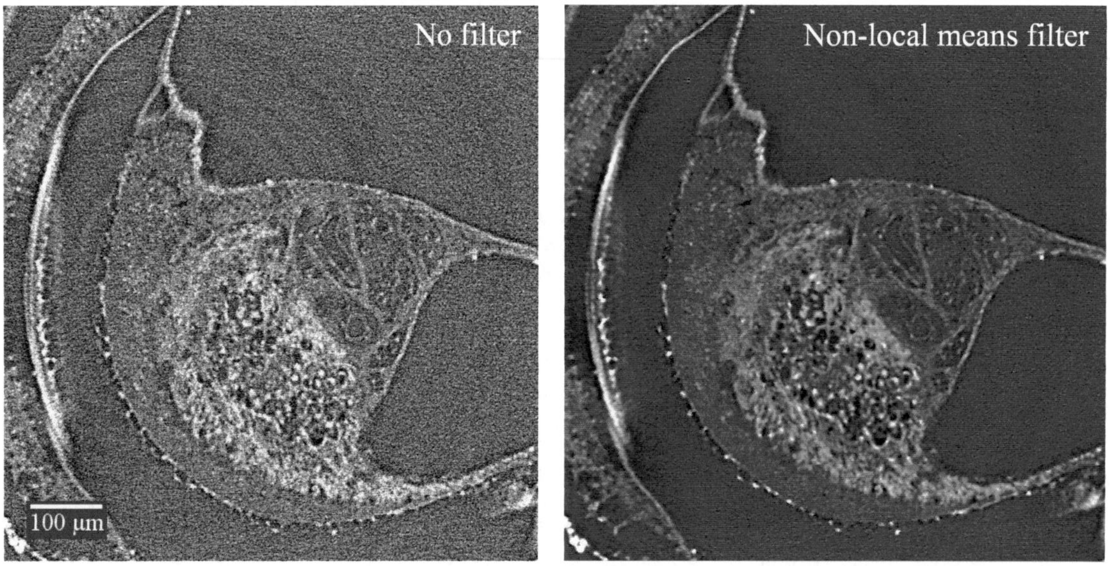

Figure 1. As this tomographic slice through a cochlea [3] demonstrates, the computationally-intensive non-local means filter can be very effective at removing noise from tomograms. Such denoising is particularly useful for samples which must be imaged with a limited dose, such as biological specimens and those involved in rapid time-resolved tomography.

For these reasons, as well as the difficulties that can otherwise be inherent in segmenting and taking measurements from tomographic data, users commonly spend many months analysing data recorded in a single beamtime. Computing power is therefore key to improving productivity and reducing the time between I13-2 beamtimes and publication. Computers that can load complete or multiple volumes into RAM are advantageous, as is sufficient processing power to accelerate computationally-intensive operations. Many of the software packages currently available for visualisation and analysis are not amenable to cluster operation and must be run on single workstations.

I13-2 users commonly lack the computing power to analyse data efficiently at their home institutions, and some even lack computers with sufficient RAM into which they can load complete tomographic volumes. Other users do not have access to expensive licenses for image-processing software. It was therefore considered useful to make powerful dedicated graphics workstations available to users. To manage access to these machines, a 'data beamline' was established.

2. Project management

Users are scheduled for visits to the data beamline in the same was as visits are made for regular beamlines, with bookings made via the User Administration System [4] and provision of financial support for travel and subsistence. Scheduling is managed by an appointed beamline staff member. Data from I13-2 beamtimes are held on a high-performance GPFS disk array for 120 days before being archived. Data beamline visits are preferentially made within this timeframe, but if visits are made after data have been archived they are restored to the high-performance array. Users are given high priority on the central DLS computing cluster for the duration of their visit.

3. Workstation specifications

Workstation specifications were determined via discussion with FEI Company about optimising performance of their software package Avizo, and are as follows: 512GB DDR4 RAM, dual Intel Xeon E5-2687W v3 CPUs (3.1GHz, 10 cores, 20 threads), dual NVIDIA Quadro K6000 GPUs with 12GB GDDR5 RAM and ECC, dual 512GB SSDs in RAID 1, 10Gbs^{-1} optic fibre connections and 30-inch monitors. Use of RAM three times larger than loaded data is recommended. A typical tomographic volume generated from pco.edge 5.5 images, once converted from 32- to 16- or 8-bit, is 28 or 14GB in size, respectively. 512GB RAM therefore allows for up to six 16-bit or twelve 8-bit tomographic volumes to be loaded and processed simultaneously.

4. Software

Tomographic reconstruction, via Dawn [5] and Savu [6], is performed on the central DLS computing cluster, although reconstructions are normally performed during I13-2 beamtimes. Various packages are available for visualisation and analysis of tomograms and radiographs, including Avizo (FEI Company, USA), Drishti [7], ImageJ [8] and Fiji [9]. Proprietary programming languages including MATLAB (Mathworks, USA) and Mathematica (Wolfram Research, USA) are available. Other software can be installed upon request.

5. Usage to date

The data beamline was launched in January 2016. In the six months that followed, thirteen user visits were made (some with multiple attendees), lasting 4.5 days on average. The system is due to be replicated on other beamlines at DLS.

6. Acknowledgements

The authors would like to express their thanks to various people at DLS who helped to set up the data beamline: K Poulter for project management support, M Basham for advice on computer specifications, A Fifield for IT acquisition, T Friedrich for cluster management, F Ferner for data storage management, R Lear for network administration, and A Strange for User Administration System support.

References

[1] Xu W W, Tzanakis I, Srirangam P, Mirihanage W U, Eskin D G, Bodey A J and Lee P D 2016 Synchrotron quantification of ultrasound cavitation and bubble dynamics in Al–10Cu melts *Ultrasonics Sonochemistry* **31**

[2] Menke H P, Andrew M G, Blunt M J and Bijeljic B 2016 Reservoir condition imaging of reactive transport in heterogeneous carbonates using fast synchrotron tomography — Effect of initial pore structure and flow conditions *Chemical Geology* **428**

[3] Rau C and Richter C-P 2015 Imaging cochlear soft tissue displacement with coherent x-rays *Physica Scripta* **90**

[4] Diamond User Administration System https://uas.diamond.ac.uk/uas/

[5] Basham M, Filik J, Wharmby M T, Chang P C Y, El Kassaby B, Gerring M, Aishima J, Levik K, Pulford B C A, Sikharulidze I, Sneddon D, Webber M, Dhesi S S, Maccherozzi F,

Svensson O, Brockhauser S, Naray G and Ashton A W 2015 Data Analysis WorkbeNch (DAWN) *J Synchrotron Radiat* **22** 853-8
[6] Atwood R C, Bodey A J, Price S W T, Basham M and Drakopoulos M 2015 A high-throughput system for high-quality tomographic reconstruction of large datasets at Diamond Light Source *Philos Trans A Math Phys Eng Sci* **373**
[7] Limaye A 2012 Drishti: a volume exploration and presentation tool *Proc SPIE 8506: Developments in X-Ray Tomography VIII*
[8] Schneider C A, Rasband W S and Eliceiri K W 2012 NIH Image to ImageJ: 25 years of image analysis *Nat Meth* **9** 671-5
[9] Schindelin J, Arganda-Carreras I, Frise E, Kaynig V, Longair M, Pietzsch T, Preibisch S, Rueden C, Saalfeld S, Schmid B, Tinevez J-Y, White D J, Hartenstein V, Eliceiri K, Tomancak P and Cardona A 2012 Fiji: an open-source platform for biological-image analysis *Nat Meth* **9** 676-82

Optimal focusing system of the Fresnel zone plates at the Synchrotron SOLEIL NanoARPES beamline

José Avila, Antoine Boury, Borja Caja-Muñoz, Chaoyu Chen, Stephane Lorcy & Maria C. Asensio*

ANTARES Beamline, Synchrotron SOLEIL, Université Paris-Saclay, L'Orme des Merisiers, Saint Aubin-BP 48, 91192 Gif sur Yvette Cedex, France.

E-mail: asensio@synchrotron-soleil.fr

Abstract. ANTARES beamline (BL), operating at very low photon energies, is a new soft X-ray scanning microscope recently built at SOLEIL Synchrotron, that offers a spectroscopic non-destructive nano-probe to study advanced materials. It combines a set of Fresnel Zone Plates (FZP) able to focalize the beam spot up to a few tenths of nanometres with a stable and precise sample nanopositioning (<1 nm). High energy-, angular- and spatial- resolution allow accurate electronic and chemical imaging combining angle-resolved photoelectron spectroscopy (NanoARPES or k-nanoscope) and core level detection by using both photoemission and X-ray absorption. Here, we report our latest results related to the optimization of the post-focusing mirrors system as well as its impact on the ultimate spatial resolution of the whole microscope.

1. Introduction

Recently, a fast progress in modern microscopic methods has been realised to image nano-objects and heterogeneous samples in nano and mesoscopic scale. In particular, the X-ray spectromicroscopy, which combines the abilities to get the spectroscopic information by X-rays excitation of the samples with the microscopy.. However, beyond the achievement of Angstrom spatial resolution, the challenge still remains in order to consolidate lateral manometric resolution with high resolution energy-resolved spectroscopic methods able to produce chemical and electronic imaging. Part of these challenging objectives have been already accomplished by the study of core levels and their chemical shifts using scanning photoelectron microscopy (SPEM) and X-ray scanning transmission microscopy (SXTM) techniques. However, these two powerful methods are no skilful in producing electronic imaging of the valence band states close to the Fermi level. Even if these electrons are essential determining the transport, magnetic and chemical properties of the investigated samples.

Even if we are strongly concerned by core level imaging, our approach has been in addition to adapt the competences of ARPES to the FZP and scanning system set-up to produce electronic imaging. The result is a new sophisticate and versatile nanoscopy named nanoARPES our k-nanoscope[1, 2, 3]. This technique is one of the most direct methods of studying the electronic structure of solids and is the only truly momentum-resolved probe, which is essential for the investigation of low dimensional and strongly anisotropic systems. By measuring momentum and kinetic energy of the electrons photoemitted from a sample illuminated with radiation of energy larger than the material work function, it is possible to gain information on both energy and the momentum of the electronic excitations close to the Fermi level, inside the solid. In this contribution we present the latest electronic and chemical imaging of nano-structured samples, as well as the repercussions of the BL focusing post-focalization on the accuracy and precision of the NanoARPES images.

2. Layout of the focusing system and results

The ANTARES BL was remarkably designed for NanoARPES experiments, core levels imaging and x-ray absorption spectromicroscopy in the photon energy range between 10 and 1000 eV using lineal and circular polarized light. The BL includes two insertions devices and the main components are: its pre-focalization optics, a plane grating monochromator (PGM) and two post-focalisation optics each of one is a Pseudo-Wolter, for more details see [1, 2, 3] and references therein. Each independent pseudo-Wolter system is composed by a set of spherical and toroidal mirrors in horizontal derivation. One pseudo-Wolter is able to focus directly on the sample while the second one its focus is coincident with position of the pinhole at the Zone Plate object plane, see Figure 1.

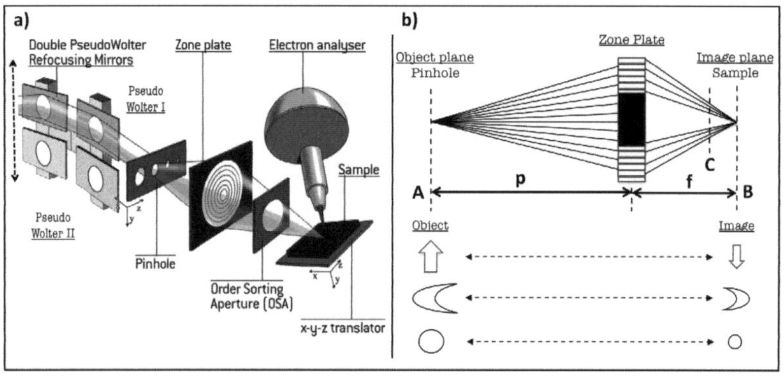

Figure 1. Schematic representations of the Pseudo-Wolter post-focusing and FZP systems. (a) The Figure presents the focalisation system of the ANTARES BL with two pseudo Wolter systems mounted in an elevator system that allow easily to select the illumination focus at the pinhole (Wolter I) or at the sample (Wolter II). Panel **(b)** shows schematically the focalization of the X-ray of the BL using the FZP

The FZP is fixed on a three-axises manipulator with high precision and stability, which can set the FZP in or out the beam path. Besides, the optical focalising system also consists of a pinhole and an Order Selection Aperture (OSA) system placed between the FZP and the sample, which suppresses unwanted diffraction orders. The pinhole is situated at the **(object)** or **(source)** position and the sample at the **(image)** position, of the optical system, espectively (see Fig. 1b). The distances between the FZP and the object and the image are called **p** and **f**, respectively. For ANTARES BL the distance **p** is close to $\approx$1.5 m, although it depends to a slight extent on the photon energy. The FZP, as a diffracted lens, focusses the X-rays at the image plane, placed at a distance **f** from de FZP. The primary aim of the BL focalization system is to define an almost perfect infinitely circular point source (or object) because if the shape of the source is distorted, the generated image will be also distorted with the same shape, as it has been schematically indicated in Fig. 1b. Most importantly, the size as well as the shape of the image in plane **B** is defined by the demagnification, which is a function of the magnitudes **p** and **f** . Effectively, the simplified expression of the lowest spatial resolution for a typical FZP is given by [4]. :

$$\delta = \frac{1.22\triangle r}{m} \quad (1)$$

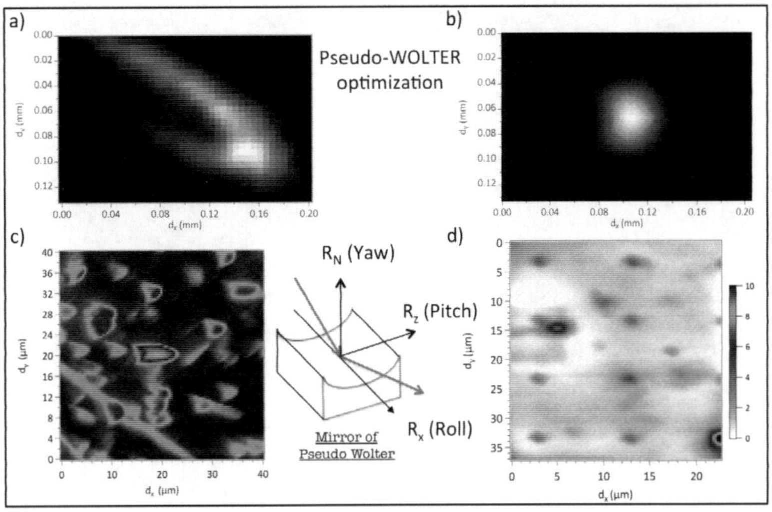

Figure 2. Optimization of the Pseudo-WOLTER Panel **(a)** shows the shape of the light point at the pinhole, plane **A** in Fig. 1b, when the positioning parameters of the pseudo-Wolter are not optimized (i.e. Roll, Pitch and Yaw positioning). Panel **(c)** image of rectangular nanowires of 1um long and 200 nm width measured with the point source illustrated in panel **(a)**. Panel **(c)** shows the shape of the beam as in panel **(a)** once the Pitch and Yaw parameters have been optimized. In panel **(d)**, shows the image of ordered gold spheres of 500 nm diameter on GaAs substrate taken with a point source as shown in panel **(b)**. In this case the imaged nano-objects do not present the coma-like shape as in panel **(c)**.

where $\triangle r$ is the outer-most zone width and m is the diffraction order of the ZP. Reaching this optimal highest spatial resolution of a FZP depends on many parameters. However, the experimental spatial resolution usually obtained by FZP systems is influenced by other extrinsic factors to be consider. Thus the experimental spatial resolution would be better calculated by [5].

$$\delta = \sqrt{\delta_i^2 + \delta_g^2 + \delta_c^2} = \sqrt{(\frac{1.22\triangle r}{m})^2 + (\sigma\frac{q}{p})^2 + (2r\frac{\triangle E}{E})^2} \quad (2)$$

where δ_i is the intrinsic resolution of the ZP (equation 1), δ_g is the demagnified source size with σ the source size and **q** and **f** the source FZP and focus distances, and δ_c is the chromatic aberration being $2r$ the diameter of the FZP. Notice that the term δ_g, particularly takes into account the impact of a deformed and enlarged point source on the spatial resolution of the obtained image, (see figure 1b).

Figure 2c displays clearly the dramatic repercussions of a not well defined shape and size of the point source on the chemical and electronic images acquired with the ANTARES k-nanoscope. Effectively, rectangular topological insulator Sb_2Te_3 nanowires on a Si substrate are shown in the image with a coma-like shape due to a defective bad-defined point source. The image display the photoemission intensity of the Te 4d core level, however, instead to present the real shape of the nano-object, it show a fully distorted nanowire, that reproduces exactly the shape of the beam at the pinhole, (plane **A** in FIg. 1b). A similar image is obtained recording the PES

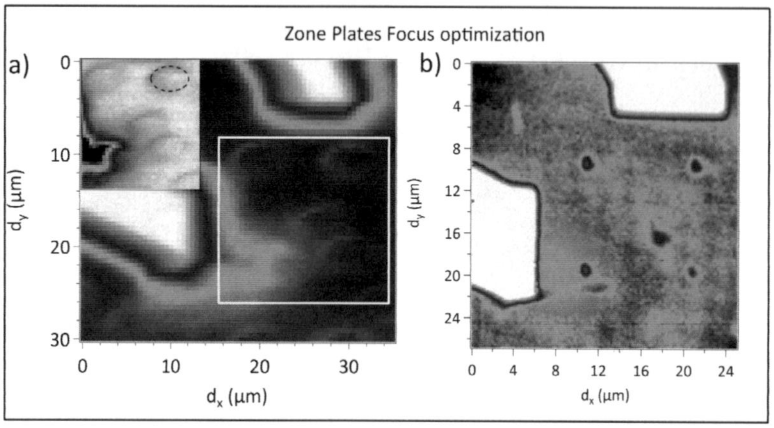

Figure 3. Focal distance optimization of the Fresnel Zone Plate Panel **(a)** shows the image gold spheres on a GaAs substrate using a defocusing position of the FZP at the **(C)** plane. In the inset, the circular shape (black dashed line) of the nano-objects resembles the shape of the light spot at the plane **(C)**, see Fig. 1b. Panel **(b)** displays the same image with the FZP located at the right focal distance at the plane **(B)**.

intensity the Au 4f core level from severals spheres of Au on GaAs substrate in Fig. 2d. The coma-like shape of the gold nano-objects is missing because the point source has been optimized to a almost perfect spherical shape represented in Fig. 2b.

Moreover, the optimal spatial resolution is obtained following the standard procedure to determine the best **(f)** value, locating the FZP and the sample at the right distance. An example of the effect of the defocusing on an image is presented in figure **3**. Both images have been obtained recording the PES intensity near Fermi level of a lithographic sample of gold spheres on a GaAs substrate. However, panel **(a)** show the distorted image obtained when the sample is placed in plane **(C)**, while the high accurate image displayed in panel **(b)** is acquired when the sample is located at the optimal work distance, in plane **(B)**.

3. Discussion

In summary, the ANTARES set-up has proved the feasibility of an innovative instrument where NanoARPES has been combined a well-focalized beam of X-rays together with a precise mechanical scanning system. The good reproducibility of the ANTARES k-nanoscope has been made possible by the optimization of sensible parameters of the BL optics.

4. Acknowledgments

The Synchrotron SOLEIL is supported by the Centre National de la Recherche Scientifique (CNRS) and the Commissariat à l'Energie Atomique et aux Energies Alternatives (CEA), France.

5. References

[1] Avila J and Asensio M C. 2014 *Synchrotron Radiation News* **27** 24-30
[2] Bostwick, A. et al. 2012 *Synchrotron Radiation News* **25** 19-25
[3] Jose, A. et al. 2013 *Journal of Physics: Conference Series* **425** 192023
[4] O.E. Myers et al. 1951 *Am. Jr. Physic.* **19** 357
[5] Yu, W et al. 199 *Review of Scientic Instruments* **70** 2238

IOP Conf. Series: Journal of Physics: Conf. Series **849** (2017) 012040 doi:10.1088/1742-6596/849/1/012040

Simple and robust synchrotron and laboratory solutions for high-resolution multimodal X-ray phase-based imaging

M. Endrizzi[1], F.A. Vittoria[1], P.C. Diemoz[1], G.K. Kallon[1], D. Basta[1], A. Zamir[1], C.K. Hagen[1], U. H. Wagner[2], C. Rau[2], I.K. Robinson[3,4], A. Olivo[1,3]

[1] Department of Medical Physics and Biomedical Engineering, University College London, London WC1E 6BT, United Kingdom
[2] Diamond Light Source, Harwell Oxford Campus, Didcot OX11 0DE, United Kingdom
[3] Research Complex at Harwell, Harwell Oxford Campus, Didcot OX11 0FA, United Kingdom
[4] London Centre for Nanotechnology, London WC1H 0AH, United Kingdom

E-mail: `m.endrizzi@ucl.ac.uk`

Abstract. Edge illumination X-ray phase contrast imaging techniques are capable of quantitative retrieval of differential phase, absorption and X-ray scattering. We have recently developed a series of approaches enabling high-resolution implementations, both using synchrotron radiation and laboratory-based set-ups. Three-dimensional reconstruction of absorption, phase and dark-field can be achieved with a simple rotation of the sample. All these approaches share a common trait which consists in the use of an absorber that shapes the radiation field, in order to make the phase modulations introduced by the sample detectable. This enables a well-defined and high-contrast structuring of the radiation field as well as an accurate modelling of the effects that are related to the simultaneous use of a wide range of energies. Moreover, it can also be adapted for use with detectors featuring large pixel sizes, which could be desirable when a high detection efficiency is important.

1. Introduction

X-ray phase-contrast imaging (XPCI) enables incorporating phase information into the image formation process and subsequent image interpretation. This enhances and complements the capabilities of conventional radiography, which is based upon the absorption of X-ray radiation within the sample under study [1–3]. Among the various methods that have been developed over the years to enable XPCI [4–19], we focus here on edge illumination (EI) [20] which can be adapted for use with synchrotron radiation, microfocus and extended laboratory sources [21,22]. EI allows quantitative retrieval of attenuation, phase and scattering information [23,24] and offers high sensitivity and robustness against mechanical and thermal instabilities [25–29]. It can also be extended from planar to three-dimensional imaging, and also in that case deliver low doses to the specimen [30,31]. We report on and compare three experimental set-ups where a significant structuring of the X-ray beam by means of an attenuating mask enables simple and robust XPCI experiments, compatible with synchrotron radiation as well as X-ray beams generated with laboratory sources.

2. Methods

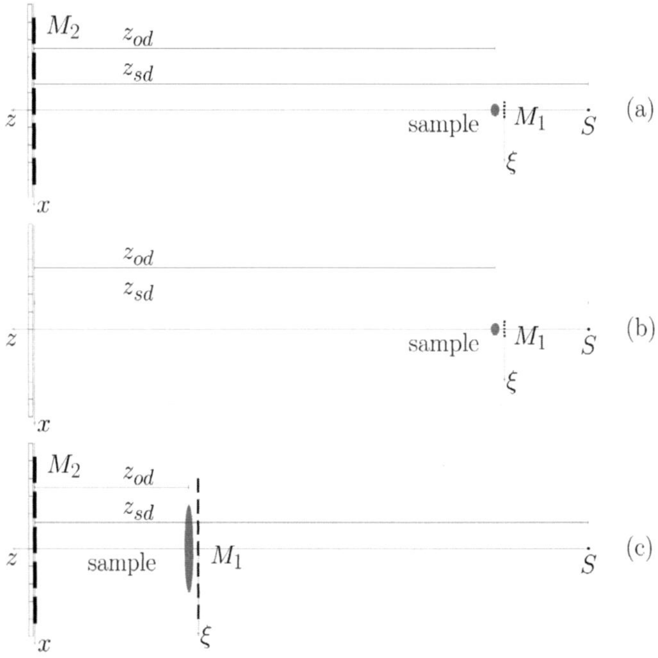

Figure 1: Sketch of the different experimental set-ups for edge-illumination XPCI, X-rays travel from right to left. (a) Laboratory high-resolution set-up: the sample mask is placed relatively close to the micrometre-sized focal spot. Magnification factor is about 10 in this configuration. Typical parameters: spot size $S = 3.5$ μm, $z_{sd} = 130$ cm and $z_{od} = 117$ cm [22]. The field of view is limited to millimitre-sized objects. (b) Laboratory beam tracking set-up: the detector mask can be removed and the modifications induced by the sample can be tracked by using groups of pixels [32]. (c) Low-magnification configuration where the sample mask is closer to the detector and an extended source can be used. Typical parameters: spot size $S = 70$ μm, $z_{sd} = 200$ cm and $z_{od} = 40$ cm [21]. The field of view allows for imaging of centimetre-sized objects.

In its synchrotron implementation, a typical EI imaging set-up consists of an absorbing slit that is placed immediately before the sample. When EI is translated to a laboratory set-up, the slit is usually replaced with a mask, a two-dimensional pattern alternating between transmitting and absorbing septa, and an extended area can be covered with a single exposure. This element spatially structures the beam and makes the phase shifts imparted by the object to the X-ray beam detectable. A second aperture, or mask, is placed before the detector with the function of analysing the X-ray beam after it has passed through the sample. This transforms the phase shift induced by sample into variations in the intensity recorded by each detector pixel.

Three examples are presented in Figure 1. The top panel (a) shows a high-resolution set-up operating with a tungsten anode laboratory source at 80 kVp [22], which enables micrometre-resolution X-ray phase imaging with a relatively compact design. The source size is about $S = 3.5$ μm and the sample mask is placed at 13 cm from it. Despite the use of a small focal spot, the coherence length $(z_{sd} - z_{od})\bar{\lambda}/\pi S \approx 0.4$ μm is small if compared to the characteristic lengths of the masks used in the experiment, which have typical periods of approximately 20 μm. The middle panel (b) presents a sketch of the beam tracking approach. This method can be implemented by removing the detector mask, and adapted to synchrotron radiation [33, 34] as well as to laboratory microfocal X-ray tubes [32]. In the third panel (c) the most commonly used configuration is depicted. In this case, the geometrical magnification G is close to unity, which allows the use of an extended focal spot, typically in the range of 70 – 100 μm. Also in this case, considering for example a molybdenum target at 40 kVp and $(z_{sd} - z_{od}) = 160$ cm, the coherence length is low, approximately 0.5 μm, and should be compared to mask periods in the range of 80 – 100 μm. In terms of energy bandwidth, all these systems are achromatic [35] and fully exploit the wide spectrum produced by conventional laboratory X-ray tubes.

The illumination function describes how the detected intensity changes as a function of the relative displacement between the sample and the detector apertures. This intensity is maximum when the apertures are perfectly aligned, and progressively decreases with increasing misalignment. The typical working points are the two positions where the intensity is about 50% of the maximum. In these configurations the refraction induced by the sample translates into an increased or decreased detected intensity, depending on the combination of sign of refraction angle and slope of the illumination function. This can be used to quantitatively separate the sample's amplitude and phase, for example by using [36]:

$$I_{L,R} = \exp\left[-\int_{\mathcal{O}} \bar{\mu} dz\right] I_F(\Delta\xi_{1,2} - z_{od}\,\bar{\phi}_x/G) \tag{1}$$

where $\bar{\mu}$ is the linear attenuation coefficient, evaluated at the effective energy of the system. This energy is obtained by weighting each monochromatic component in the beam according to the source spectrum and the detector response [37, 38]. $\bar{\phi}_x = \partial\bar{\phi}/\partial x$ is the partial derivative of ϕ evaluated at the system's effective energy and I_F is the value of the illumination function the image has been acquired at. The integration is carried out over the spatial extension of the sample $\mathcal{O}$. The following two images are also defined $I_\Sigma = I_L + I_R$ and $I_\Delta = I_L - I_R$. It is important to note that the image I_Δ/I_Σ depends only on $\bar{\phi}_x$ and the experimental parameters, while I_Σ depends only on the sample's absorption [36].

In the beam tracking approach the effects of absorption T, refraction δx and scattering s are formulated as [33]:

$$I(x) = TI_0(x - \delta x) * s(x) \tag{2}$$

where I and I_0 are the intensity distributions acquired with and without the sample in the beam. With a high-resolution detector it is possible to resolve the intensity distributions corresponding to each pre-sample aperture that fits within the field of view. With this data, an over-constrained problem is formulated and the sample parameters are extracted through a non-linear fitting procedure, by comparing the shape of the intensity distributions recorded with and without the sample in the beam.

Equation 2 can also be used in the case of Figure 1(c), where the detector mask is present and a small geometrical magnification is used. By indicating the illumination and object function by L and O, respectively, and by representing them both through a multi-Gaussian model $L(x) = \sum_{n=1}^{N}(A_n/\sqrt{2\pi\sigma_n^2})\exp\left[-(x-\mu_n)^2/2\sigma_n^2\right]$ and $O(x) = \sum_{m=1}^{M}(A_m/\sqrt{2\pi\sigma_m^2})\exp\left[-(x-\mu_m)^2/2\sigma_m^2\right]$, the intensity recorded at the detector can be expressed as follows [24]:

$$\frac{I(x)}{I_0} = t\sum_m\sum_n A_{mn}\exp\left[-\frac{(x-\mu_{mn})^2}{2\sigma_{mn}^2}\right] \tag{3}$$

where $\mu_{mn} = \mu_m + \mu_n$, $\sigma_{mn}^2 = \sigma_m^2 + \sigma_n^2$ and $A_{mn} = A_m A_n(1/\sqrt{2\pi\sigma_{mn}^2})$. In many practical cases, a single Gaussian term is enough to accurately represent both L and O, and the system of Equations 3 can be analytically inverted. Three input images are required and they have to be recorded at specific relative displacements between the two masks. One image is acquired

IOP Conf. Series: Journal of Physics: Conf. Series **849** (2017) 012040 doi:10.1088/1742-6596/849/1/012040

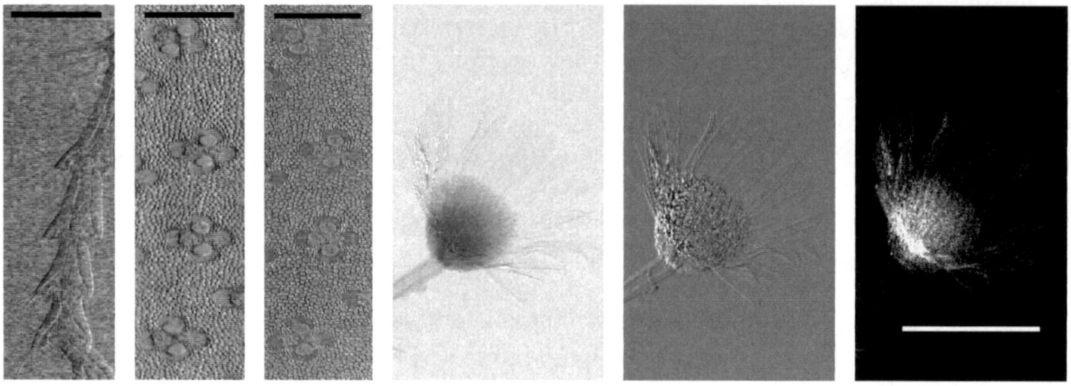

(a) (b) (c) (d) (e) (f)

Figure 2: Experimental images. (a) microscopic details of a ground beetle leg, intensity projection acquired at $\Delta\xi = 2\ \mu$m. Differential phase (b) and transmission (c) images of bamboo wood micro-structure retrieved by using Equation 1. Scale bar is 500 μm in panels (a)-(c). Transmission (d), differential phase (e) and scattering (f) images of a flower (Chrysanthemum leucanthemum) obtained with the low magnification system of Figure 1(c). Scale bar is 1 cm for panels (d)-(f).

with the masks perfectly aligned ($\Delta\xi_2 = 0$) and the other two images are acquired with a symmetric displacement: ($\Delta\xi_1 = -\Delta\xi_3$). This produces the same reduction of intensity in the two symmetric positions, and it is achieved by displacing the masks by the same amount in opposite directions. The optimal value for $\Delta\xi_{1,3}$ typically depends on the specific imaging system design, a common choice is to displace the mask until the detected intensity reaches approximately 50% of the maximum (obtained at $\Delta\xi = 0$). If this simplified approximation is not valid for a particular experiment, more terms need to be retained in Equation 3 and the problem can still be solved by means of a numerical procedure as, for example, non-linear fitting.

3. Results

Figure 2 reports examples of experimental images acquired with laboratory implementations of EI. Panel (a) shows microscopic details of an insect leg as visualised by a single intensity projection image (I_L image, $\Delta\xi = 2\ \mu$m). Panels (b) and (c) are the differential phase and transmission images of the micro-structure of bamboo wood (I_Δ/I_Σ and I_Σ). In this case, the $I_{L,R}$ images were recorded with a masks displacement of $\Delta\xi_1 = -\Delta\xi_2 = 2\ \mu$m. This set-up featured a high X-ray beam energy (80 kVp on a tungsten target) and the insect leg can be effectively considered a pure phase object. Panels (d)-(f) show the transmission, differential phase and scattering images of a flower, respectively. This set-up used a lower energy: the X-ray spectrum generated by a molybdenum target operated at 40 kVp. The scale bar is 500 μm for panels (a)-(c) and 1 cm for panels (d)-(f).

Finally we present a comparison of a detail extracted from a breast tissue image as an example of a potential medical application. The specimen had a thickness of approximately 2.5 cm and was fixed in formalin. It was obtained from a mastectomy after informed consent and the study was approved by the local ethical regulatory bodies. Figure 3(a) shows the conventional, absorption-contrast based radiography while Figure 3(b) shows the EI one. The scale bar is 5 mm. It is apparent how the fibrous structure is more clearly represented in the EI image with respect to the conventional one. It should be noted that the EI image was obtained with a single exposure and both amplitude and phase effects are contributing to the image contrast. It

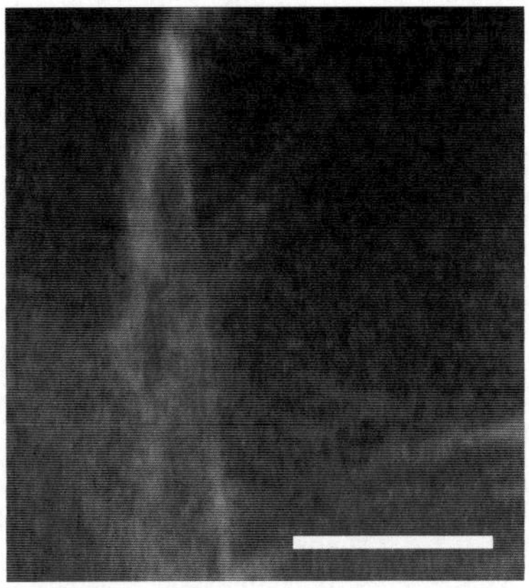

(a)

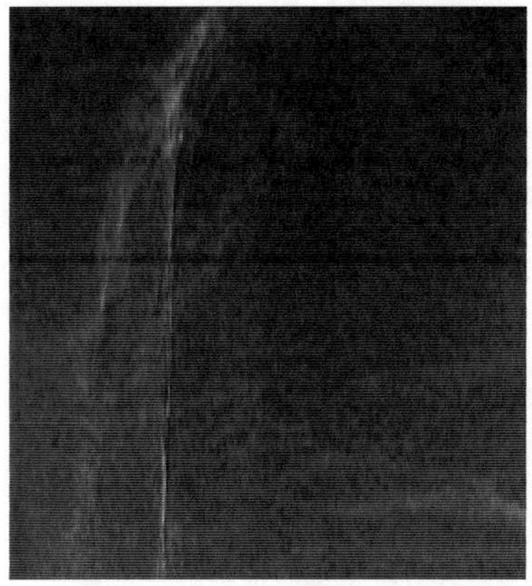

(b)

Figure 3: Comparison between the standard (a) absorption-contrast based radiography and the (b) EI XPCI one. The scale bar is 5 mm. The specimen is breast tissue, approximately 2.5 cm thick. The detail of interest makes apparent how the fibrous structure is visible with much higher detail and contrast in the image obtained by means of the EI method, in comparison to conventional, absorption-based imaging.

was shown that with a laboratory EI system used in this configuration it is possible to perform low-dose mammography [30].

4. Conclusion

We reported and compared different implementations of the edge illumination for synchrotron- and laboratory-based X-ray phase contrast imaging applications. All these approaches are based on the use of an amplitude modulator that is placed before the sample. This is typically achieved by means of an absorbing slit, or mask, that structures the radiation field making the phase distortions introduced by the sample detectable. Different implementation regimes are available, resulting in widely ranging resolution and X-ray energy values, which can be tailored to specific applications in materials study, medical imaging and more.

Acknowledgements

This project was supported by the UK Engineering and Physical Sciences Research Council Grants EP/I022562/1 and EP/I021884/1. ME was supported by the Royal Academy of Engineering under the RAEng Research Fellowships scheme. PCD was supported by Marie Curie Career Integration Grant No. PCIG12-GA-2012-333990 within the Seventh Framework Programme of the European Union.

References

[1] Fitzgerald R 2000 *Physics Today* **53** 23–26
[2] Wilkins S, Nesterets Y I, Gureyev T, Mayo S, Pogany A and Stevenson A 2014 *Philosophical Transactions of the Royal Society of London A: Mathematical, Physical and Engineering Sciences* **372** 20130021
[3] Bravin A, Coan P and Suortti P 2013 *Physics in Medicine and Biology* **58** R1
[4] Bonse U and Hart M 1965 *Applied Physics Letters* **6** 155 –156
[5] Goetz K, Foerster E, Zaumseil P, Kalashnikov M P, Mikhailov I A, Sklizkov G V and Fedotov S I 1979 *Kvantovaia Elektronika Moscow* **6** 1037–1042
[6] Davis T J, Gao D, Gureyev T E, Stevenson A W and Wilkins S W 1995 *Nature* **373** 595–598

[7] Snigirev A, Snigireva I, Kohn V, Kuznetsov S and Schelokov I 1995 *Review of Scientific Instruments* **66** 5486–5492
[8] Ingal V N and Beliaevskaya E A 1995 *Journal of Physics D Applied Physics* **28** 2314–2317
[9] Wilkins S W, Gureyev T E, Gao D, Pogany A and Stevenson A W 1996 *Nature* **384** 335–338
[10] Chapman D, Thomlinson W, Johnston R E, Washburn D, Pisano E, Gmür N, Zhong Z, Menk R, Arfelli F and Sayers D 1997 *Physics in Medicine and Biology* **42** 2015–2025
[11] Clauser J F 1998 Us patent 5,812,629
[12] David C, Nohammer B, Solak H H and Ziegler E 2002 *Applied Physics Letters* **81** 3287–3289
[13] Momose A, Kawamoto S, Koyama I, Hamaishi Y, Takai K and Suzuki Y 2003 *Japanese Journal of Applied Physics* **42** L866
[14] Mayo S C and Sexton B 2004 *Optics Letters* **29** 866–868
[15] Pfeiffer F, Weitkamp T, Bunk O and David C 2006 *Nature Physics* **2** 258–261
[16] Wen H, Bennett E E, Hegedus M M and Rapacchi S 2009 *Radiology* **251** 910–918
[17] Morgan K S, Paganin D M and Siu K K W 2012 *Applied Physics Letters* **100** 124102
[18] Bérujon S, Ziegler E, Cerbino R and Peverini L 2012 *Physical Review Letters* **108** 158102
[19] Wen H, Gomella A A, Patel A, Lynch S K, Morgan N Y, Anderson S A, Bennett E E, Xiao X, Liu C and Wolfe D E 2013 *Nature Communications* **4**
[20] Olivo A, Arfelli F, Cantatore G, Longo R, Menk R H, Pani S, Prest M, Poropat P, Rigon L, Tromba G, Vallazza E and Castelli E 2001 *Medical Physics* **28** 1610–1619
[21] Olivo A and Speller R 2007 *Applied Physics Letters* **91** 074106 (pages 3)
[22] Endrizzi M, Vittoria F A, Diemoz P C, Lorenzo R, Speller R D, Wagner U H, Rau C, Robinson I K and Olivo A 2014 *Optics Letters* **39** 3332–3335
[23] Munro P R, Ignatyev K, Speller R D and Olivo A 2012 *Proc. Natl. Acad. Sci. USA* **109** 13922–13927
[24] Endrizzi M, Diemoz P C, Millard T P, Jones J L, Speller R D, Robinson I K and Olivo A 2014 *Applied Physics Letters* **104** 024106
[25] Marenzana M, Hagen C K, Borges P D N, Endrizzi M, Szafraniec M B, Ignatyev K and Olivo A 2012 *Physics in Medicine and Biology* **57** 8173
[26] Diemoz P C, Endrizzi M, Zapata C E, Pešić Z D, Rau C, Bravin A, Robinson I K and Olivo A 2013 *Physical Review Letters* **110**(13) 138105
[27] Diemoz P, Hagen C, Endrizzi M and Olivo A 2013 *Applied Physics Letters* **103** 244104
[28] Millard T P, Endrizzi M, Ignatyev K, Hagen C K, Munro P R T, Speller R D and Olivo A 2013 *Review of Scientific Instruments* **84** 083702
[29] Endrizzi M, Basta D and Olivo A 2015 *Applied Physics Letters* **107** 124103
[30] Olivo A, Gkoumas S, Endrizzi M, Hagen C K, Szafraniec M B, Diemoz P C, Munro P R T, Ignatyev K, Johnson B, Horrocks J A, Vinnicombe S J, Jones J L and Speller R D 2013 *Medical Physics* **40** 090701
[31] Hagen C, Munro P, Endrizzi M, Diemoz P and Olivo A 2014 *Medical Physics* **41** 070701
[32] Vittoria F A, Kallon G K, Basta D, Diemoz P C, Robinson I K, Olivo A and Endrizzi M 2015 *Applied Physics Letters* **106** 224102
[33] Vittoria F A, Endrizzi M, Diemoz P C, Wagner U H, Rau C, Robinson I K and Olivo A 2014 *Applied Physics Letters* **104** 134102
[34] Vittoria F A, Endrizzi M, Diemoz P C, Zamir A, Wagner U H, Rau C, Robinson I K and Olivo A 2015 *Scientific reports* **5**
[35] Endrizzi M, Vittoria F A, Kallon G, Basta D, Diemoz P C, Vincenzi A, Delogu P, Bellazzini R and Olivo A 2015 *Optics Express* **23** 16473–16480
[36] Munro P R, Hagen C K, Szafraniec M B and Olivo A 2013 *Optics Express* **21** 11187–11201
[37] Munro P R T and Olivo A 2013 *Phys. Rev. A* **87**(5) 053838
[38] Endrizzi M, Oliva P, Golosio B and Delogu P 2013 *Nucl. Instr. Meth. Phys. Res. A* **703** 26 – 32 ISSN 0168-9002

IOP Conf. Series: Journal of Physics: Conf. Series **849** (2017) 012041 doi:10.1088/1742-6596/849/1/012041

Super-resolution x-ray imaging using interaction between periodic structure of object and standing wave generated with total-reflection-mirror interferometer

Yoshio Suzuki

Department of Advanced Materials Science, Graduate School of Frontier Sciences, University of Tokyo, Kashiwanoha, Kashiwa, Chiba, 277-8561, Japan

yoshio2081@gmail.com

Abstract. A super-resolution method in projection-type x-ray imaging is proposed. In this method, interference fringes generated with a two-beam interferometer are used for detecting the fine periodic structure of the object. When the sample has a fine periodic structure, the structure can be detected as interaction between the periodic structure of object and the standing wave formed by the two-beam interferometer. Feasibility studies have been carried out using wavefront-division interferometer with total-reflection-mirror optics and a resolution test chart as a model sample. The fine structures with a period up to 100 nm were detected as modulation of transmitting x-ray intensity at 11.5 keV.

1. Introduction

Spatial resolution of projection-type imaging in the x-ray region is generally limited by the spatial resolution of the imaging detector. Owing to the technical limitations, the spatial resolution of x-ray imaging detector is presently limited to around 1 μm. The higher spatial resolution is, of course, achievable by using x-ray image-forming optics such as Fresnel zone plate, total reflection mirrors, and compound refractive lens. However, the fabrication of high-resolution optical elements for hard x-rays is still a difficult task at present.

Recently, we have proposed a method for overcoming the spatial resolution limit in the x-ray projection imaging, in which interaction between periodic structure of object and a standing wave generated by a two-beam interferometer is used [1]. The fine periodic structures beyond the spatial resolution of imaging system can be detected as modulation of transmitting beam intensity. This method is a kind of super resolution, i.e. spatial resolution beyond the resolution limit of imaging system. Many types of optical systems for two-beam interferometer have ever been developed in the x-ray region: Fresnel biprism optics for hard x-rays [2, 3], total-reflection-mirror system for soft x-rays [4], mono-prism optics [5], total-reflection-mirror system for hard x-rays [6], and bilens optics [7]. In our previous experiment, a two-beam interferometer with single prism optics, which is an analogue of Fresnel's biprism optics in the visible light region, was used for generation of periodic standing wave, and it was confirmed that the fine periodic structures can be detected as the modulation of transmitting beam intensity as well as the modulation of fluorescent x-ray yield.

However, in our previous experiment, there is a problem due to the absorption loss at the prism. It is difficult to produce uniform fine periodic fringes in a wide field of view, and it is also difficult to apply it to soft x-rays. In the present experiment, a total-reflection-mirror interferometer is used instead of the refractive prism, and a standing wave with a period up to 100 nm can be generated. The feasibility study for super-resolution imaging is carried out using a resolution test pattern having fine periodic structures as a model object.

2. Experimental setup

The principle of interaction between x-ray interference fringes and object is schematically shown in Fig. 1. When two parallel waves are crossed on the object plane, a sinusoidal periodic field is generated by the interference between two plane waves. The period, d, of the standing wave is expressed as $d = \lambda /(2\sin(\phi /2)$, where λ is wavelength of x-ray and ϕ is intersecting angle between two beams. The d is approximately rewritten as $d \sim \lambda/\phi$ for small ϕ. When an object having a periodic structure is placed in the standing wave field, modulation in intensity of transmitting beam, corresponding to the interaction between the periodic structure of specimen and the standing wave, is observed by scanning the interference fringe position.

The maximum modulation in the transmitting beam intensity is observed when the period of the structure in the object is equal to the interference fringe pitch. If the fringe period is slightly mismatched with the object structure, a moiré pattern is observed in the measured projection image. Thus, the existence of fine periodic structures can be detected as a modulation of transmitting beam intensity that is independent on the spatial resolution of the imaging detector.

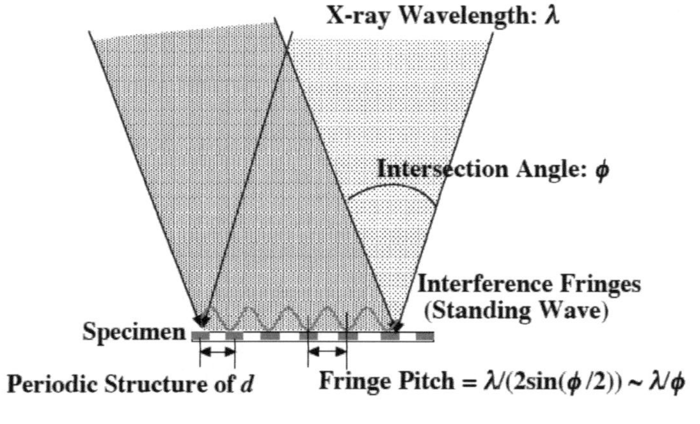

Figure 1. Schematic diagram of standing wave generated with two-beam interferometer and its interaction with periodic structure of object

The experiments have been performed at the BL20XU of SPring-8 [8]. Schematic diagram of experimental setup is shown in Fig. 2. The undulator radiation from 173-pole planar undulator is monochromatized by passing through a liquid-nitrogen-cooling Si 111 double-crystal monochromator. As shown in the figure, the coherent intersecting beam can be generated by a wavefront-division interferometer with a plane total-reflection mirror, i.e. Lloyd's mirror configuration with plane wave illumination. A part of plane wave is deflected by the total-reflection mirror and superimposed on the direct beam. The standing wave is generated in the beam-overlapping area. The deflection angle can be changed by the glancing angle to the mirror. The plane mirror used in the experiment is 150 mm-long Pyrex mirror without any surface coating. The surface figure accuracy of the mirror is better than $\lambda/10$ for He-Ne laser wavelength (632.8 nm). The spatial coherence of incoming wave is defined by a slit placed at about 196 m from the interferometer. In the present experiment, the spatial coherence is required only in the vertical direction, because the vertical beam deflection geometry is employed. Therefore, the coherence is defined only in the vertical direction with a slit of a width of 10 μm. Therefore, the spatial coherence area defined by the fringe visibility of 0.88 is calculated to be larger than 0.5 mm at an x-ray wavelength of 1 Å. The object is placed at about 20 mm from the rear end of the mirror.

A resolution test chart (XRESO-50HC, NTT-AT) is used as a model sample of periodic structures. The test chart made of 0.5 μm-thick-tantalum has stripe patterns whose periods are 200 nm and 100 nm, respectively. A negative pattern is drawn in a square region of 19.5 μm x 19.5 μm. Transmission image of object is measured by an indirect sensing x-ray camera consisting of a phosphor screen (20 μm-thick single crystal of $Lu_2SiO_3:Ce^+$), optical lens system, and a cooled CMOS image sensor. The effective pixel size of the detector is 0.244 μm, and the distance from object and imaging detector is about 2 mm. X-ray energy of 11.5 keV (wavelength of 1.078 Å) is chosen in the experiment to maximize the absorption contrast of the test pattern made of Ta.

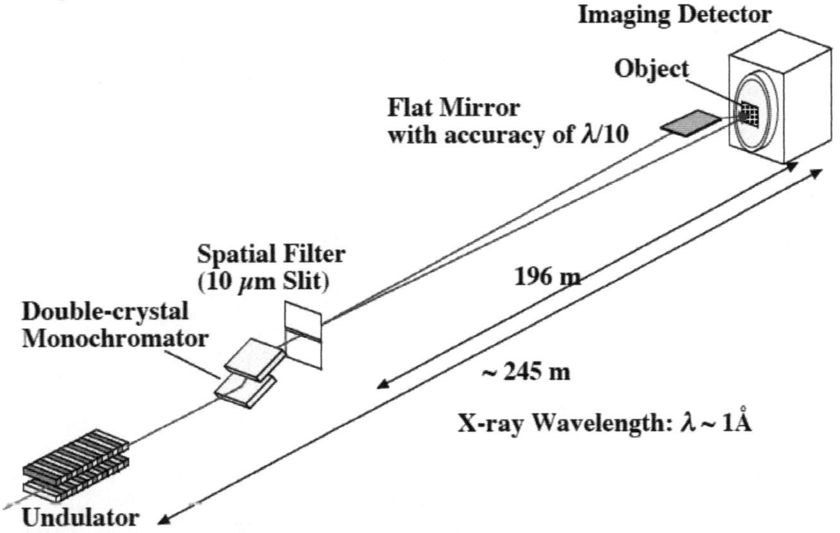

Figure 2. Schematic diagram of experimental setup at BL20XU of SPring-8.

3. Results and Discussions

Typical transmission images measured at an intersecting angle that is slightly (0.02 mrad) detuned from the optimum angle are shown in Fig. 3. Clear moirê fringes of about 4 periods (5 μm pitch) are visible both in the 200 nm-period pattern image (Fig. 3a) and in the 100 nm-period pattern image (Fig. 3b). The quantitative measurements have been performed by means of scanning the position of the standing wave. The sequential images were acquired by step-by-step translational scan of the mirror position. The qualitative intensity modulation by fringe scan of standing wave is shown in Fig. 4. Here, the beam intensity is integrated in a 20 μm x 20 μm square region of the periodic patterns. Clear sinusoidal modulation with a period of 200 nm is observed in the transmitting beam intensity as shown in Fig. 4a. The 100 nm-period pattern is also measured by the similar manor. The result is shown in Fig. 4b. The sinusoidal intensity modulation with a period of 100 nm is also clearly observed.

The exact periodic length of standing wave can be determined by the intersection angle of two plane waves that can be measured by far-field image combining with a spatial pinhole placed in front of the object [1]. The measured pitches are 199.3 nm and 100.8 nm for the 200 nm-pattern and 100 nm-pattern, respectively. As a result, it is confirmed that nanometer periodic structures can be detected by the standing-wave super-resolution technique with the total-reflection-mirror interferometer, and the quantitative measurement of periodic length is also capable.

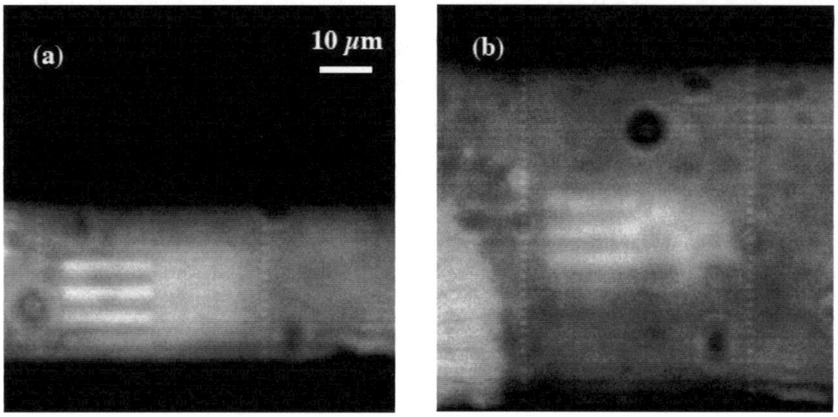

Figure 3. Transmission image of resolution test patter with standing wave illumination. (a) Line and space patterns with a period of 200 nm at ϕ = 0.522 mrad (offset angle of 0.02 mrad). (b) 100 nm-period patterns at ϕ = 1.086 mrad (offset angle of 0.02 mrad). X-ray energy is 11.5 keV (1.078 Å). Exposure time is 10 s.

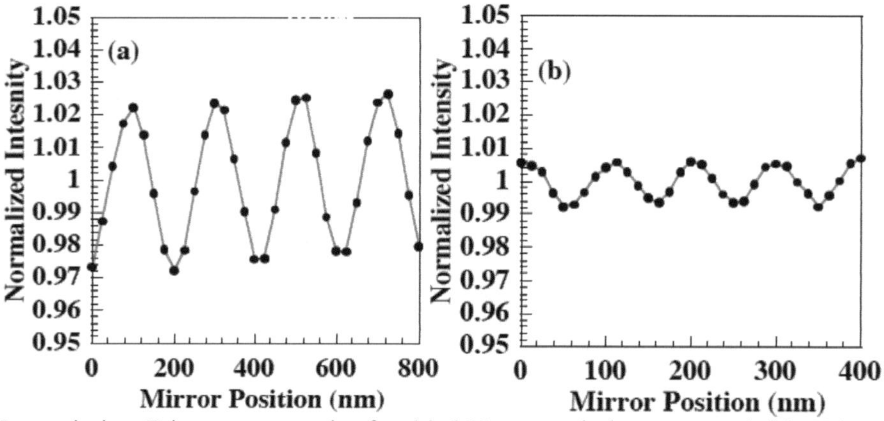

Figure 4. Transmission Fringe scan results for (a) 200 nm-period pattern and (b) 100 nm-period pattern. X-ray energy of 11.5 keV (1.078 Å). Solid circles are measured data. Integration time for each point is 10 s.

4. Acknowledgement

The synchrotron radiation experiments were performed at BL20XU of Spring-8with the approval of Japan Synchrotron Radiation Research Institute (JASRI) (Proposal No. 2013B1255).

References

[1] Y. Suzuki, Rev. Sci. Instrum. **86**, 043701 (2015).
[2] A. R. Lang and A. P. W. Makepeace, J. Synchrotron Radiat. **6**, 59 (1999).
[3] A. F. Isakovic, A. Stein, J. B. Warren, A. R. Sandy, S. Narayanan, M. Sprung, J. M. Ablett, D. P. Siddons, M. Metzlerc, and K. Evans-Lutterod, J. Synchrotron Radiat. **17**, 451 (2010).
[4] W. Cash, A. Shipley, S. Osterman, and M. Joy, Nature **407**, 160 (2000).
[5] Y. Suzuki, Jpn. J. Appl. Phys. **41**, L1019 (2002).
[6] Y. Suzuki and A. Takeuchi, Jpn. J. Appl. Phys. **47**, 8595 (2008).
[7] A. Snigirev, I. Snigireva, V. Kohn, V. Yunkin, S. Kuznetsov, M. B. Grigoriev, T. Roth, G. Vaughan, and C. Detlefs, Phys. Rev. Lett. **103**, 064801 (2009).
[8] Y. Suzuki, et al, AIP Conf. Proc. **705**, 344 (2004).

Estimating the resolution of real images

R Mizutani[1]*, R Saiga[1], S Takekoshi[2], C Inomoto[2], N Nakamura[2], M Arai[3], K Oshima[3], M Itokawa[3], A Takeuchi[4], K Uesugi[4], Y Terada[4] and Y Suzuki[5]

[1] Department of Applied Biochemistry, Tokai University, Kanagawa 259-1292, Japan
[2] Tokai University School of Medicine, Kanagawa 259-1193, Japan
[3] Tokyo Metropolitan Institute of Medical Science, Tokyo 156-8506, Japan
[4] JASRI/SPring-8, Sayo, Hyogo 679-5198, Japan
[5] Graduate School of Frontier Sciences, University of Tokyo, Chiba 277-8561, Japan
*Correspodence: ryuta@tokai-u.jp

Abstract. Image resolvability is the primary concern in imaging. This paper reports an estimation of the full width at half maximum of the point spread function from a Fourier domain plot of real sample images by neither using test objects, nor defining a threshold criterion. We suggest that this method can be applied to any type of image, independently of the imaging modality.

1. Introduction

The resolutions of X-ray microtomographic images have been estimated by using square-wave patterns [1]. Such patterns should be prepared with a precision much finer than the target resolution. Another method of resolution estimation is to observe edge profiles [2]. However, edge profiles can be distorted with the refraction effect, resulting in overestimation of image resolvability. A modulation transfer function (MTF) representing the frequency domain response can be determined from square-wave patterns or edge profiles. The MTF has been used to estimate the resolution, though a threshold must be artificially defined in order to specify the MTF cutoff.

The resolution can also be estimated from the high frequency limit of the Fourier domain profile. It has been proposed that five times the noise level in amplitude (Rose criterion) or twice the noise level in the power spectrum [3] should be used to define the high frequency limit. The resultant estimates largely depend on the threshold criterion [4] and hence involve arbitrariness.

This paper reports an estimation of the full width at half maximum (FWHM) of the point spread function (PSF) from a Fourier domain plot of real sample images by neither using test objects, nor defining a threshold criterion. The resolutions of micro/nano-tomographic sections were also estimated. We suggest that this method can be applied to any image, independently of the imaging modality.

2. Resolution estimation from a Fourier domain plot

The FWHM of the PSF can be estimated from a Fourier domain plot, as reported by Mizutani *et al.* [4]. This method is based on the principle that a Gaussian at an arbitrary position in the real domain is transformed into a Gaussian at the origin in the Fourier domain. The FWHM of the transformed Gaussian is then determined by plotting the logarithm of the squared norm of the Fourier transform against the squared distance from the origin. This method is only effective when the PSF can be approximated with a Gaussian, though the Gaussian approximation is applicable in most cases. Figure 1 shows example plots calculated from test images.

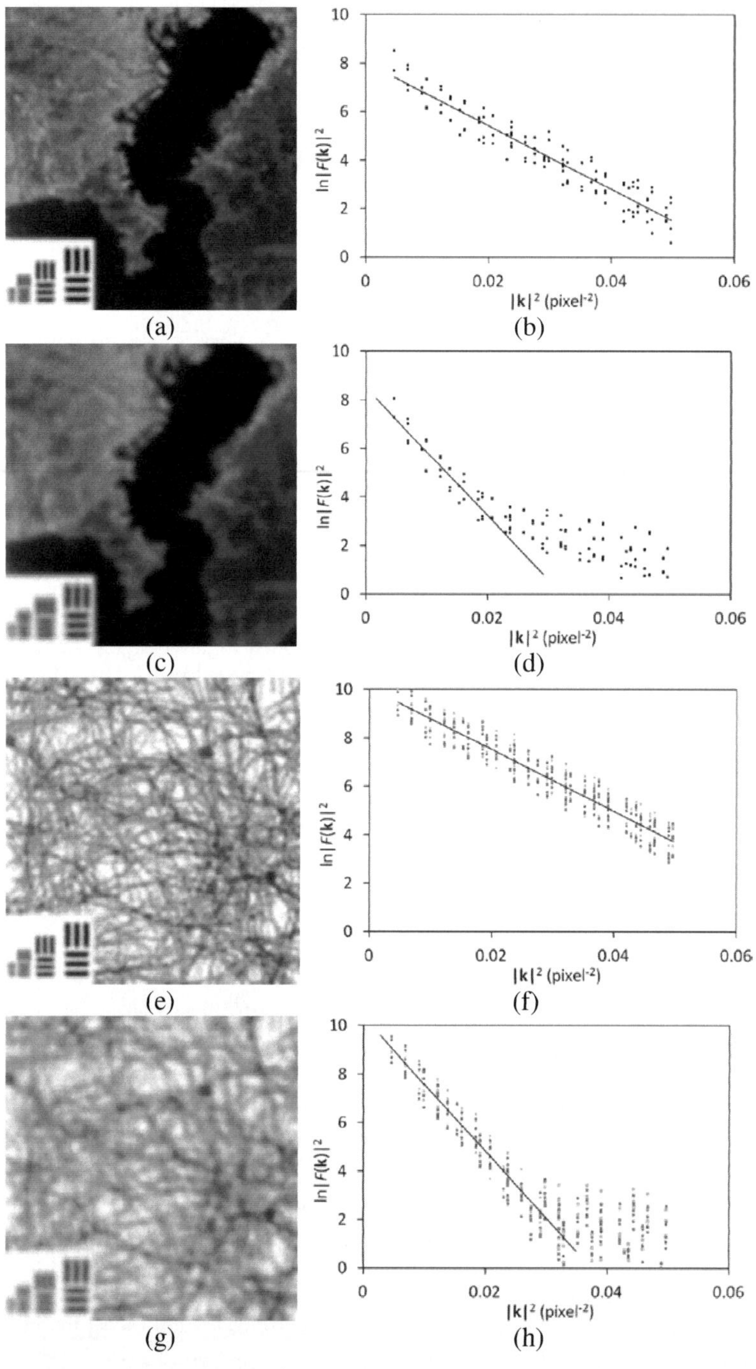

Figure 1. A satellite image of Tokyo Bay (a, c) and a color drawing of a human brain network (e, g) were convolved with Gaussians having FWHMs of 4 pixels (a, e) and 6 pixels (c, g). The lower left inset showing square wave patterns with 2, 4, 6 and 8-pixel pitches indicates that blurring comparable to the Gaussian FWHM occurred. Fourier plots of Mizutani *et al.* [4] were calculated from these test images (b, d, f, h). Red, green or blue dots indicate plots of each color plane. Linear correlations representing Gaussian PSFs are indicated with thin lines.

A satellite image of Tokyo Bay (JAXA) was converted into gray-scale and subjected to 2 × 2 binning in order to eliminate intrinsic blurring. A color drawing was taken from a figure of the human brain network [5]. Square wave patterns with pitches of 2, 4, 6 and 8 pixels were embedded in these original images. Then, test images were generated by convolving the original images with Gaussians. Red, green and blue planes of the network drawing were individually processed.

The resolution estimation procedure was implemented in the RecView software (available from https://mizutanilab.github.io/). Regions within 10 pixels from each Fourier coordinate axis were not used for the plot, since streaks were observed along the axes in some images, presumably due to the readout properties of the detector. Then, pixels in each 5 × 5 bin were averaged in order to reduce data and simplify the plot. In this plot, the Gaussian PSF should be observed as a linear correlation [4]:

$$\ln\left|F(\mathbf{k})\right|^2 \cong -4\pi^2\sigma^2\left|\mathbf{k}\right|^2 + \text{constant},$$

where $F(\mathbf{k})$ is the Fourier transform of the image, $\mathbf{k}$ the coordinate in the Fourier domain, and σ the standard deviation of the Gaussian PSF. The equality is true when the images can be regarded as random, though most images seem to meet this requirement. Indeed, plots generated from test images showed linear correlations corresponding to the FWHM of the Gaussian PSF (Figure 1), independently of the image types. These results indicate that this method can be used to estimate the image resolution.

3. Resolution estimation of X-ray micro/nano-tomographic sections

The resolution of an X-ray micro/nano-tomographic section depends not only on the optics performance but also on a number of factors, including rotation stage wobbling, component drift, vibration, temperature fluctuation, and sample deformation during the data acquisition. Therefore, the resolution must be determined using tomographic sections and not from a single raw image. Figure 2a shows a microtomographic section of test patterns carved on an aluminum wire [2]. This image was reconstructed from a dataset acquired at the BL20XU beamline of SPring-8 with a simple projection geometry using a LuAG scintillator screen. The pixel width was 0.50 μm. The pattern with a pitch of 1.60 μm is clearly resolved, while the pattern with a 1.20 μm pitch is not visible.

Figure 2b shows the Fourier domain plot generated from the section shown in Figure 2a. In this case, the PSF slope was estimated from the linear regression between 0 and 0.05 pixel^{-2}, giving a slope of -48.4 pixel^{-2}. The FWHM of the PSF was calculated as $2\sqrt{2\ln 2} \times \sqrt{-\text{slope}}/2\pi$, giving a FWHM of 2.61 pixels. By multiplying the FWHM with the pixel width, the resolution was estimated to be 1.30 μm. This coincides with the appearance of the test patterns (Figure 2a).

Figure 2c shows a nanotomographic section of a human neuron [4]. The brain tissues were collected with informed consent from the legal next of kin using protocols approved by ethical committees of the related organizations. The image was reconstructed from a dataset acquired at the BL37XU beamline of SPring-8 using FZP optics. The effective pixel width was 23 nm. The plot of this section (Figure 2d) showed a typical profile that can be divided into three parts: the origin spike, successive PSF slope, and the right remainder of shallow descent.

The origin spike is ascribable to the Fourier profile of the overall structure. The region near the origin in the Fourier domain corresponds to large structures in the real domain. If a sample occupies only small part of the viewing field, the spike becomes more prevailing and distorts the PSF slope. In such cases, the region of interest instead of the whole image should be used to estimate the resolution.

Although the right shallow slope is distinct in Figure 2d, this slope should not be used to estimate the resolution, as discussed below. In this case, the PSF slope should be taken from the left quarter of the plot. The resolution estimated from the linear regression between 0 and 0.01 pixel^{-2} was 149 nm, while the resolution of the same nanotomograph evaluated using a test object was 100-120 nm [4]. The difference would originate from tissue deformation or sample drift during the data acquisition.

The right shallow descent represents the FWHM of 2 pixels and corresponds to the Nyquist limit of digitization. Therefore, the shallow descent must not be used to estimate the resolution. The signal-to-noise ratio of the neuron section (Figure 2c) is much lower than that of the test object section (Figure 2a). This is mainly because the pixel width of Figure 2c is approximately 22 times smaller than that in

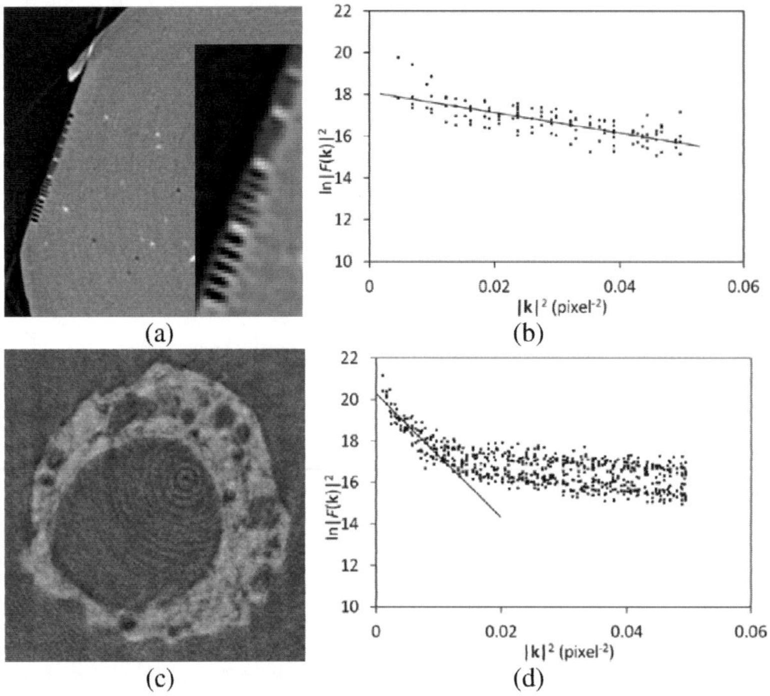

Figure 2. (a) Microtomographic section of an aluminum test object. The right inset shows magnified patterns. Patterns with pitches of 2.0 and 1.6 μm were clearly resolved, while those of 1.2 μm and lower were not. (b) Fourier domain plot of the test object section. The thin line indicates the possible PSF slope. (c) A nanotomographic section of human brain neuron. Image width: 10 μm. (d) Fourier domain plot of the neuron section. The thin line indicates the possible PSF slope. The right remainder slope corresponds to the Nyquist limit of digitization.

Figure 2a. Noise in the image raises the Nyquist slope, making it difficult to identify the PSF slope, as can be seen in Figure 2d. Hence, raw data should be acquired with a sufficient signal-to-noise ratio to visualize fine structures, and in turn, to allow the detection of the PSF slope.

The results obtained in this study indicated that the FWHM of PSF can be estimated from the sample image itself. We suggest that this method can be applied to any type of image, independently of the imaging modality.

Acknowledgments

This work was supported in part by Grants-in-Aid for Scientific Research from JSPS (nos. 21611009, 25282250 and 25610126). The synchrotron radiation experiments were performed with the approval of JASRI (proposal nos. 2013A1384, 2014B1083, 2015A1160, 2015B1101 and 2016B1041).

References

[1] Mizutani R, Takeuchi A, Uesugi K and Suzuki Y 2008 *J. Synchrotron Radiat.* **15** 648
[2] Mizutani R, Taguchi K ,Takeuchi A, Uesugi K and Suzuki Y 2010 *Nucl. Instrum. Meth. A* **621** 615
[3] Modregger P, Lübbert D, Schäfer P and Köhler R 2007 *Physica Status Solidi (a)* **204** 2746
[4] Mizutani R *et al* 2016 *J. Microsc.* **261** 57
[5] Mizutani R, Takeuchi A, Uesugi K, Takekoshi S, Osamura R Y and Suzuki Y 2010 *Cerebral Cortex* **20** 1739

IOP Conf. Series: Journal of Physics: Conf. Series **849** (2017) 012043 doi:10.1088/1742-6596/849/1/012043

Synchrotron X-ray micro-tomography at the Advanced Light Source: Developments in high-temperature in-situ mechanical testing

Harold S. Barnard[1]*, A.A.MacDowell[1], D.Y.Parkinson[1], P.Mandal[1], M.Czabaj[2], Y.Gao[3], E.Maillet[3], B.Blank[4], N.M.Larson[5], R.O.Ritchie[6], B.Gludovatz[6], C.Acevedo[6], D.Liu[7]

[1]Advanced Light Source, Lawrence Berkeley Laboratory, Berkeley CA, USA
[2]University of Utah, Dept. of Mechanical Engineering, Salt Lake City, UT, USA
[3]GE Global Research, Niskayuna, NY, USA
[4]Pulse Ray Inc. Beaver Dams, NY, USA
[5]University of California, Dept. of Materials, Santa Barbara, CA, USA
[6]Materials Sciences Division, Lawrence Berkeley Laboratory, Berkeley CA, USA
[7]Dept. of Materials, University of Oxford, Oxford, Oxfordshire, UK

E-mail: hbar@lbl.gov

Abstract. At the Advanced Light Source (ALS), Beamline 8.3.2 performs hard X-ray micro-tomography under conditions of high temperature, pressure, mechanical loading, and other realistic conditions using environmental test cells. With scan times of 10s–100s of seconds, the microstructural evolution of materials can be directly observed over multiple time steps spanning prescribed changes in the sample environment. This capability enables in-situ quasi-static mechanical testing of materials. We present an overview of our in-situ mechanical testing capabilities and recent hardware developments that enable flexural testing at high temperature and in combination with acoustic emission analysis.

1. Introduction

As the use of advanced structural materials, such as carbon fiber composites and ceramic matrix composites (CMC), expand in aerospace, turbine, nuclear, and other engineering disciplines, there is a growing need for understanding the response of these materials' microstructure to their operational environments through high resolution imaging and in-situ testing. Beamline 8.3.2 at the Advanced Light Source (ALS) specializes in synchrotron hard X-ray micro-computed-tomography (μCT) [1, 2]. With μCT, transmission radiographs of an object are taken from multiple angles and are then used to computationally reconstruct a 3D image of the object's internal structure. The high X-ray flux from synchrotron sources like the ALS uniquely enable high-speed micro-scale X-ray imaging and tomography. At 8.3.2, μCT scans typically require 10s–100s of seconds to acquire high-resolution (0.6 μm/pixel) data sets. With short acquisition timescales, relative to other lab-based μCT instruments, synchrotron tomography is well suited for in-situ mechanical testing materials. Analysis of advanced structural materials exposed to incrementally increasing loads can provide detailed knowledge of how material microstructure evolves under realistic loading and failure conditions.

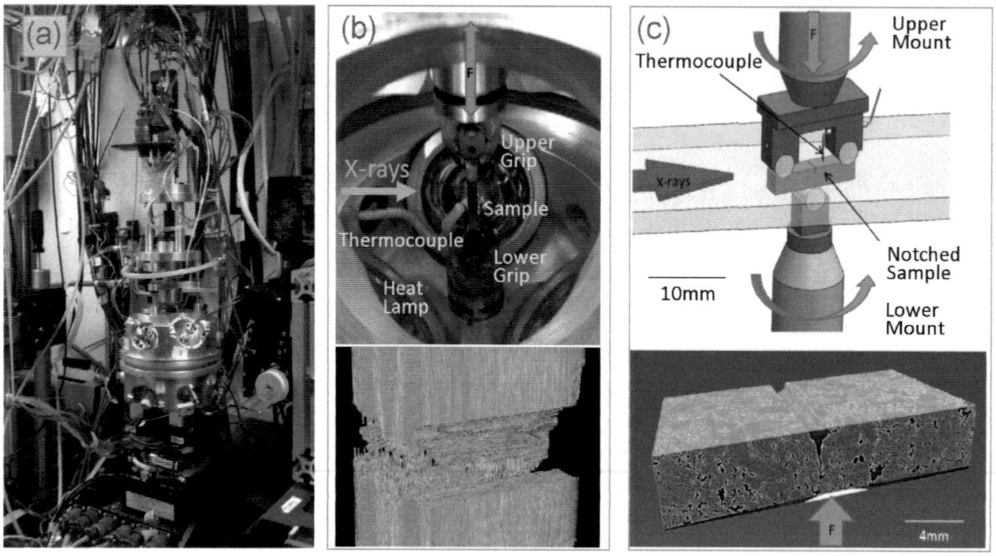

Figure 1. (a) Photo of environmental cell for high temperature mechanical testing. (b) Uniaxial loading grippers and volume rendering of analyzed CMC sample. (c) Three Point bend loading system and volume rendering of analyzed graphite sample.

In-situ environmental cells are a critical part of the beamline 8.3.2. program. Various cells are available for studies spanning fields such as aerospace, plant physiology, geology, and materials science. Each of these cells is used to impose unique controlled environmental conditions on samples. Studying materials under loads, especially at high temperature (800° to > 1600°C), is critical for understanding properties and failure modes of advanced structural materials. Uniaxial loading tests of such materials, particularly tensile testing of CMCs has been and continues to be a highly successful part of the 8.3.2. user program [3]. Over the past several years, the majority of mechanical testing on 8.3.2 has been performed using a high temperature loading cell described in detail in reference [4] and shown in Figure 1. To address growing scientific needs, we are developing new in-situ mechanical testing capabilities to include in-situ tensile testing with acoustic emission analysis in addition to flexural testing with both 3-point and 4-point bending.

2. Acoustic Emission Analysis

Acoustic emission (AE) analysis is an established technique used in industry for monitoring how a material changes under load, particularly the formation of fractures and damage. AE analysis refers to the recording of elastic waves produced by damage within a material with high frequency transducers. For loaded samples, analysis of characteristics of the detected waves such as propagation time, amplitude, energy, and characteristic shape can indicate damage location and mechanism. AE has been shown to be beneficial for analysis of fiber-reinforced composites including CMCs [5, 6, 7, 8, 9] and can therefore be an important tool to couple with μCT.

Preliminary AE analysis has been performed on data collected during room temperature in-situ tensile tests. These measurements successfully detected the initiation and propagation of cracks in loaded CMC samples and have therefore demonstrated that AE analysis can be performed with in-situ tomography (Figure 2). AE was monitored using two sensors clamped in place 20 mm apart on either side of the gauge section of CMC sample. The AE system used was composed of Pico HF-1.2 sensors connected to an IL-LP-WS, 26 dB gain preamplifier, and

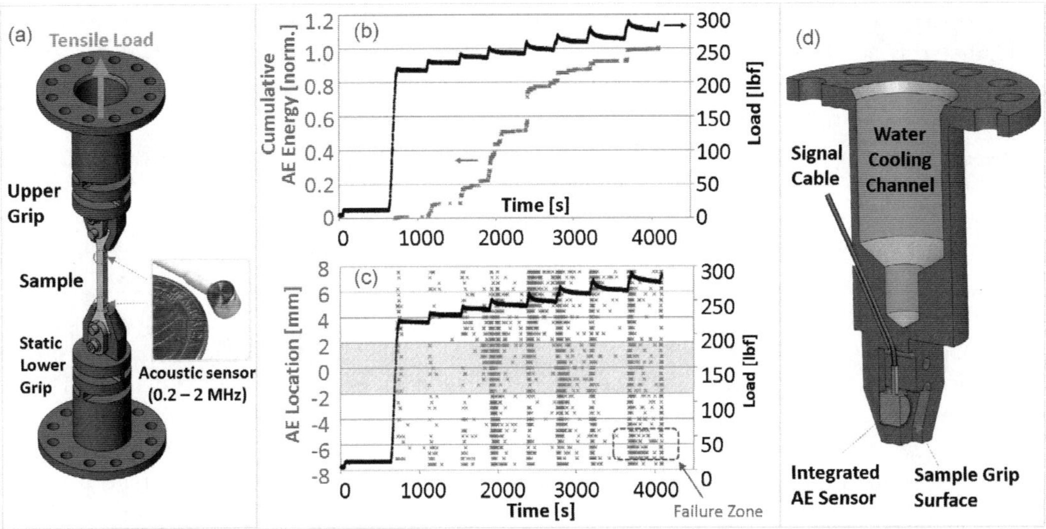

Figure 2. (a) Schematic of tensile loading system: sample, grippers, and AE sensor placement. (b) Cumulative AE energy vs. time and (c) location of AE events vs. time for CMC sample. In (c) the highlighted region corresponds to the scanned length and the dashed-line box indicates the failure zone. (d) Design concept for water-cooled sample gripper with integrated AE sensor.

a 1283 USB AE Node from Mistras Group.

Cumulative AE signal energy and signal amplitudes were monitored continuously during tensile loading until the sample failed, and the AE signals were used to guide the load increase. Load was increased until AE activity became significant. Further load increments were determined based on the amount of increased accumulated AE energy. μCT scans were taken between each loading step and are thus closely associated with development of cracks in the samples during continued load increase. Figure 2b shows cumulative AE energy as a function of time. Cumulative AE energy is a direct measure of damage progression; the energy of an AE signal being directly related to the surface area created by fracture. From the difference in times of arrival of signals at the top and bottom sensors, the location of each AE event can also be determined. Figure 2c shows that significant damage occurred along the entire gauge and that failure location corresponds to a zone of high AE activity prior to failure.

Further work will focus on correlating the locations and signal characteristics of AE events in the scanned volume with observations from the tomography scans. In addition, since the sensors' thermal limits are typically $\sim 200^{\circ}$C, water-cooled sample grippers with integrated AE sensors are in development to make high temperature AE analysis possible (Figure 2d).

3. Flexural Testing

A common method for determining mechanical properties of materials is by flexural testing using 3-point or 4-point bend configurations [10, 11, 12], where the number of points refers to the number of contacts points used to impart a bending moments on the sample. Flexure testing enables analysis of the fracture mechanics of materials. Broadly speaking, the 3-point-bend test is best suited for studying initiation and propagation of cracks due to concentration of tensile stresses in the center of sample. The 4-point-bend test is well suited for determining the bulk properties of the sample under pure flexure, which occurs between the two inner loading points. Flexural tests are relatively straightforward in the engineering laboratory setting because

IOP Conf. Series: Journal of Physics: Conf. Series **849** (2017) 012043 doi:10.1088/1742-6596/849/1/012043

sample preparation is simple and bending instruments are well established. These tests become more challenging for in-situ tomography environments due to geometric constraints but can provide critical understanding of the mechanisms for material failure and fracture resistance. Development of flexure systems at beamline 8.3.2 are described in the following sections.

3.1. Horizontal Three-Point Bending

A horizontally oriented 3-point bending apparatus was developed for the high temperature loading cell used on 8.3.2 [13]. This mechanism, shown in Figure 1c, consists of two 4.8 mm (3/16 inch) diameter rollers that load a horizontal beam-like sample from the top, and a stationary 4.8 mm diameter roller that supports the sample from the bottom. This configuration was chosen because stationary support roller allows the middle of the sample – where crack propagation is typically observed – to remain in the field of view with minimal motion as the sample is loaded. To accommodate high temperature experiments the loading and support rollers are made from alumina (Al_2O_3) and are supported by a stainless steel structure mounted in water cooled supports. A thermocouple is also integrated into the design to directly measure the sample temperature. 10 mm and 16 mm span was chosen for samples of 2–4 mm in width w and height h.

The sample's large aspect ratio L/w – longest dimension L versus shortest dimension w – and its horizontal orientation, can be problematic for X-ray attenuation in L and contrast. Therefore this 3-point bend method works best for samples that have sufficient transmission in L and has been demonstrated with low-Z materials such as graphite [13].

3.2. Vertical Four-Point Bending

A vertically oriented 4-point bend apparatus is currently in development. The prototype for the 4-point bend mechanism that is compatible with our high temperature uniaxial loading cell is shown in Figure 3. This mechanism converts the vertical linear motion of the loading stage into a symmetric bending moment applied to both ends of the sample. The vertical bending configuration is advantageous because it eliminates the contrast issues that arise with large aspect ratio samples mounted horizontally (as in section 3.1).

The bending grips are adjustable to accommodate samples of different thickness and are made from super-alloy MAR M247 because they operate at high temperature with limited conductive cooling. The support arms are made from phosphor bronze and are conductively cooled through the water-cooled support posts.

With 4-point bend measurements, it is desirable to load the sample purely with bending moments applied to the sample through the four contact points. With this mechanism, however, axial compressive loads cannot entirely be avoided. To minimize compression, the bending arms are made to the maximum length allowed by the internal geometry of the environmental cell. Based on the geometry of the current design, the axial compression force should be fairly minimal, 20 times less than the bend force applied at the contact points.

Preliminary 4-point bend measurements were performed on CMC samples demonstrating that samples can be loaded to failure in-situ between tomography scans. Digital image correlation (DIC) [14] analysis was also performed ex-situ to characterize the performance of the 4-point bend mechanism, as shown in Figure 3b. In this analysis, the motion of the load stage, grippers, and sample were tracked to determine the relationship between the strain distribution across the sample and the motion of the mechanism. Figure 3d shows the strain distribution as a function of applied displacement Δ_{applied} by the chamber's vertical uniaxial loading stage. The strain measurements on both tensile and compressive sides of the sample were within $\sim 5\%$ of each other, indicating relatively symmetric loading and minimal influence of the compressive load. Figure 3e shows the motion of the sample's centroid – typically near the region of interest – versus the applied displacement Δ_{applied}. As expected, this shows that the centroid $C_{\text{vert,horiz}}$

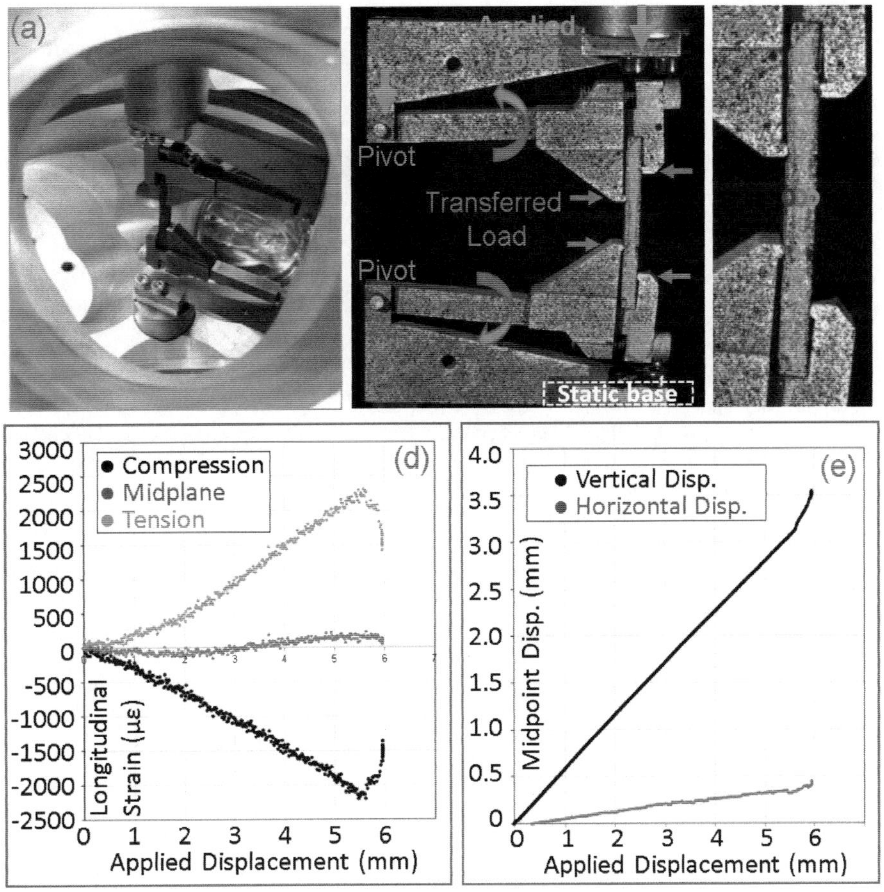

Figure 3. (a) Photo of 4 point bend mechanism. (b) Digital image correlation (DIC) analysis of 4-point bend system. Arrows indicate motion of and forces generated by components. (c) Zoomed-in view of DIC analysis showing strain distribution in the sample regions highlighted by circles correspond to (d) the plot measured strain vs applied displacement. (e) Shows the vertical and horizontal displacement of the midpoint of the sample.

moves substantially in the vertical direction with $\Delta C_{\text{vert}} \approx \Delta_{\text{applied}}/2$. Motion of the sample between μCT scans can problematic however it can be managed with a $>$ 5 mm tall x-ray window, appropriate positioning of the sample and registration algorithms to correct for the sample motion in the analysis. Development of this system is ongoing and looks promising for in-situ 4-point bending.

4. Conclusions

Coupling μCT with high temperature mechanical testing provides powerful tools for understanding the microstructural behavior of advanced materials. Due to increasing scientific demand, at beamline 8.3.2, we have expanded our capabilities to include acoustic emission (AE) and flexural testing. Preliminary analysis with AE has been performed, with high temperature AE experiments planned. Three-point bending has been successfully implemented and used for brittle materials such as graphite. Four-point bending has been demonstrated with CMCs and continues to be developed.

References

[1] Barnard H S, MacDowell A, Parkinson D, Venkatakrishnan S, Panerai F and Mansour N 2016 Developments in synchrotron x-ray micro-tomography for in-situ materials analysis at the advanced light source *SPIE Optical Engineering and Applications, vol. 9967* (International Society for Optics and Photonics)

[2] MacDowell A, Parkinson D, Haboub A, Schaible E, Nasiatka J, Yee C, Jameson J, Ajo-Franklin J, Brodersen C and McElrone A 2012 X-ray micro-tomography at the advanced light source *SPIE Optical Engineering and Applications, vol. 8506* (International Society for Optics and Photonics)

[3] Bale H A, Haboub A, MacDowell A A, Nasiatka J R, Parkinson D Y, Cox B N, Marshall D B and Ritchie R O 2013 *Nature Materials* **12** 40–46

[4] Haboub A, Bale H A, Nasiatka J R, Cox B N, Marshall D B, Ritchie R O and MacDowell A A 2014 *Review of Scientific Instruments* **85** 083702

[5] Maillet E and Morscher G N 2015 *Mechanical Systems and Signal Processing* **52** 217–227

[6] Morscher G N and Godin N 2015 *Ceramic Matrix Composites: Materials, Modeling and Technology* 569–590

[7] Maillet E, Baker C, Morscher G N, Pujar V V and Lemanski J R 2015 *Composites Part A: Applied Science and Manufacturing* **75** 77–83

[8] Baker C, Morscher G N, Pujar V V and Lemanski J R 2015 *Composites Science and Technology* **116** 26–32

[9] Maillet E, Godin N, RMili M, Reynaud P, Fantozzi G and Lamon J 2014 *Composites Part A: Applied Science and Manufacturing* **57** 8–15

[10] Kuhn H and Medlin D 2000 *ASM International, Member/Customer Service Center, Materials Park, OH 44073-0002, USA, 2000. 998*

[11] ASTM 2016 Standard test method for flexural strength of manufactured carbon and graphite articles using three-point loading at room temperature: D7972 14 Tech. rep. ASTM International, West Conshohocken

[12] ASTM 2016 Standard test method for flexural properties of continuous fiber-reinforced advanced ceramic composites: C134113 Tech. rep. ASTM International, West Conshohocken

[13] MacDowell A A, Barnard H, Parkinson D Y, Haboub A, Larson N, Zok F, Parerai F, Mansour N N, Bale H, Gludovatz B *et al.* 2016 High temperature x-ray micro-tomography *AIP Conference Proceedings* vol 1741

[14] Hild F and Roux S 2012 *Digital image correlation* (Wiley-VCH, Weinheim)

IOP Conf. Series: Journal of Physics: Conf. Series **849** (2017) 012044 doi:10.1088/1742-6596/849/1/012044

Investigation of Measurement Condition for 3-Dimensional Spectroscopy by Scanning Transmission X-ray Microscopy

T. Ohigashi[1], Y. Inagaki[1], A. Ito[2], K. Shinohara[2] and N. Kosugi[1]

[1]UVSOR Synchrotron, Institute for Molecular Science, NINS, Okazaki, Aichi 444-8585 Japan
[2]School of Engineering, Tokai University, Kitakaname, Hiratsuka, Kanagawa 258-1292, Japan

ohigashi@ims.ac.jp

Abstract. A sample cell for performing computed tomography (CT) was developed. The 3-dimensional (3D) structure of polystyrene spheres was observed and the fluctuation of reconstructed linear absorption coefficients (LAC) was 9.3%. To improve the quality of data in 3D spectro-microscopy, required measurement condition is discussed.

1. Introduction
Combination of X-ray microscopy and computed tomography (CT) is one of the advanced techniques to be realized in forth-coming diffraction limit synchrotron-based light sources, considering the high transmittance of X-rays and the high spatial resolution of the X-ray microscopy. This combination allows us to analyze 3-dimentional (3D) structures without any destructive process of samples. Among several CT techniques, the full-field microscopy [1-5] may be more suitable methods than the scanning one from the viewpoint of data acquisition time. The scanning transmission X-ray microscope (STXM) in the soft X-ray region has been established as a powerful tool for the 2D chemical analysis based on near edge X-ray absorption fine structure (NEXAFS) and also been applied to 3D chemical state mappings [6-8]. In the soft X-ray region, the thickness of typical samples is around one micron; therefore, the spatial resolution of microscopy, a few tens nm, is suitable for performing CT. In most cases, tomographic data by using scanning type microscopy have been acquired under the condition of restricted angle projection due to the limitations of the measurement time and the shape and mounting of samples. In order to realize more quantitative analysis, the full projection (180° or 360°) measurement is highly required [9]. In this report, an experimental result of CT measurements is shown. The feasibility and the measurement condition to perform 3D spectroscopic measurements in STXM are also discussed.

2. Development of a sample cell for the CT measurement
A sample cell for the CT measurement is shown in Fig.1. This cell consists of a base plate, a small two phase stepper motor (AM-1020, Faulhaber) with a spur gearhead (12/5) and a sample clamp. Samples can be attached on a tip of a needle or in a glass capillary tube, and then be held by the sample clamp.

Fig. 1: The sample cell for CT

IOP Conf. Series: Journal of Physics: Conf. Series **849** (2017) 012044 doi:10.1088/1742-6596/849/1/012044

The stepper motor is driven by a programmable stepping driver (AD PM 00, Faulhaber) via a feedthrough. The controllable accuracy of the rotating angle is approximately 0.1°/full step. Design of the base plate is compatible with ALS-based STXM systems.

3. Experiments and results

3.1. STXM beamline, BL4U

The present measurement was performed at STXM beamline, BL4U, in UVSOR Synchrotron (Okazaki, Japan) [10]. In this experiment, a Fresnel zone plate (FZP) with a long focal length was used for the long working distance and focal depth. Parameters of the FZP is as follows; diameter of 240 μm, outermost zone width of 35 nm, diameter of a 0th order stop of 90 μm, pattern material of nickel on a silicon nitride membrane (100 nm thick) support. Diameter of an order select aperture (OSA) is 65 μm. In this experiment, the photon energy was 280 eV, below the carbon K-edge, and the focal length of the FZP is 1.90 mm. Size of a fixed exit slit used as a virtual source was set to 50×50 μm^2. In this geometry of the STXM, the working distance is ~400 μm, the spot size of the focused beam is 73×73 nm^2 and the calculated focal depth is 2.2 μm.

3.2. Sample and data acquisition

A cross sectional image of polystyrene spheres with diameter of 5 μm was obtained by CT. The spheres were fixed on a tip of tungsten needle with glue. In this experiment, the sample was rotated by 180° in total to get quantitative reconstructed values. 50 STXM images were totally acquired with rotating the sample by 3.6° each. The dwell time was 3 ms, the scanning step size was 150 nm, and the size of the image was 100×100 pixels. Due to wobbling of the sample position and refocusing, it took 2.5 hour to acquire all images. One of the STXM images is shown in Fig. 2. The rotational axis of the CT sample cell is along the vertical direction of the image.

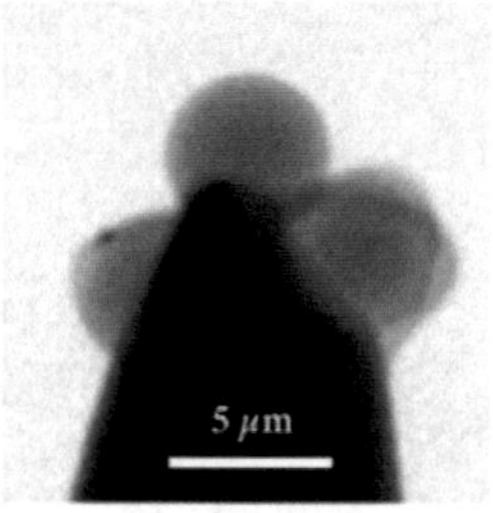

Fig. 2: STXM image of a sample

3.3. CT image reconstruction

The alignment of a series of 50 STXM images was done manually to extract sinograms. Cross sectional images were reconstructed by a home-made filtered-back projection algorithm with assuming the incident X-rays are parallel beams. One of the reconstructed cross sectional images (CT image) and a 3D volume image are shown in Fig. 3 (a) and (b), respectively. A profile along the radial direction of Fig. 3(a) is shown in Fig. 3(c). Judging from 10-90% of the edge of the profile, the spatial resolution is estimated as >300 nm. In Fig. 3(d), a histogram of the CT image is shown and a peak of the linear attenuation coefficient (LAC) at 0.173 μm^{-1} implies polystyrene (*cf.* 0.185 μm^{-1} at 280 eV [11]). An FWHM of that peak is 0.016 μm^{-1}, 9.3% of the peak LAC.

4. Discussion

In this experiment, structure of the sample is homogeneous so that the fluctuation of reconstructed LACs (i.e. the FWHM of the peak) will become an error in spectroscopy by CT. The fluctuation of LACs was 9.3% in this experiment and should be suppressed. There are several reasons for this fluctuation; (a) resolving power of the monochromator, (b) beam hardening, (c) photon flux, (d) degradation of the spatial resolution and (e) error by reconstruction algorithm, where (a), (b) and (c) can be negligible. The resolving power of the monochromator at BL4U is >3,000 in this configuration [10] so that the fluctuation of the LAC is less than 0.01%. The effect of beam hardening due to the higher order incident X-rays, such as bright edges on the CT image, is not clearly seen in Fig. 3(a) and 3(c) [12]. The photon flux onto one pixel was 3.5×10^4 [photons/dwell] and the fluctuation is estimated as 0.5%.

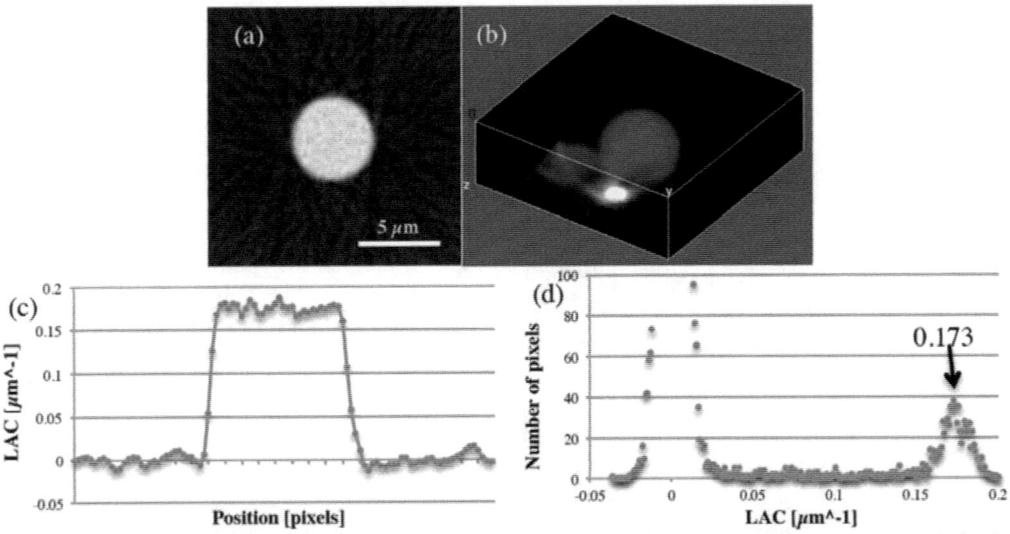

Fig. 3: Reconstructed images of the polystyrene spheres; (a) a cross sectional image and (b) 3-dimensional volume image. (c) A profile along radial direction and (d) a histogram of (a).

To estimate (d) error by the reconstruction algorithm and the required number of rotations, the reconstruction of a phantom with the same size as in the experiment, 100×100 pixels, is simulated. The distributions of LACs of the phantom is shown in Fig. 4(a). Sinograms are extracted from the phantom with 25, 50, 100, 150, 200 and 250 rotations per 180°, respectively. The phantoms are reconstructed and their FWHMs at the peak at 0.185 μm^{-1} are plotted in 4(c). In this plot, the FWHM saturates at 0.0021 μm^{-1} above 100 rotations and the fluctuation is 1.1%. Therefore, we can improve the fluctuation caused by reconstruction algorithm from 2.2% to 1.1% by increasing the number of rotation from 50 to 100.

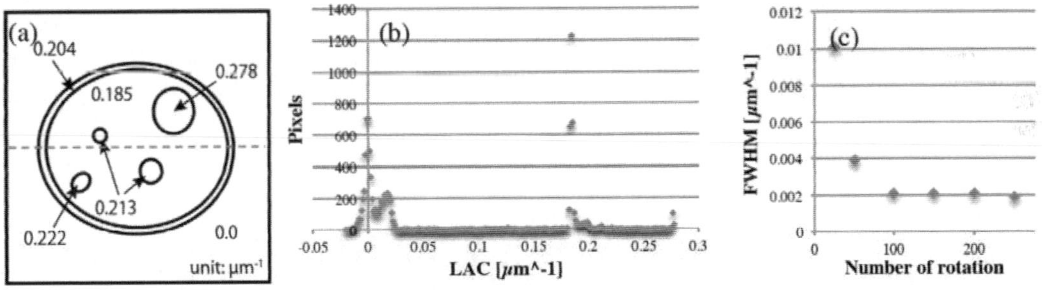

Fig. 4: (a) Distribution of LAC of the phantom and (c) a histogram of the reconstructed sinogram with 100 rotations. (c) The FWHMs of the peak at 0.185 μm^{-1} are plotted against the number of rotations.

In the experiment, the expected spatial resolution was around 150 nm, similar to the scanning step size, but >300 nm in fact. This degradation of the spatial resolution was caused by mechanical issues, such as insufficient focal depth of the FZP, instability of the sample and wobbling of a rotational axis of the stepper motor. Effect of the degraded resolution is simulated by reconstructing the phantom which is filtered by Gaussian with radius of 1 pixel. The histograms of the reconstructed normal and filtered phantoms are shown in Fig. 5(a) and the detail of the peaks around 0.185 μm^{-1} is shown in inset. The FWHM of the filtered phantom becomes slightly larger than the normal one, 0.0024 and 0.0020, respectively. Furthermore, the peak of the filtered phantom has a tail to the higher LAC and its peak shifts toward the higher LAC. Profiles of the reconstructed phantoms are shown in Fig. 5(b). The profiles are extracted from center of the phantom, shown in Fig. 4(a) as a dashed line. Even though integrated intensities of normal and filtered phantoms are same, reconstructed LACs of the filtered one are slightly higher than normal one. This result is due to the smaller diameter of the profile of the filtered phantom in Fig. 5(b). Therefore, density of the reconstructed LACs of the filtered phantom becomes higher than the normal one.

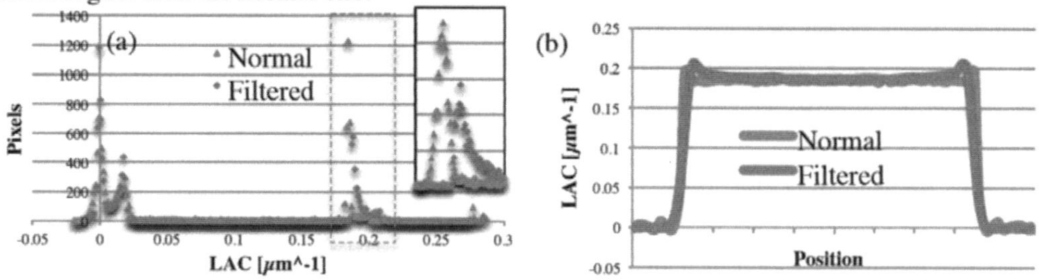

Fig. 5: (a) Histograms of the CT image of the normal and Gaussian filtered phantoms and inset shows detail of the peak around 0.185 μm^{-1}, (b) profiles of reconstructed images

5. Conclusion

The first trial of CT by STXM was performed and the fluctuation of the reconstructed LAC of 9.3% was obtained. This fluctuation should be suppressed to perform spectroscopic measurement by CT. Then, we investigated the measurement condition by simulation to obtain more reliable reconstructed LACs and 100 rotations per 180° is necessary for data acquisition. The spatial resolution of STXM is also effective to obtain reliable LACs so that the FZP with suitable parameters, spatial resolution and sufficient focal depth, has to be improved. For performing actual 3D spectro-microscopy, the new CT sample cell should be developed for high stability and low wobbling. That reduces the measurement time remarkably (e.g. 17.5 hour is estimated in current condition).

Acknowledgement

This work was partly supported by Nanotechnology Platform Program (Molecule and Material Synthesis) of the Ministry of Education, Culture, Sports, Science and Technology (MEXT), Japan, and also by JSPS Grants-in-Aid for Scientific Research (JSPS KAKENHI Grant Nos. 16K05022 and 26248010).

References

[1] Larabell C A and Le Gros M A 2004 *Mol. Biol. Cell* **15** 957
[2] Hoshino M and Aoki S 2008 *Appl. Phys. Express* **1** 067005
[3] Ohigashi T, Fujii H, Usui K, Namba H, Mizutani H, Takemoto K and Kihara H 2011 *AIP Conf. Proc.* **1365** 124-127
[4] Schneider G, Guttmann P, Rahbein S, Werner S and Follath R 2012 *J. Struct. Biol.* **177** 212-223
[5] Sasov A and Van Dyck D 1998 *J. Microsc.* **191** 151-158
[6] Obst M, Wang J and Hitchcock A P 2009 *Geobiol.* **7** 577-591
[7] Berejnov V, Susac D, Stumper J and Hitchcock A P 2012 *ECS Transactions* **50** 361-368
[8] Schmid G, Zeitvogel F, Hao L, Ingino P, Kuerner W, Dynes J J, Karunakaran C, Wang J, Lu Y, Ayers T, Schietinger C, Hitchcock A P and Obst M 2014 *Microsc. Microanal.* **20** 531-536
[9] Johansson G A, Tyliszczak T, Mitchell G E, Keefe M H and Hitchcock A P 2007 *J. Synchrotron Rad.* **14** 395-402
[10] Ohigashi T, Arai H, Araki T, Kondo N, Shigemasa E, Ito A, Kosugi N and Katoh M 2013 *J. Phys: Conf. Ser.* **463** 012006
[11] Henke B L, Gullikson E M and Davis J C 1993 *Atomic Data and Nuclear Tables* **54** 181-341
[12] Tsuchiyama A, Uesugi K, Nakano T and Ikeda S 2005 *Am. Mineral.* **90** 132-142

Low background, UHV compatible scintillator detector for the CLS cryo scanning soft X-ray microscope

A F G Leontowich[1], D M Taylor[1], J Wang[1], C N Regier[1], T Z Regier[1], R Berg[1], D Beauregard[1], J J Dynes[1], C Senger[2], J Swirsky[1], C Karunakaran[1], A P Hitchcock[3] and S G Urquhart[2]

[1] Canadian Light Source Inc., 44 Innovation Boulevard, Saskatoon, SK, S7N 2V3
[2] Department of Chemistry, 110 Science Place, University of Saskatchewan, Saskatoon, SK, S7N 5C9, Canada
[3] Brockhouse Institute for Materials Research, 1280 Main Street West, McMaster University, Hamilton, ON, L8S 4M1, Canada

adam.leontowich@lightsource.ca

Abstract. A new soft X-ray scanning transmission X-ray microscope (STXM) optimized for cryo spectro-tomography was designed and commissioned at the Canadian Light Source (CLS). The instrument was required to achieve ultra high vacuum and be compatible with in-situ plasma cleaning. It also required a scintillator detector, and the design of this detector had to evolve to meet these environmental requirements. The scintillator deposition technique, and the suppression of background by introduction of an edge filter are also presented.

1. Introduction

Soft X-ray (100 – 2000 eV) scanning transmission X-ray microscopy (STXM) is an established technique with many instruments operating around the world and more under construction [1]. One advantage of STXM over complimentary techniques such as scanning electron microscopy (SEM) or electron energy loss in transmission electron microscopes (TEM-EELS) is relatively low dose interrogation [2], which has enabled investigations of radiation sensitive samples such as organics. One of the earliest detectors developed for STXM was a scintillator-photomultiplier tube (PMT) [3-5] placed behind the sample to measure transmitted X-rays. Although many other detectors have been implemented in STXM over the years, the PMT detector remains in wide use because it still has the highest counting efficiency of any transmission detector below about 600 eV, fully covering the water window. The PMT detector is and has been the default transmission detector for the ambient-STXM at the Canadian Light Source (CLS) Spectromicroscopy beamline across the full photon energy range.

A new STXM, optimized for cryo spectro-tomography (cryo-STXM), was recently designed and commissioned at the CLS. Buildup of ice and carbon contamination on the cryo-cooled sample would negatively affect the experiments, so the instrument was designed to achieve ultra high vacuum (UHV), and be compatible with in-situ plasma cleaning. The cryo-STXM required a PMT detector capable of the same environmental requirements. Modification of an existing PMT detector design [5] was necessary to achieve UHV and plasma compatibility, without sacrificing efficiency or signal to noise performance.

2. Materials selection for UHV and plasma environment

The overall mechanical and optoelectronic design is centered on the proven Hamamatsu R647P PMT, with new emphasis on materials to meet the goal of UHV and plasma compatibility (Figure 1). A Hamamatsu C9744 photon counting unit, C9727 power supply and a National Instruments PCIe-6351 data acquisition card complete the counting circuit.

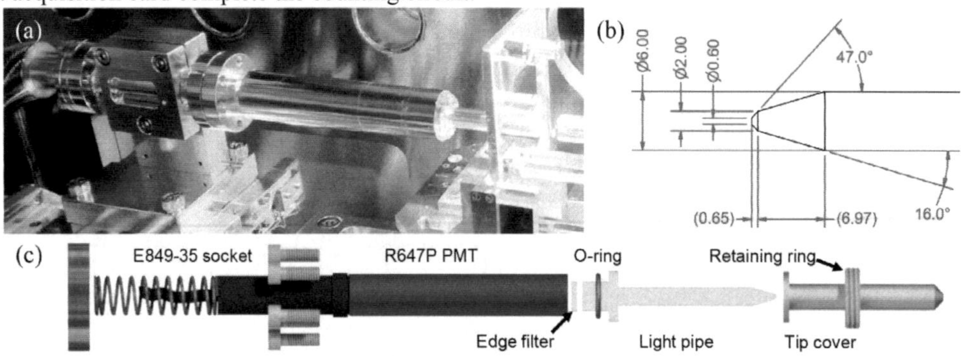

Figure 1. (a) Photograph of the assembled PMT detector installed in the cryo-STXM. Overall length from tip cover to bellows is 221 mm. (b) Dimensions of the light pipe tip in millimeters. (c) Exploded view drawing depicting the internal components.

The PMT is in air, inside a re-entrant housing to protect it from chamber process gasses, especially He which causes premature failure. The housing weldments are electropolished 304L stainless steel, while the tip cover and threaded retaining ring are electropolished 6061 Al. A stainless steel welded bellows is used, and the ConFlat® screws are Au rather than Ag coated, as Ag rapidly develops a black particulate Ag_2O powder in an oxidizing plasma. The light pipe material is borosilicate glass (Hilgenberg GmbH) with all optically active surfaces fire polished. Compared to poly(methyl methacrylate) (PMMA) which was used in the past, glass has superior vacuum properties. In addition, PMMA is etched by oxidizing [6] and reducing plasmas [7], whereas glass is practically inert. A small amount of DuPont Krytox® LVP lubricant was applied to the baked out (120°C for 48 h) brown Viton® o-ring before setting it in the vented groove. A thin coat of Dow Corning High Vacuum Grease was applied to both surfaces of the optical components, which are out of vacuum.

In this configuration, following CLS UHV cleaning and handling procedures, the cryo-STXM instrument with the assembled PMT detector installed has achieved pressures in the 10^{-9} Torr range without baking. It has also endured several 10 minute exposures to hydrogen plasma (GV10x, ibss Group) without any noticeable effects or changes to PMT performance. This PMT detector cannot be baked in the current configuration as the R647P tube cannot exceed 50°C. However, we plan to add air cooling of the tube housing weldment to permit baking. A glass-to-metal seal for the light pipe could further improve vacuum performance.

3. Scintillator deposition

P43 phosphor powder (2.1 μm median particle diameter, NP-1043-01, Nichia) was selected as the scintillator material as it exhibits good all-around performance and high efficiency [5,8], and has proven reliable in the ambient-STXM over a decade. To deposit the loose white P43 powder, a vanishingly thin layer of Dow Corning High Vacuum Grease is transferred to the light pipe tip. The tip is then plunged into a vial of desiccated P43. Puffs of compressed air are used to remove excess powder while periodically inspecting the uniformity and thickness of the layer using an optical microscope. The layers produced by this method are inconsistent. The quality of the scintillator layer has a very significant impact on the detector efficiency, so we explored many other methods including dipping, settlement and evaporation from suspension [9], and thermal evaporation. After review, we find the best method is still plunging the greased tip into desiccated scintillator powder. We compared optical micrographs to raster scanned detector images to determine the characteristics of a well performing layer (Figure 2).

IOP Conf. Series: Journal of Physics: Conf. Series **849** (2017) 012045　doi:10.1088/1742-6596/849/1/012045

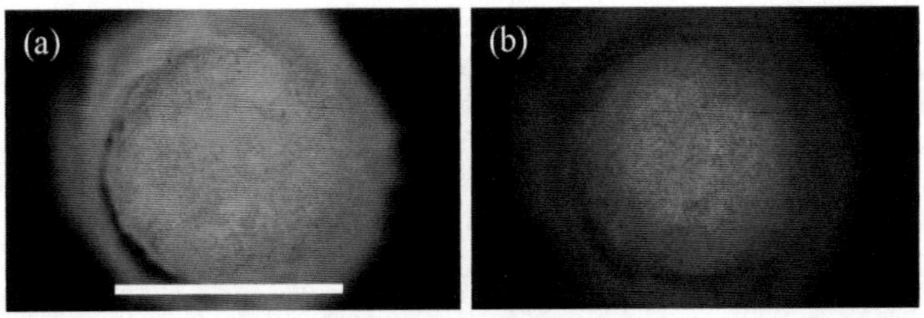

Figure 2. Optical micrographs of a well performing P43 layer in (a) reflection and (b) transmission, 200× magnification, scale bar = 600 μm. A uniform layer a few grains thick was found to be optimal.

After optimizing the deposition method, we compared the efficiency to the ambient-STXM PMT detector. These experiments were performed in the ambient-STXM using the same counting circuit and settings for both detectors. The efficiency of the cryo-STXM PMT detector relative to the ambient-STXM PMT detector was measured to be 77% at 320 eV (absolute efficiency (AE) ~30%), 85% at 395 eV (AE ~50%), and 104% at 700 eV (AE ~60%). A further study of the absolute efficiency is planned.

4. Suppressing background with an edge filter

The sensitivity curve of the R647P PMT spans 300 – 650 nm. It is sensitive to the visible light from the scintillator (signal), as well as any other light present inside the cryo-STXM chamber (background), including scattered laser light from the interferometer. This background has a negative effect on the limit of detection for optical density measurements and spectroscopy. The tip cover partially intercepts background light, and the sample and order sorting aperture also act as light baffles. Purpose-made light baffles are implemented in other STXMs. It is not practical to add a visible light filter over the tip cover aperture (*i.e.*, in the X-ray path) because the soft X-rays would be strongly attenuated and the signal would suffer. However, we hypothesized that it would be possible to suppress the laser light portion of the background by inserting an edge filter between the light pipe and the PMT (Figure 1).

The wavelength of the cryo-STXM interferometer is 632.8 nm (HeNe). The calculated transmission spectrum of the edge filter (SP01-633RU-9.7-D, Semrock) is presented in figure 3a. The cut off wavelength is dependent on the angle of incidence (AOI) and polarization, but the transmission of 632.8 nm light at AOIs from 0 – 28° is practically zero. The X-ray excited optical luminescence (XEOL) spectrum of P43 was measured at the CLS Spherical Grating Monochromator (SGM) beamline and is presented in figure 3b.

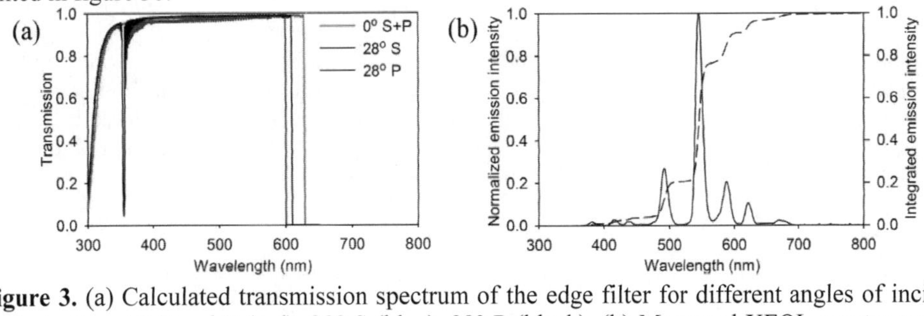

Figure 3. (a) Calculated transmission spectrum of the edge filter for different angles of incidence and polarization: 0° S and P (red), 28° S (blue), 28° P (black). (b) Measured XEOL spectrum of P43 phosphor (solid line), and the integrated spectrum (dashed line).

We measured the XEOL spectrum using excitation energies of 300, 500, 800, 1200, and 1800 eV. All of these spectra were identical. Visually the emission appears bright green with λ_{max} at 546 nm, and 91% of intensity is below 600 nm. Knowing the spectral properties of the edge filter and of P43, the geometry of the light pipe was designed so that 1) light emitted into the light pipe will totally internally reflect down to the PMT, and 2) this light will reach the edge filter with an AOI less than 28°.

Experiments were performed to quantify the effect of the edge filter on the signal and the background (Table 1). First, the PMT power supply voltage and the discriminator were optimized following manufacturers procedures. The counting card accuracy was found to be perfect using a pulse generator. Internal components including the zone plate, order sorting aperture, and sample plate were placed in actual measurement positions. There are no purpose-built light baffles in the instrument.

Table 1. Effect of the edge filter on background and signal measured in the cryo-STXM.

Conditions	Background (Hz, /10)	Signal + background (Hz, /10)
Filter out, laser on	940	2.31×10^5
Filter out, laser off	14	2.31×10^5
Filter in, laser on	60	2.04×10^5
Filter in, laser off	14	2.04×10^5
Filter in, laser off, tip aperture blocked	14	N/A

Some counts were observed when the tip cover aperture was physically blocked by two layers of Al foil. We attribute these to electronic noise (cables, thermal electrons within the PMT, etc.). When the lasers were switched off, the count rate was identical to the physically blocked trial; it appears that the lasers are responsible for all light-based detector background in the cryo-STXM. The edge filter reduced background counts by a factor of 15, but it did not eliminate all laser light. This may be due to fabrication tolerances of the light pipe or the edge filter. The edge filter reduced the real signal by 13%.

The edge filter is a coating which could be directly deposited onto the light pipe or PMT. This would ease assembly and could improve the detector efficiency. It may also be beneficial to employ an interferometer with a laser wavelength outside the sensitivity curve of the PMT in future STXMs.

5. Conclusion

A scintillator-PMT detector was fabricated for the CLS cryo-STXM. The instrument including the detector has achieved pressures in the 10^{-9} Torr range without baking, and endured multiple exposures to hydrogen plasma. The detector efficiency was found to be similar to the ambient-STXM PMT detector. Introduction of an edge filter was found to reduce the detector background by a factor of 15.

Acknowledgements

The CLS cryo-STXM is a component of a CFI Leading Edge fund project, "Enhancing the Spectro-microscopy Beamline and Endstations at the Canadian Light Source" (U. Sask; S G Urquhart, Project Leader), with partial financial support from AFCC Automotive Fuel Cell Cooperation Corp. The CLS is supported by the Canada Foundation for Innovation, Natural Sciences and Engineering Research Council of Canada, the University of Saskatchewan, the Government of Saskatchewan, Western Economic Diversification Canada, the National Research Council Canada, and the Canadian Institutes of Health Research.

References

[1] Hitchcock A P 2012 Soft X-ray imaging and spectromicroscopy *Handbook on Nanoscopy (*vol 2*)*, ed G Van Tendeloo *et al*, (Wiley) chapter 22 pp 745-791
[2] Hitchcock A P, Dynes J J, Johansson G, Wang J and Botton G 2008 *Micron* **39**, 741
[3] Curran S C and Baker W R 1944 *United States Atomic Energy Commission Report* MDDC 1296
[4] Everhart T E and Thornley R F M 1960 *J. Sci. Instrum.* **37**, 246
[5] Fakra S, Kilcoyne A L D and Tyliszczak T 2004 *AIP Conf. Proc.* **705**, 973
[6] Vourdas N, Boudouvis A G and Gogolides E 2005 *J. Phys.: Conf. Series* **10**, 405
[7] Yoshimura S *et al* 2013 *Japan. J. Appl. Phys.* **52**, 090201
[8] Kilcoyne A L D *et al* 2003 *J. Synchrotron Rad.* **10**, 125
[9] Sasaki K Y and Talbot J B 1999 *Adv. Mat.* **11**, 91

Performance of the HERMES beamline at the carbon K-edge

S Swaraj[1*], R Belkhou[1], S Stanescu[1], M Rioult[1], A Besson[1] and A P Hitchcock[2]
[1] Synchrotron SOLEIL, L'Orme des Merisiers, Saint-Aubin - BP 48, F-91192 Gif-sur-Yvette cedex, FRANCE
[2] McMaster University, Hamilton, Ont. L8S 4M1 CANADA

*sufal.swaraj@synchrotron-soleil.fr

Abstract. Contamination of soft X-rays beamline optics due to carbon cracking and deposition under X- ray irradiation is especially critical for spectromicroscopy operations near the carbon K-absorption edge from organic materials, polymers and nanoparticles. In this paper we present the strategy and procedure followed on the HERMES beamline (Synchrotron SOLEIL) to minimize carbon contamination of the beamline optics. Measurements on a complex organic test sample are reported to demonstrate the performance of the beamline at the carbon K-edge in imaging, spectroscopy and spectromicroscopy modes.

1. Introduction

With the recent advances in brilliance, coherence and stability of synchrotron sources, it is becoming exceedingly important that the gained advantages are not lost downstream in the beamline due to contaminated optics. Carbon contamination of beamline optics is a well-known phenomenon that can significantly undermine the performance of a beamline, especially for soft X-rays. It not only affects the spectral and imaging performance near the carbon K-edge but can also result in reflectivity losses at higher energies [1]. Basic steps/conditions leading to the formation of these contamination layers are well established by various theoretical and experimental studies [2,3]. These include, (a) the presence of carbon radical precursor molecules, (b) the adsorption of precursor molecules onto the optics surface, and (c) cracking of the precursor molecules primarily by the electrons photo-emitted from the optics and contaminant at the positions where the X-ray beam hits the optics. Since the 1970's various procedures have been proposed for cleaning beamline optics. These methods include UV-ozone [1], DC glow discharge [4-8], RF plasma [9], pulsed laser [10] and zero-order SR (synchrotron radiation) activated *in-situ* cleaning [11-15]. While all of these techniques have proven to be effective to various degrees, the choice of technique is often limited by ease of applicability, risk of optics damage and down time. From this point of view, the zero-order SR activated *in-situ* cleaning appears to be the most appropriate and effective [12]. This technique involves the use of SR generated, activated oxygen radicals as a chemically reactive cleaning agent that reacts with carbon to form CO and CO_2, which are then pumped away. Warburton et al. [15] were the first to suggest that exposing beamline optical elements to continuous flow of pure O_2, and not only during a specific cleaning procedure that involves zero-order SR, can suppress carbon contamination completely. Risterucci et al. [11] have experimentally shown that the use of intense monochromatic soft x-ray photon beams can have a cleaning effect on a Pt(111) surface and is highly effective at cryogenic temperatures. They also suggested that this procedure could have benefits for optics at room temperature as well.

However, it is important to state that preventive measures should take precedence to strategies for decontamination. This was followed at HERMES beamline by minimizing any contaminant source in the beamline. Wherever possible, all the motors, encoders and cables have been placed outside vacuum optical chambers. All the vacuum vessels of the beamline are made exclusively of Al and have undergone a reactive pre-treatment to have a ca. 300 nm thick Al_2O_3 passivated layer that acts as a barrier and prevents any absorption/desorption of carbon from the chamber walls [16,17]. Finally, all the optics vacuum chambers have been conditioned for several weeks under radiation with dummy mirrors. The real mirrors were installed only after achieving UHV with the X-ray beam present. These steps were aimed at improving the performance of the beamline in the so-called water-window energy domain. In this paper we present the monitoring of beamline contamination over a period of 4 years and demonstrate the effectiveness of these steps.

2. Experimental

Experimental observations were performed at the HERMES beamline with a ring current of 450 mA corresponding to the multi-bunch operation mode. A residual gas analyzer (RGA, *MKS* instruments) was used for partial gas pressure measurements. Carbon K-edge spectra were obtained either using a Si photodiode, installed downstream of the monochromator or using a photomultiplier tube at the Scanning Transmission X-ray Microscope (STXM) end-station. ~200 nm thin sections of a previously studied polyurethane test sample [18] embedded with Poly(Styrene-co-Acrylonitrile) (SAN) and Poly-Isocyanate Poly-Addition (PIPA) nanoparticles were prepared and transferred on to silicon nitride window for STXM measurements. The Fresnel zone plate (FZP – the STXM focusing element) used for this measurement has an outer ring width of 30 nm. Data analysis software *aXis2000* was used for the analysis of the presented STXM data [19].

3. Results and discussion

3.1. HERMES beamline

HERMES is an undulator beamline (2 APPLE-II type undulators of 42 & 64 mm period) covering an overall energy range of 70-2500 eV. It is a dual-branch beamline that has XPEEM (X-ray Photoemission Electron Microscope) and STXM end-stations. Full details are given in Belkhou et al. [20]. Briefly, the first M1 optics presents three mirrors in chicane mode working in pairs depending on the energy range (M1A/B for low energy at 2.5° incidence; M1A/C for high energy, 1.2° incidence). M1A is a plane mirror (SiC substrate, with Ni and Rh stripes) while M1B (Si substrate, Ni coating) and M1C (Si substrate, Rh coating) are toroidal mirrors focusing the beam horizontally 0.5 m after the gratings and collimating the beam vertically for parallel incidence on the gratings. The PGM (*Plane Grating Mirror*) monochromator has two VLS-VGD (*Variable Line Spacing – Variable Groove Depth*) gratings (450 l/mm and 600 l/mm) coupled with a M2 plane mirror (Si substrate, Ni and Rh coatings). An additional M3 toroidal mirror (Si substrate, Au coating) is used inside the monochromator optics vessel to focus the beam to the exit slits. The deflection chamber further downstream has M5 cylindrical mirror (Si substrate, Au coating) and M4 toroidal mirror (Si substrate, Au coating). When inserted in the optical path, the M5 mirror deflects the beam to the KB focusing optics on the XPEEM branch. On the STXM branch the M4 mirror focuses the beam to a set of secondary slits that acts as a source for the FZP. The M1, monochromator, and the deflection chamber are all made of Al and have been reactively pretreated to obtain ~300 nm of Al_2O_3 coating internally. A partial pressure of pure oxygen, $P(O_2)$ of ~$4x10^{-8}$ mbar is maintained during beam operation in all these chambers.

3.2. Discussion on the evolution of carbon contamination

An RGA is installed permanently on the M1 chamber since its installation in 2012. The partial pressures in the chamber are regularly monitored in order to follow the hydrocarbon contamination and the overall effects of continuous O_2 flow. Figure 1a presents the evolution of C, CO and CO_2

content over a period of 4 years. These measurements were obtained in the absence of X-rays and O_2 partial pressure. The partial pressures presented in Fig. 1a were normalized to the residual O_2 partial pressures. The plot indicates clearly that the carbonaceous content in the M1 chamber has decreased by at least an order of magnitude. A visual inspection of the M1 mirrors also confirms the absence of any carbon trace. In addition, reflectivity measurements on the M1 mirrors were performed after tuning the undulator harmonic to the carbon edge. No significant changes in reflectivity profile across the length of the M1 mirrors were observed, confirming the absence of any contaminating layer. This cleaning effect is expected since the source of precursor hydrocarbon molecules, initially present in the form of external contaminants, are likely to be significantly reduced during this extended period of time (48 months). This chamber has no in-vacuum motors or encoders installed that could be an additional contamination source. On the other hand, while the monochromator M3 optics receives much less monochromatic power load (<10 mW) compared to the M1 optics (c.a. 300W pink beam), the presence of in-vacuum encoders act as a major source of contamination. These encoders are unavoidable since high precision movements are necessary in the monochromator. Due to this proximity to a major source of contaminants, the monochromator optics were found to have a visually prominent carbon trace. Absorption by this carbon trace (which only existed on M3) resulted in a significant drop at the carbon K-edge (85%) near 285 eV as shown in figure 1b. It has to be noted that the visual carbon trace on the optics was already observed within the first two months of operation. Although the in-vacuum encoders and cables are claimed by manufacturers to be UHV compatible in terms of out-gassing rate, the manufacturers do not guarantee that these components will not desorb hydrocarbons when hit by X-rays. While the in-vacuum source of contaminants is significant in this chamber, it is nevertheless limited. On visual inspection after another two years of continuous operation, it was found that the carbon trace was completely removed in the region of the mirror M3 continuously exposed to X-rays for more than 2 years (Fig. 1c). Figure 1c shows a photo of the contaminated region that was created by the exposure of the mirror to X-rays during commissioning and initial alignment. The decontaminated part on the mirror is the region that was exposed to X-rays for more than two years. This was also observed as a reduction in the carbon dip from 85% to 15% (Fig. 1b). It has been shown recently that the presence of atomic oxygen on Au surfaces dramatically increases the dissociation rate of oxygen [21] at higher temperatures. Thus we speculate that although M3 receives only a small power load, it does lead to some temperature change, and even a few degree increase in temperature may aid the decontamination process.

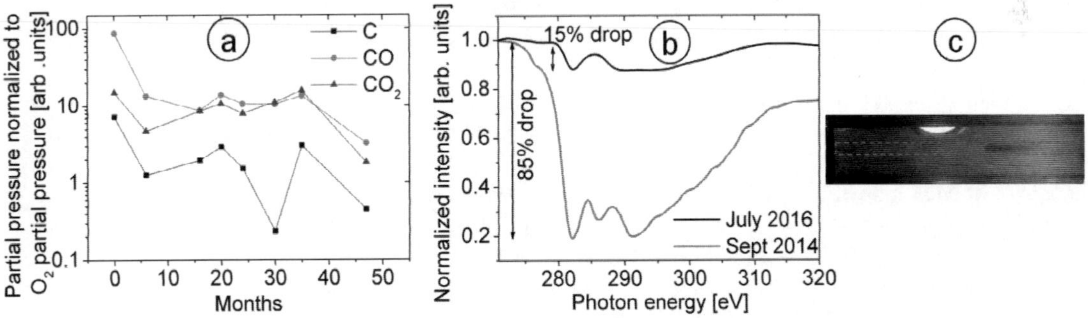

Figure 1. (a) Normalized RGA data showing the evolution of C, CO and CO_2 content with time, (b) I0 carbon dip over a period of two years measured after the monochromator exit slit. The Sep 2014 trace was after initial commissioning while the July 2016 trace was after *in situ* cleaning with O_2 and the X-ray beam (c) Contaminated and 'X-ray beam decontaminated' region of the M3 mirror

3.3. Carbon K-edge measurements on a polyurethane test sample

In order to demonstrate the performance of the beamline at the carbon K-edge we performed STXM analysis of polyurethane samples embedded with SAN and PIPA nanoparticles. This sample was chosen because of the experimental challenges it presents due to the similarity in the C K-edge NEXAFS spectra of the components and the facile radiation damage of some of the components [18].

Figure 2(a-c) present individual chemical component maps extracted by fitting STXM stacks to reference spectra of the components. The SAN and PIPA particles were reliably differentiated and their average sizes, 50 - 300 nm, are consistent with previous results [18]. NEXAFS spectra of individual components extracted from the STXM stacks are presented in figure 2d. The sharp C1s→π* transitions observed in the NEXAFS spectra for all three components span a range of 2 eV (285.1 to 287.2 eV). It is evident from the presented results that the beamline optics are sufficiently clean to allow differentiation of chemically similar components near the carbon K-edge.

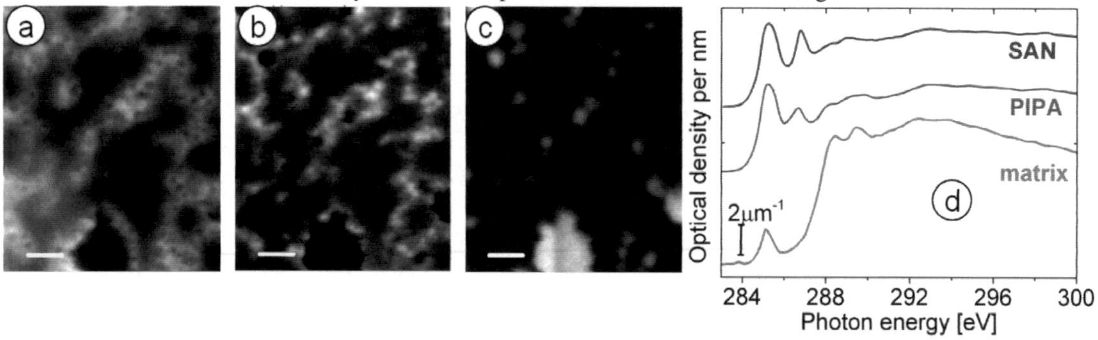

Figure 2. Individual chemical component maps of (a) polyurethane matrix, (b) SAN and (c) PIPA (scale bar: 1 μm) and (d) NEXAFS spectra of each component extracted from the carbon K-edge STXM stack, and converted to an absolute scale (OD/nm).

4. Conclusion

The high quality images and spectra at the carbon K-edge are a consequence of *in-situ* suppression and reversal of carbon contamination of the HERMES beamline optics. Avoiding in-vacuum sources of contaminants and maintaining a continuous flow of oxygen in the optics chambers are favorable conditions for an almost effortless and efficient cleaning of the contaminated optics. We stress that the *in-situ* cleaning of the optics can be achieved even by monochromatic beam at room temperature.

References

[1] C. Chauvet et al., 2011, *J. Synchrotron Rad.* **18** 761.
[2] K. Boller et al., 1983, *Nucl. Instr. and Meth.* **208** 273.
[3] J. Hollenshead & L. Klebanoff, 2006, *J. Vac. Sci. Technol. B* **24** 64.
[4] T. Koide et al. 1986, *Nucl. Instr. and Meth. in Phys. Res.* A **246** 215.
[5] R. B. Gillette & B. A. Kenyon, 1971, *Appl. Opt.* **10** 545.
[6] W. D. Beverly et al. 1973, *MIT press, Cambridge, Mass.* **31** 159.
[7] W. R. hunter et al., 1973, *Appl. Opt.* **12** 2800.
[8] W. R. McKinney et al., 1982, *Nucl. Instr. and Meth.* **195** 371.
[9] E. Pellegrin et al., 2014, *J. Synchrotron Rad.* **21** 300.
[10] A. Singh et al. 2013, *Applied Surface Science* **283** 612.
[11] P. Risterucci et al., 2012, *J. Synchrotron Rad.* **19** 570.
[12] A. Toyoshima et al., 2015, *J. Synchrotron Rad.* **22** 1359.
[13] A. Toyoshima et al., 2012, *J. Synchrotron Rad.* **19** 722.
[14] Roger W.C. Hansen et al.,1994, *Nucl. Instr. and Meth. in Phys. Res.* A **347** 249.
[15] W. K. Warburton et al., 1992, *Nucl. Instr. and Meth. in Phys. Res.* A **319** 240.
[16] H. Ishimaru, 1989, *J. Vac. Sci. Technol. A* **7** 2439.
[17] H. F. Dylla, 1993, *J. Vac. Sci. Technol. A* **11** 2623.
[18] A. P. Hitchcock et al., 2001, *Ultramicroscopy*, **88** 33.
[19] A. P. Hitchcock, aXis2000 is freely available at http://unicorn.mcmaster.ca/aXis2000.html
[20] R. Belkhou et al., 2015, *J. Synchrotron Rad.* **22**, 968.
[21] X. Deng et al., 2005, *J. Am. Chem. Soc.* **127** 9267.

NANOSURVEYOR 2: A COMPACT INSTRUMENT FOR NANO-TOMOGRAPHY AT THE ADVANCED LIGHT SOURCE

Richard Celestre[1], Kasra Nowrouzi[1,2], David A. Shapiro[1], Peter Denes[1], John M. Joseph[1], Andreas Schmid[3], Howard A. Padmore[1]

[1]Advanced Light Source, Lawrence Berkeley National Laboratory, Berkeley, CA 94720
[2]University of California, Berkeley, CA 94720
[3]Molecular Foundry, Lawrence Berkeley National Laboratory, Berkeley, CA 94720

E-mail: **rscelestre@lbl.gov**

Abstract. The Advanced Light Source has developed a compact tomographic microscope based on soft x-ray ptychography for the study of nanoscale materials [1,2]. The microscope utilizes the sample manipulator mechanism from a commercial TEM coupled with laser interferometric feedback for zone plate positioning and a fast frame rate charge-coupled device detector for soft x-ray diffraction measurements. The microscope has achieved point to point (25 nm steps) scan rates of greater than 120 Hz with a positioning accuracy of better than 1 nm RMS. The instrument will enable the use of commercially available sample holders compatible with FEI transmission electron microscopes thus also allowing *in-situ* measurement of samples using both soft x-rays and electrons. This instrument is a refinement of a currently commissioned instrument called The Nanosurveyor, which has demonstrated resolution of better than 10 nm in two dimensions using 750 eV x-rays. Once moved to the new Coherent Scattering and Microscopy beamline it will enable spectromicroscopy and tomography of nano-materials with wavelength limited spatial resolution.

1. Introduction

X-ray ptychography holds the promise of overcoming the resolution, depth of field and efficiency limitations of imaging with conventional x-ray optics [1,2]. It instead relies upon phase retrieval from coherent diffraction data which can be measured to very high numerical aperture with high efficiency. The use of soft x-rays for microscopy, though limiting sample thickness to a few microns, provides very high contrast and exquisite sensitivity to electronic and magnetic states of matter. This, coupled with the high coherent flux available in the soft x-ray range at modern synchrotron sources allows for high spatial resolution with very short exposure times. The ability to image materials with near wavelength limited spatial resolution and functional systems with high time and chemical resolution has previously been demonstrated [2,3]. As x-ray source brightness increases, there is an ever increasing demand for high performance imaging systems. We present a prototype design for an ultra-stable, high speed scanning x-ray microscope which is fully compatible with many electron microscopy sample holders. The microscope, called Nanosurveyor 2, will enable wavelength limited x-ray microscopy and facilitate correlative imaging using x-rays and electrons.

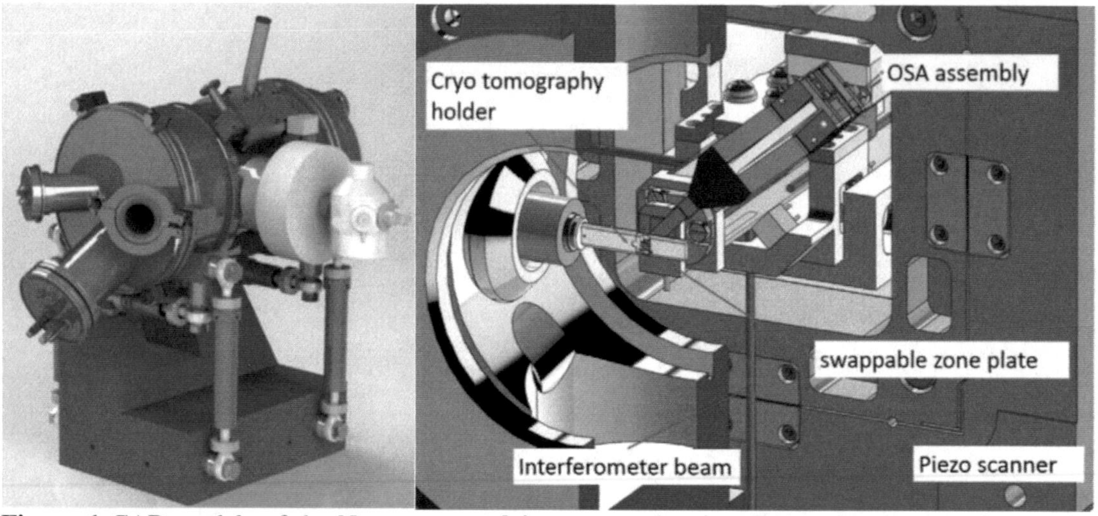

Figure 1 CAD models of the Nanosurveyor 2 instrument showing Gatan cryo tomography sample holder and internal zone plate optic scanning hardware with reference to TEM sample holder.

2. Mechanical and Optical Design

The prototype nanosurveyor 2 instrument is based on the sample area hardware of an FEI CM200 series TEM. We have modified the octagon sample area hardware removed from an FEI TEM to create a larger vacuum chamber incorporating both upstream and downstream sections in order to house the required beam conditioning hardware and the LBNL designed fast CCD [4]. This approach was taken in order to leverage the existing design of the TEM for high stability, which also utilizes air-side motor hardware for thermal stability, and to provide for ease of measurement of samples going between EM and XRM instruments. The sample is held and positioned by a standard FEI Compustage sample manipulator which has been upgraded to higher resolution optical encoders. Most functionality present in a standard TEM installation remains in our implementation of the system. This allows users to mount samples in commercially available sample holders designed for tomography, cryo-tomography and those holders specially designed for *in-situ* experiments.

The basic configuration of the system is similar to other scanning zone plate systems in use at the ALS in that the focusing optic is mounted into a high stiffness scanning piezo flexure stage with X and Y axes (orthogonal to the x-ray beam direction) [5]. The interferometer's fiber-optics heads are rigidly mounted to the fixed assembly which supports the sample goniometer such that mechanical path lengths to the sample are minimized. These interferometer beams are returned by a polished and gold coated surface on the zone plate mounting assembly in order to track position of the zone plate in XY while scanning. A compact, three axis order sorting aperture assembly is mounted within the frame of the piezo scanner in order to facilitate tracking of the focused x-ray beam with the order sorting aperture (OSA). This enables us to utilize the entire 120 μm range of our XY scanning system rather than being limited to a fraction of the OSA diameter (typically 50 μm).

The focusing optic travels along the beam axis on a novel stage that integrates the motion stage with the interferometric mirror surfaces and the mounting system for the zone plate optic. This system is mounted within the open frame of the XY piezo scanner. The design of this linear travel stage is centered around an octagonal piece of sapphire which has been polished and had designated reflecting surfaces coated for use as mirrors for the interferometric tracking system. Other faces act as the bearing surfaces for stick/slip piezo actuators which provide the motive force to translate the optical elements along the beampath. This assembly is both lightweight and very stiff, leading to a high resonant frequency in the scanning system. The design specification for the system was to allow for positioning the beam on the sample at a rate matching the maximum frame rate of the fast CCD

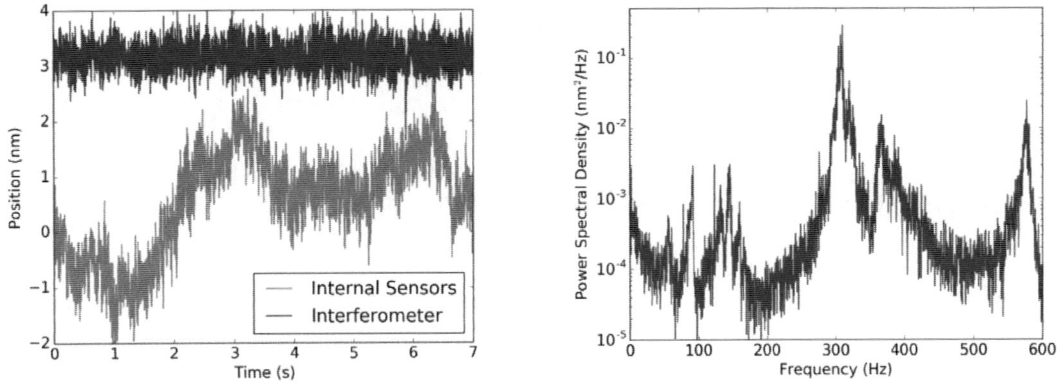

Figure 2 Left: position of the zone plate as measured by the interferometer when the zone plate is in closed loop control of the internal capacitive sensors (red line) and the interferometer (blue line). The position is measured at 100 kHz but the data is displayed at 1 kHz. Right: power spectrum of the zone plate vibrations when in closed loop control of the interferometer. The first resonance is at 300 Hz and over the full 7 second measurement has an RMS of 1 nm.

(200 frames per second) which resides in a light tight vacuum section immediately downstream of the sample.

3. Stability Analysis

Two mechanisms are available for closing the loop in the zone-plate XY stage: feedback from internal capacitive sensors of the stage, or an external interferometer measurement. Although the interferometer heads have the advantage of directly measuring the position of the final moving part (as opposed to the capacitive sensors, which measure the motor movement,) their accuracy depends on the stability/stiffness of the structure in between the optical head mounts and the moving stage. Figure 2 shows interferometer measurements of the zone plate when under closed loop control. The capacitive sensors stabilize the stage to high precision but there is still some residual relative motion of the zone plate at the few nanometer scale. The first resonant frequency for these vibrations is at 300 Hz and over the full 7 second measurement has a Root Mean Square (RMS) amplitude of 1 nm.

The interferometer can output position information either in incremental form (quadrature) or absolute (High Speed Serial Link - HSSL). While quadrature benefits from faster update rates, limited by a 40ns clock, it has the disadvantage of suffering from a tradeoff between communicated precision and maximum speed with which the stage can move. This means that when the feedback parameters and the stage tuning is configured for optimal scanning performance, the quadrature precision cannot be set below 1 nm. This is especially critical for a new beamline called COSMIC at the ALS, dedicated to coherent imaging and scattering. Due to the high coherent flux, motor moves will be the speed bottleneck. An alternative is to use HSSL, which is immune to lossy communication, and has virtually no limit on communicated precision. The downside of HSSL is a longer clock cycle of 80 ns, and the need for 48 cycles to communicate a single position. This still amounts to only a few milliseconds per position update, which is sufficient, given motor moves take milliseconds.

4. Conclusions

We present the conceptual design and preliminary performance analysis of a new compact scanning transmission x-ray microscope dedicated to ptychographic imaging at the ALS. The microscope is based upon the sample goniometer and transfer system of a commercial TEM which has been customized to also house a high performance zone plate scanning mechanism. This mechanism can perform point to point scans at well over 100 Hz which is fast enough to keep pace with the short exposure times possible at new high brightness beamlines and the short readout time of a custom fast frame rate CCD. We have also demonstrated that the stability of the instrument is suitable to reach the microscope design goal of soft x-ray wavelength limited imaging.

Acknowledgments
The Advanced Light Source is supported by the Director, Office of Science, Office of Basic Energy Sciences, of the U.S. Department of Energy under Contract No. DE-AC02- 05CH11231.

References

[1] Thibault P, Dierolf M, Menzel A, Bunk O, David C and Pfeiffer F 2008 Science 321 379–382
[2] Shapiro D A, Yu Y S, Tyliszczak T, Cabana J, Celestre R, Chao W, Kaznatcheev K, Kilcoyne A L D, Maia F, Marchesini S, Meng Y S, Warwick T, Yang L L and Padmore H A 2014 Nature Photonics 8 765–769 ISSN 1749-4885 URL http://www.nature.com/nphoton/journal/v8/n10/full/nphoton.2014.207.html
[3] Denes P, Doering D, Padmore H A, Walder J P and Weizeorick J 2009 Re- view of Scientific Instruments 80 083302 ISSN 0034-6748, 1089-7623 URL http://scitation.aip.org/content/aip/journal/rsi/80/8/10.1063/1.3187222
[4] Kilcoyne A L D, Tyliszczak T, Steele W F, Fakra S, Hitchcock P, Franck K, Anderson E, Harteneck B, Rightor E G, Mitchell G E, Hitchcock A P, Yang L, Warwick T and Ade H 2003 Journal of Synchrotron Radiation 10 125–136 URL http://dx.doi.org/10.1107/S0909049502017739
[5] Lim J, Li Y, Alsem D H, So H, Lee S C, Bai P, Cogswell D A, Liu X, Jin N, Yu Ys, Salmon N J, Shapiro D A, Bazant M Z, Tyliszczak T and Chueh W C 2016 Science 353 566–571 ISSN 0036-8075, 1095-9203 URL http://science.sciencemag.org/content/353/6299/566

IOP Conf. Series: Journal of Physics: Conf. Series **849** (2017) 012048 doi:10.1088/1742-6596/849/1/012048

Coherence Length and Vibrations of the Coherence Beamline I13 at the Diamond Light Source

U.H. Wagner[1], **A. Parson**[1], **C. Rau**[1,2]

[1] Diamond Light Source, Chilton, Didcot, Oxon OX11 0DE

[2] Northwestern University, Chicago (IL), USA, 60611-3008

ulrich.wagner@diamond.ac.uk

Abstract. I13 is a 250 m long hard x-ray beamline for imaging and coherent diffraction at the Diamond Light Source. The beamline (6 keV to 35 keV) comprises two independent experimental endstations: one for imaging in direct space using x-ray microscopy and one for imaging in reciprocal space using coherent diffraction based imaging techniques [1]. In particular the coherence experiments pose very high demands on the performance on the beamline instrumentation, requiring extensive testing and optimisation of each component, even during the assembly phase. Various aspects like the quality of optical components, the mechanical design concept, vibrations, drifts, thermal influences and the performance of motion systems are of particular importance. In this paper we study the impact of the front-end slit size (FE slit size), which determines the horizontal source size, onto the coherence length and the detrimental impact of monochromator vibrations using in-situ x-ray metrology in conjunction with fringe visibility measurements and vibration measurements, based on centroid tracking of an x-ray pencil beam with a photon-counting detector.

1. Beamline Layout and Coherence Length adjustment via Front-end Slits

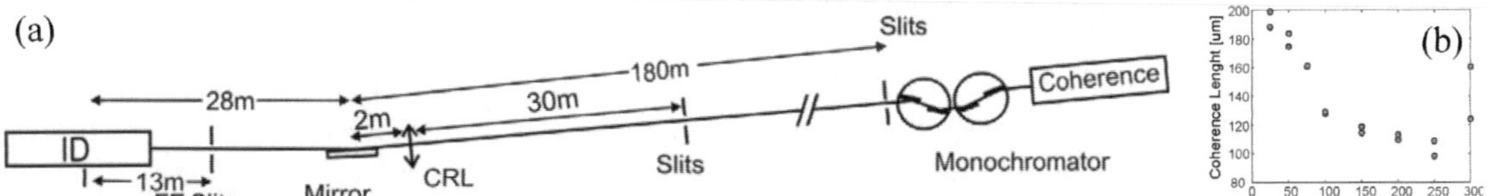

Figure 1. (a) Beamline Layout (b) Variation of the coherence length with FE slit size at 8.0 KeV.

The beamline layout is shown in figure 1(a). The three outstanding features of the beamline are: The front-end slits (FE slits), which are located close to the horizontal source point, thanks to the double mini-beta design [2], and allow to directly adjust the horizontal source size and subsequently the coherence length. Secondly all optical components are horizontally deflecting to minimise the loss of coherent flux [3], arising from the optics performance and the component's stability. Thirdly the monochromator is located close to the end-station to allow for water cooling and thus avoids the technical challenge to use a liquid nitrogen cooling system, which does not induce vibrations and drifts.

Figure 1(b) shows how opening and closing the FE slits adjusts the coherence length, which has been measured using the fringe visibility of a boron-fibre diffraction pattern [4]. In [5] details about the measurements at I13 are given: The vertical coherence length has been found to be larger than 350 μm whilst the horizontal coherence length appears to be limited to 200 μm, even if the FE slits are closed to approximately 25 μm. This might be due to limitations within the FE slit design, in particular the slit blades, or vibrations within the horizontally deflecting beamline optics.

As shown in [3] any vibrations of the monochromator will have the most profound impact onto the beamline performance, since it is the most downstream optical component. Simply speaking a rotational vibration between the first and second crystal of the monochromator with an rms-amplitude α_{rms} at a distance $r_m = 197.5$ m from the FE slits, which are opened to d_{Slit}, will lead to an increase of the apparent FWHM of the virtual source $w_s{}^2 \approx d_{Slit}{}^2 + \left(2\sqrt{8\ln(2)}\,\alpha_{rms}\, r_m.\right)^2$. Using the van Citter-Zernicke theorem and the apparent source size one can derive the coherence length $l_{tc} = \frac{\sqrt{2}\lambda}{\pi}\frac{r_s}{w_s}$, whereby $2l_{tc}$ corresponds to the FWHM of the mutual intensity function and $r_s =$ 201 m is the distance from the FE slits to the boron-fiber. Vice versa, if one measures the coherence length one can derive the apparent source size and subsequently, if the FE slit size d_{Slit} is known, an upper estimate of the monochromator vibrations.

Table 1. Coherence length, apparent source size and vibrations versus FE slit size: Slit sizes beyond 100 μm have not been considered, since the Gaussian profile of the x-ray beam increasingly determines the photon-beam size rather than the FE slits for larger slit openings.

FE Slit Size d_{Slit}[μm]	25	50	75	100
Coherence Length l_{tc} [μm]	195	180	160	129
Apparent Source Size w_s [μrad(FWHM)]	77	83	93	115
Angular Vibration α_{rms} [nrad(rms)]	57	55	66	73

Table 1 is derived from figure 1(b) and shows that the monochromator vibrations vary between 57 nrad(rms) and 73 nrad(rms). The data for the fringe visibility analysis were collected over 40 seconds per point, whilst the Bragg axis control loop was switched off. The variation of the vibration- level arises most likely from the relatively short data acquisition time. Reducing the overall amplitude of vibrations, in particular transient excitations, will yield the largest benefits in increasing the horizontal coherence length, but requires more detailed in-situ spectral analysis.

2. In-situ vibration measurements using a Photon Counting Detector

It is very beneficial to study vibrations already during the assembly of the instrument using optical or mechanical metrology equipment [6] to ensure a correct design and assembly and to mitigate any glitches. Nevertheless the final and ultimate tests have to be performed using in-situ x-ray based techniques. The most simple and direct method to measure vibrations is to use an x-ray pencil beam and track the motion of its centroid with a high frame-rate and high resolution x-ray detector.

Detectors achieving a resolution of less than 1 μm tend to have lower quantum efficiency (typically around 0.44 counts/x-ray Photon) and higher noise levels (>2.2 counts(rms)), since they are detecting the x-rays indirectly via a scintillation screen and a high resolution microscope objective. Directly detecting detectors can have quantum efficiencies approaching 1.0 and are virtually noise free, but have a larger pixel size and thus less resolution. This drawback can often be compensated by better sub-pixel interpolation:

For simplicity, let us assume an x-ray pencil beam, which is precisely one pixel wide and centered between two pixels, so that it is split equally between the two pixels and the total number of counts per pixel is c_p. If the pixel size is d_P and the beam moves towards one pixel by δx, then its counts will increase to $c_p(1 + 2\delta x/d_P)$ and for the other pixel, it will decrease to $c_p(1 - 2\delta x/d_P)$, thus converting position information into a difference Δc in counts between the two pixels. The position move is then given by $\delta x = \Delta c(d_P/4c_p)$ and thus limited by the smallest difference Δc, which one can measure. In the case of a noise-free detector the shot-noise of the incident x-ray photon beam, will eventually determine the smallest observable centroid movement: The shot-noise corresponds to the square-root of the number of photons hitting a pixel and with Gaussian error propagation as an approximation, one obtains $\Delta c^2 = 2c_p$. Hence the position resolution is limited by $\Delta x^2 = {d_P}^2/8c_p$.

The photon counting detectors Merlin [7] is based on the MediPix Chip, has a pixel size of 55 µm, a frame-rate just over 1.2 KHz and provides back-to-back readout, i.e. it continuously samples the impinging photon beam without any interruption due to the readout process, thus implementing a lowpass filter with a rectangular impulse response and maximized spectral resolution. An intrinsic property of a photon-counting detector is the so called pile-up, i.e. two photons arriving in such a short period of time that the pulse counting circuity cannot discriminate between them and only counts one event. This will lead to a flattening relationship between the incident number of photons and measured counts. The pile-up becomes really significant for count-rates >200 KPhotons/sec, though it can be corrected for higher count-rates until the detector starts saturating.

3. Experimental Setup

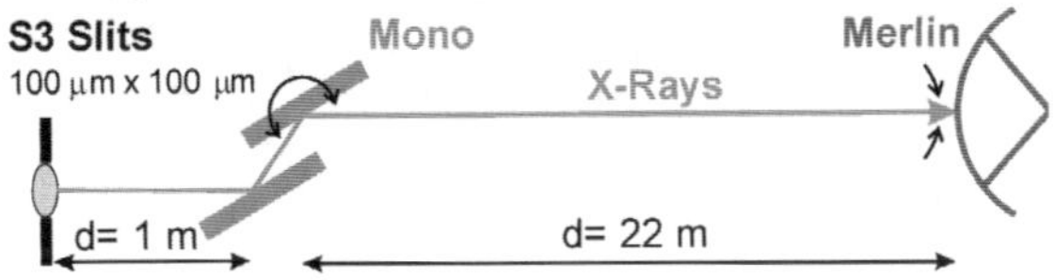

Figure 2. Setup to measure the angular vibration between the two monochromator crystals, using an x-ray energy of 9.1 KeV and Merlin, a photon-counting detector.

The experimental setup is shown in figure 2. The flux impinging onto the detector is ~200 KPhotons/Pixel/sec and the frame rate is 1.2 KHz using back to back readout. Hence there are 170 Photons/Pixel/Frame, providing a position resolution around 1.5 µm (37x sub-pixel interpolation), respectively 35 nrad pitch rotation of the monochromator crystals. In addition we could demonstrate that by illuminating a n_p pixels long row perpendicular to the measurement direction and binning the data along that row, one can improve the position sensitivity by ${n_p}^{1/2}$. The centroid positions are calculated using the EPICS ArrayDetector STAT plugin [8]. To obtain the vibration spectra a time-series with 8000 points is acquired and Fourier transformed using a Hamming-window. The statistical variation is reduced by averaging over 50 spectra. In addition the instantaneous amplitude, which is the modulus of the analytic signal and provides details about amplitude variations over time, is calculated and the different data sets are plotted in a sequence to provide an overview, how the vibrations change during the whole measurement. A Python program is used for automatizing the data acquisition and undertaking in situ data analysis.

4. Results

Table 2. Pitch vibration amplitudes of the monochromator under different operating conditions.

Bragg Axis	Chiller	Horizontal [nrad(rms)]	Vertical [nrad(rms)]
Noise Floor (Mono removed)	n.a.	39	41
Parked	Off	61	50
Parked	On	100	61
Floating	On	73	48
Normal Operation	On	166	61
Vibrating	On	280	93

The measured rms-vibration amplitudes for the monochromator under different operating conditions are summarised in table 2. To determine, if a vibration is caused by the monochromator, the monochromator is moved out of the beam and the noise floor is measured. The observed vibration level of 40 nrad(rms) corresponds well with the theoretical resolution limit of 35 nrad for the given experimental conditions and thus vibrations from other beamline-components or the machine itself are not dominant. When the monochromator is inserted, but the control system, the air supply to the

Bragg-axis and the chiller remain switched off, one observes a small increase in horizontal vibrations to 60 nrad(rms). Once the chiller is switched on this increase to 100 nrad(rms) and is mainly due to a 100Hz vibration (see inset in figure 3(a)). Switching on the air supply to the Bragg axis reduces vibrations to 73 nrad(rms), due to the reduced coupling between bearing and shaft. The most significant increase to 166 nrad(rms) is observed, when the motion control system is switched on too. Occasionally vibrations up to 280 nrad(rms) have been measured and are due to the system exciting the axis every few seconds as shown in figure 3(b).

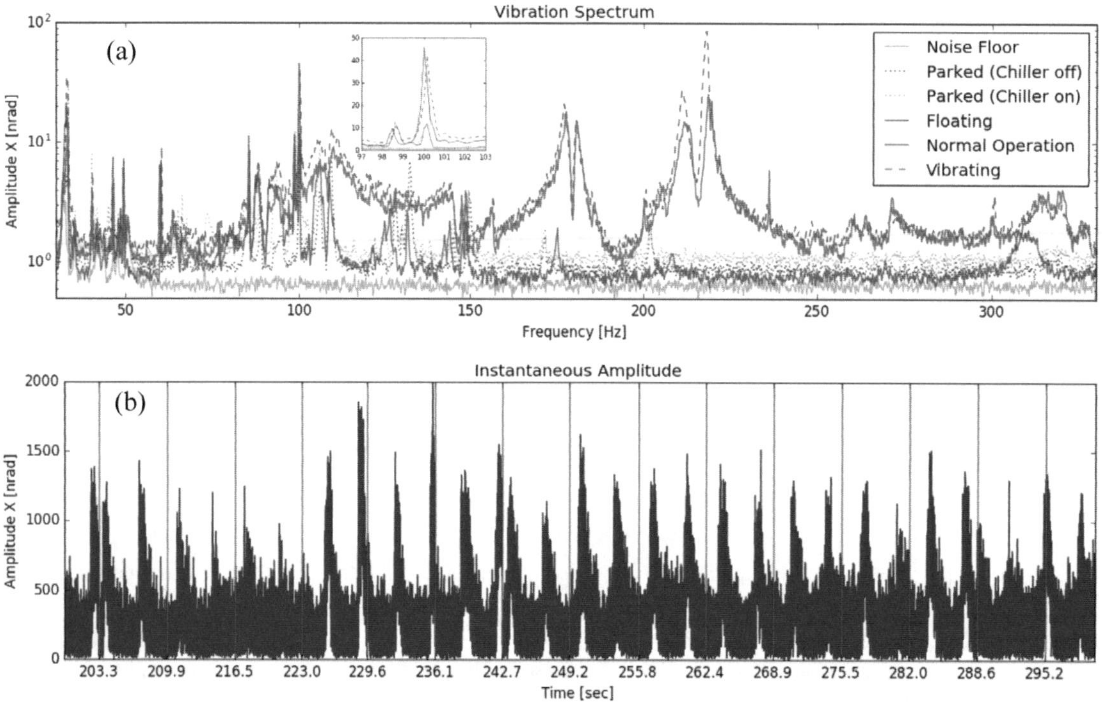

Figure 3. (a) Vibration spectra under different operating conditions. No significant vibrations are observed above 350 Hz. (b) Series of instantaneous amplitude measurements, indicating a regular excitation of the Bragg axis. The vertical lines separate the different data sets.

5. Summary

Fringe visibility measurements were used to show, how the coherence length can be adjusted with the frontend slits and an upper limit of ~200 μm was observed due to vibrations. Then vibration measurements with a resolution of 35 nrad in the spectral and time domain using an x-ray pencil beam and a photon counting detector were discussed and monochromator vibrations of ~73 nrad(rms) were measured (control system switched off), agreeing with the fringe visibility data. Finally the motion control system of the Bragg axis was identified as the main vibration source.

References

[1] C. Rau, U. Wagner, Z. Pešić, A. De Fanis 2011 *Phys. Status Solidi* **208** 2522
[2] B. Singh et al. 2009 *PAC Conference Proceedings*
[3] U.H. Wagner and C. Rau 2010 *AIP Conf. Proc.* **1234** 461
[4] V. Kohn, I. Snigireva, A. Snigirev 2000 *Pys. Rev. Lett.* **85** 2745
[5] U. H. Wagner, A. Parsons, J. Rahomaki, U. Vogt, C. Rau 2016 *AIP Conf. Proc.* **1741** 050022
[6] U. H. Wagner, S. Alcock, G. Ludbrook, J. Wiatryzk, C. Rau 2012 *AIP Conf. Proc.* **1437** 9
[7] R. Plackett, I. Horswell, E. N. Gimenez, J. Marchal et. al. 2013 *J. Instr.* **8** C01038
[8] M. L. Rivers 2010 *AIP Conf. Proc.* **1234** 51

IOP Conf. Series: Journal of Physics: Conf. Series **849** (2017) 012049 doi:10.1088/1742-6596/849/1/012049

Preparing for hard x-ray microscopy with Multilayer Zone Plates

Markus Osterhoff[1], Christian Eberl[2], Jakob Soltau[1], Hans-Ulrich Krebs[2]

[1]Institute for X-ray physics; [2] Institute for material physics, University of Göttingen, Friedrich-Hund-Platz 1, D–37077 Göttingen

E-mail: `mosterh1@gwdg.de`

Abstract. With Pulsed Laser Deposition, Multilayer Zone Plates can be fabricated to focus hard x-ray beams into 2D spots smaller than 10 nm. To put these optics into use for imaging applications, we have commissioned a new dedicated sample tower as a high-resolution module for the GINIX instrument, stationed at the P10 beamline at PETRA III. Here we summarise the motorisation and show first imaging benchmark results obtained with a "traditional" Fresnel Zone Plate. The first 2D continuous STXM scan using the new EigerX 4M detector at full 750 Hz speed is shown: a field of view of roughly about 1 µm squared has been recorded with 255 × 255 images within 96 seconds.

1. Introduction

Multilayer Zone Plates (MZPs) are basically an "optically thick" version of Fresnel Zone Plates (FZPs) [5, 6]; the fabrication using a sputter–slice technique – in our case pulsed laser deposition and slicing with a focused ion beam – allow for smooth outer most layers of width well below 10 nm and an optical thickness of several micro metre. Thus, the aspect ratio compared to FZPs based on lithography can be orders of magnitude larger, increasing efficiency even at high x-ray energies above 12.4 keV [3].

In recent years we have achieved sub-5 nm focusing in two dimensions at 7.9 keV [7]; also first imaging experiments have successfully been carried out [8]. Despite the focusing potential of outer most zones below 10 nm, resolution is limited to around 50 nm so far due to vibrations in the used versatile GINIX setup ("Göttingen Instrument for Nano-Imaging with X-rays"): This instrument has been designed for holography measurements with a resolution aim of about 100 nm [1, 2], with an emphasise on flexibility on that scale, not stability on the sub-10 nm scale.

Famous for its vivid and versatile construction, the GINIX setup will keep its basic design, but is currently upgraded for higher resolutions. On one end, sources of vibrations are located and removed; on the other end, the general purpose and tomography sample tower is exchanged with high resolution towers bringing optic (MZP or waveguide) and sample motorisation on a common base. In the next sections we introduce the new MZP tower and report on its commissioning. Current status of MZP fabrication is only touched, see [4] for more details. The imaging capabilities of the GINIX setup are summarised in section 5, before a short summary and outlook.

Figure 1. Photograph of the combined MZP–sample tower.

Table 1. Travel ranges and accuracies of MZP and sample motorisation.

purpose	motor	axes	travel range	accuracy
lens, lateral	SLC–1730	x,y,z	21 mm	few nm
lens, rotation	STT–12.7	yrot,zrot	±5°	≤mrad
sample, align	M–686	x,y	25 mm	sub-µm
sample, align	N–765	z	6.5 mm	few nm
sample, scan	P–733	y,z	30 µm	sub-nm

2. Sample Tower

The new MZP sample tower combines motorisation for optic and sample on a common optical bread board; the beam height is 125 mm above the breadboard. The sample stage consists of an x-y-stage (horizontal) M–686 by Physik Instrumente Karlsruhe (PI) and a z-stage (vertical) N–765 for alignment in the primary and focussed MZP beam; on top, a piezo scanner (P–733) allows for fast and accurate 2D scanning perpendicular to the MZP beam. The scan range is limited to 30 µm, but encoder values show excellent position accuracy of rms less than 0.2 nm for 1 µm scans faster than 1 Hz.

Alignment of the MZP is facilitated by three orthogonal stick-slip positioners by SmarAct (Oldenburg) for translations, and a piezo-driven Gimbal mount for rotations. See Tab. 1 and Fig. 1 for travel ranges and a photograph. Note that in this first iteration, robustness is prioritised over degrees of freedom; especially there are no sample rotations available, yet.

The piezo scanner allows for continuous 2D-movements with a servo cycle of 50 µs and is configured for horizontal forward–backward movements with intermediate vertical steps to fill the 2D area. During linear movements, the Eiger detector (Dectris AG, Dättwill) is triggered

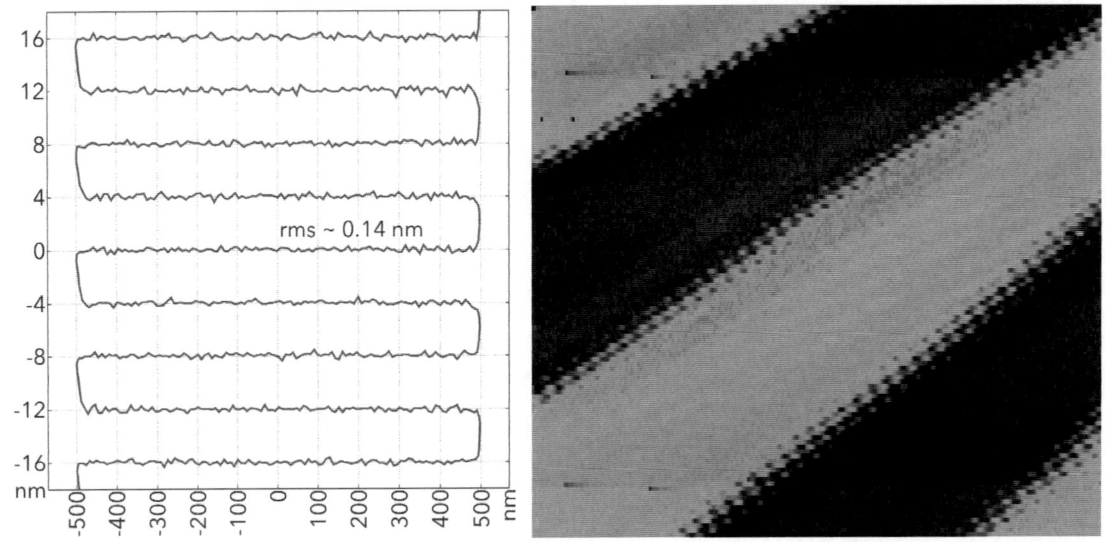

Figure 2. Continuous 2D scan: (left) encoder positions (vertical ROI) of a 1 μm/340 ms fast scan, showing a vertical rms accuracy better than 0.2 nm at 4 nm step size; (right) final STXM image of a Siemens star test pattern, taken with Eiger detector with frame frequency of 750 Hz. The field of view is about 1 μm.

for 750 Hz acquisitions. During commissioning (see following section), the field of view was set to a square of side length 1 μm, with 255 × 255 acquisitions during 96 s. The rather conservative overhead for acceleration and due to short exposure time of 1 ms was about 48 % but could be reduced.

3. Commissioning Results

First, the sample tower itself was tested with the P10 beamline's compound refractive lenses and a micro-focus at 13.8 keV. Gold colloids and semiconductor nanowires (InP) have been used as test samples. Afterwards, a Fresnel Zone Plate (FZP) with outermost zone widths of 30 nm – a courtesy of Christian David (Swiss Light Source) – was used to commission the MZP tower with a "gold standard" optic. Due to the large working distance of 34.5 mm, these experiments are less prone to collisions than with short-f MZPs. Also, on order sorting aperture (OSA) has been used to absorb higher diffraction orders.

2D raster scans in discrete stepping mode have been carried out on a Siemens star test pattern, alternating with CRL beam and FZP beam to find the overlap, and to find the FZP's focal plane. These step-by-step scans have been complemented with the full continuous measurements, that allowed for the first time 255 × 255 raster positions to be taken within less than two minutes at the GINIX setup. Parts of the piezo scanner's encoder positions and the Siemens star STXM image are shown in Fig. 2. Due to limited time during commissioning, the setup's full potential could not be tapped, yet. The z-encoder shows an rms accuracy of better than 0.14 nm during continuous scans of 1 μm/340 ms per line, which is less than four per cent of the step size (4 nm in vertical direction). Complementary interferometric distance measurements between FZP and sample holder are not yet implemented, but under progress.

4. Current Multilayer Zone Plates

Recent optimisation in MZP fabrication, especially in laser parameters and material choice have enabled apertures of up to 15 μm, an increase by one order of magnitude compared to first lenses, with a ML area coverage of about 96 per cent. The focal length at 8 keV has been increased to 0.5 mm, also a factor of ten larger than the first lens that was used as proof of concept for focusing. For more details on fabrication, see [3, 4].

One advantage of MZPs over lithographic FZP is the virtually unrestricted optical thickness. The initial glass fibres that are pulled to thicknesses of one to two micron posses a usable length of hundreds of micron; together with fine outer most zones of few nano metres, aspect ratios of several tens of thousand should be possible. This makes MZPs especially interesting for high energy applications: Neglecting absorption and volume effects, the ideal phase shift of π between the two neighbouring layers of Ta_2O_1 and ZrO_2 is obtained for an optical thickness of $0.5\ \mu m/keV^2$.

5. Hard X-ray imaging at GINIX

The MZP–sample tower is a new high-resolution module for the flexible GINIX setup (Göttingen Instrument for Nano-Imaging with X-rays), operated by the AG Salditt at the coherence beamline P10 at PETRA III. GINIX is open for user proposals and highly flexible – nevertheless, a couple of routine experiments are offered, like nano-SAXS (scanning SAXS), propagation based (waveguide holographic) imaging extended to holographic tomography, and also WAXS and fluorescence measurements are supported. For more details, see [1, 2][1].

6. Summary and Outlook

A new dedicated MZP–sample tower has been commissioned; continuous scanning and triggering of the Eiger detector has been carried out with a frame rate of 750 Hz, allowing STXM measurements with 255×255 images to be taken in less than two minutes. The MZP tower is mobile and will be further evaluated at the GINIX and other instruments.

Acknowledgements

We gratefully thank Michael Sprung for excellent support and granting commissioning beam time, Christian David for the FZP. We thank Bastian Hartmann, Peter Luley, Peter Nieschalk (IRP), Florian Döring and Volker Radisch (IMP) for outstanding support in construction and fabrication, and Tim Salditt for stimulating and encouraging discussions. We acknowledge funding by Deutsche Forschungsgemeinschaft through SFB 755 projects C12 and INF, and BMBF through grants 05K13MG4, 05K16MG2.

References

[1] Kalbfleisch S et al 2010 *AIP Conf. Proc* **1234**, 433–6
[2] Kalbfleisch S et al 2011 *AIP Conf. Proc* **1365**, 96–9
[3] Eberl C et al 2014 *Appl. Surf. Sci* **307**, 638–44
[4] Eberl C et al 2015 *XRM Proc.* (this volume).
[5] Schmahl G, Rudolph D and Niemann B 1980 *Quarterly Reviews of Biophysics*
[6] David C et al 20xx *Journal*
[7] Döring F et al 2013 *Optics Express* **21** 19311–23
[8] Osterhoff M, Eberl C, Döring F, Wilke RN, Wallentin J, Krebs HU, Sprung M, Salditt T 2015 *Journal of Applied Crystallography* **48**, 116–24

[1] and `http://photon-science.desy.de/facilities/petra_iii/beamlines/p10_coherence_applications/infrastructure/eh2_standard_micro_focus_setup_and_gini_x_setup/gini_x_setup/index_eng.html`.

Nanoscale imaging applications of soft X-ray microscope based on a gas-puff target source

P. W. Wachulak*[a], A. Torrisi[a], A. Bartnik[a], L. Wegrzynski[a], T. Fok[a], H. Fiedorowicz[a]

[a] Institute of Optoelectronics, Military University of Technology, Kaliskiego 2 Str., 00-908 Warsaw, Poland

* wachulak@gmail.com

Abstract. A compact microscope based on nitrogen double stream gas puff target soft X-ray source, which emits radiation in the water-window spectral range, at the wavelength of $\lambda = 2.88$ nm, is presented. The microscope, employing ellipsoidal grazing incidence condenser and a Fresnel zone plate objective, is capable of capturing images with a 60 nm spatial resolution and exposure time as low as a few seconds. Examples of different applications of the SXR microscopy, and its applicability for various fields of science, are presented and discussed.

1. Introduction

The water-window spectral range ($\lambda = 2.3$ - 4.4 nm) [1] is particularly suitable for biological imaging (of cells, membranes, lipids, DNA plasmids, etc.) because it is possible to obtain high contrast images due to a significant difference in absorption of water (oxygen) and carbon - biological sample constituents. Soft X-ray (SXR) microscopy has been successfully employed with different techniques, mainly in transmission mode, using diffractive optics, such as zone-plates [2], raster scanning of the sample by focused SXR beam [3] or as a contact microscopy [4]. Short wavelength sources, such as synchrotrons, allow for imaging in the water-window with spatial resolution of 10 nm [5]. However, for a worldwide spread and possible commercial applications, the development and use of a compact high resolution imaging systems is of major importance. In such systems the tradeoff between the exposure time and the size and complexity of the system is still challenging [6–8]. In this work we try to partially overcome presented limitations showing a SXR microscopy system, based on a double stream nitrogen gas puff target [9], capable of resolving 60 nm features, with a desk-top footprint.

2. Description of the SXR microscope

The experimental setup can be seen in figure 2. A nitrogen plasma was produced by focusing Nd:YAG laser pulses 4ns/ 0.74J (NL303HT, from EKSPLA) with an f = 25 mm focal length lens, onto a double stream gas puff target. The gaseous target is formed by two circularly symmetric nozzles. The inner and outer nozzles inject nitrogen (at pressure of 8 bar) and helium (6 bar), respectively. The nozzle timing and position in respect to the laser focal spot were optimized for highest photon flux at the sample plane [10]. The SXR radiation from the nitrogen plasma is collected and focused by an ellipsoidal, axi-symmetrical nickel coated condenser mirror (Rigaku Innovative Technologies Europe s.r.o.) that can efficiently reflect radiation from the SXR region. The nitrogen plasma emission was monochromatized using a 200 nm thick, free-standing titanium filter (Lebow). The sample is imaged using a Fresnel zone plate objective, Δr=30 nm, (Zoneplates Ltd., UK) onto a SXR sensitive back-illuminated CCD camera (1 Mpix, i-Kon, DO-934N, from Andor). The geometrical magnification of the system is ~220x (sample to CCD). The source photon flux at the sample plane was equal to $(7.9 \pm 0.2)\cdot 10^9$ photons/pulse at 2.88 nm wavelength, theoretically $(8 \pm 0.2)\cdot 10^{11}$ photons/(sec·0.1%BW). This high value allows to achieve short exposure times and high spatial resolution. The half-pitch resolution of the microscope was assessed by a well established knife edge (KE) test resulting equal to 60 nm. More details can be found in [10,11].

Figure 1. Photograph of the SXR microscope.

3. Imaging results and applications

In case of microscopy, the most important factor is the quality of images of real samples and its possible applications. Some examples of water-window images of various objects are depicted in figs. 2 and 3. Exposures of 1 to 600 SXR pulses were required, at 10 Hz source repetition rate, to acquire the SXR images. During the image acquisition, the CCD camera was cooled down to -20 °C to decrease its intrinsic thermal noise. Below we present some recent applications of this system.

3.1 Material science and nanotechnology applications

A silicon nitride membrane, 50 nm thick, was used as a support for a thin layer of saccharose ($C_{12}H_{22}O_{11}$), as depicted in figure 2a). Because of the use of monochromatic radiation, it was possible to measure the transmittance of a single (71.8 %) and double (52.3 %) saccharose layers, yielding the layer thickness of ~160 nm. More details about this work can be found in [10].

A SXR image of gold microcracks is depicted in figure 2b). A high resolution image was obtained with features visible beyond the diffraction limit of the ordinary visible light microscope. A small, contrast-enhanced inset shows a branched nanocrack in more detail due to higher spatial resolution achievable with the SXR radiation. The width of the crack was ~100 nm. More details is in [11].

3.2 Biological applications

Images of dehydrated adherent CT 26 fibroblasts, derived from colon carcinoma Mus musculus (strain BALB/c) are depicted in figure 3a-c). The cells, 20-30 μm in size, were imaged using a visible light microscope a) and are directly compared to the SXR image b). Their internal structure is visible due to a phase contrast in figure 3a) and as a modulation of the sample absorption in the SXR image –b). figure 3c) shows comparison of small areas of figure 3a) and 3b). The SXR image shows enhanced spatial resolution due to much shorter wavelength, beyond the capabilities of visible light microscopes. The internal and external structures of the cells, of the order of ~60 nm, are clearly visible in the SXR image, as indicated with arrows in figure 3c), while in the visible light micrograph they cannot be resolved. More details can be found in [11].

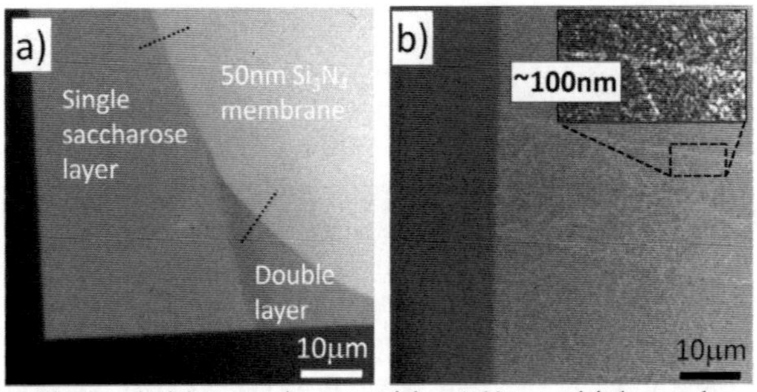

Figure 2. A 50 nm thick silicon nitride membrane with ~160 nm thick saccharose layer – a). A 100 nm thick silicon nitride membrane with 100 nm thick gold layer with microcracks - b).

CT 26 fibroblast, prepared without fixing procedure, are depicted in figure 3d-e). The first image was obtained with a traditional optical microscope (NA=0.7, 40x) and the second one shows a SXR image, acquired with 200 SXR pulses (22 sec exposure). The sample was prepared with a gradual dehydration in ethanol series (final concentration 70% EtOH), without any fixation procedure. Also in this case it is possible distinguish internal and external structures of the cells. The dehydration procedure by a carbon-rich ethanol washes causes, however, the sample to absorb too much SXR radiation.

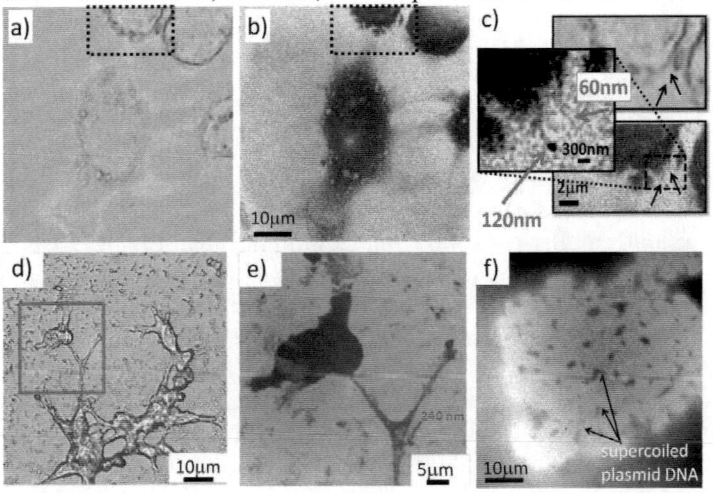

Figure 3. Comparison of visible light microscopy image – a, d) and SXR images – b, e) of dehydrated CT 26 fibroblasts, with and without fixing procedure, respectively. Image c) shows comparison of small areas and smallest features in the visible light microscopy image and the SXR image. The SXR image e) was obtained in the region indicated by a red square in figure d). A supercoiled pBR322 plasmid DNA is imaged in – f). All images, except c), the field of view (FOV) is 60x60 μm^2.

A supercoiled pBR322 plasmid DNA (4361bp), circular double-stranded DNA from Inspiralis, UK, was deposited on top of 50 nm thick Si_3N_4 membrane, from 100 ng / μl solution, dried in nitrogen atmosphere for 20 minutes and imaged - figure 3f). The aggregations are clearly visible; some of them are separated from the "DNA bulk" and vary in size from a few microns down to hundreds of nanometers. More information about those measurements is presented in [10].

4. Conclusions

In conclusion, we presented a water-window compact, desk-top microscope, based on nitrogen double stream gas puff target SXR source and Fresnel zone plate objective. The microscope allows capturing magnified images of the objects, with 60 nm half-pitch spatial resolution, exposure time as low as a few seconds, desk-top footprint, easy accessibility and simple operation. Such system does not require sample preparation, such as gold coating in case of SEM, exploits natural, optical absorption contrast in the water-window spectral range, is adequate for biological imaging and may be considered as a complementary imaging tool to the already well established techniques. It was also shown that this compact SXR microscope is capable of imaging variety of different samples from various fields of science and technology, including, but not limited to, biology, material science and nanotechnology. The versatility of such microscope may open the possibility of widespread and commercialization of such systems in the near future.

Acknowledgments

The research was supported by the Laserlab Europe Project (EU-H2020 654148), the EMJD Programme EXTATIC (No 2012-0033), the National Science Centre; award numbers UMO-2015/17/B/ST7/03718, UMO-2015/19/B/ST3/00435 and the National Centre for Research and Development, LIDER programme, award # LIDER/004/410/L-4/12/NCBR/2013. We would like thank to Dr. Daniel Adjei, Laboratoire d'Optique Appliquée, ENSTA Paris Tech, Ecole Polytechnique, CNRS, for preparing the supercoiled pBR322 plasmid DNA (4361bp). We thank also to Šárka Vondrová, Jana Turňová and Prof. Miroslava Vrbová from Czech Technical University in Prague, Czech Republic, for preparing the cell samples investigated with the SXR full field microscope.

References

[1] Da Silva L B, Trebes J E, Balhorn R, Mrowka S, Anderson E, Attwood D T, Barbee T W, Brase J, Corzett M and Gray J 1992 X-ray laser microscopy of rat sperm nuclei. *Science (New York, N.Y.)* **258** 269–71

[2] Schneider G, Niemann B, Guttmann P, Rudolph D and Schmahl G 1995 Cryo X-ray microscopy *Synchrotron Radiation News* **8** 19–28

[3] Kirz J, Jacobsen C, Lindaas S, Williams S, Zhang X A E and H M 1994 *Synchrotron Radiation in the Biosciences* (Oxford: Oxford University Press)

[4] Poletti G, Orsini F and Batani D 2005 Study of Multicellular Living Organisms by SXCM (Soft X-Ray Contact Microscopy) *Solid State Phenomena* **107** 7–10

[5] Chao W, Fischer P, Tyliszczak T, Rekawa S, Anderson E and Naulleau P 2012 Real space soft x-ray imaging at 10 nm spatial resolution *Optics Express* **20** 9777

[6] Takman P A C, Stollberg H, Johansson G A, Holmberg A, Lindblom M and Hertz H M 2007 High-resolution compact X-ray microscopy *Journal of Microscopy* **226** 175–81

[7] Johansson G A, Holmberg A, Hertz H M and Berglund M 2002 Design and performance of a laser-plasma-based compact soft x-ray microscope *Review of Scientific Instruments* **73** 1193

[8] Legall H, Blobel G, Stiel H, Sandner W, Seim C, Takman P, Martz D H, Selin M, Vogt U, Hertz H M, Esser D, Sipma H, Luttmann J, Hofer M, Hoffmann H D, Yulin S, Feigl T, Rehbein S, Guttmann P, Schneider G, Wiesemann U, Wirtz M and Diete W 2012 Compact x-ray microscope for the water window based on a high brightness laser plasma source *Optics Express* **20** 18362–9

[9] Wachulak P W, Bartnik A, Skorupka M, Kostecki J, Jarocki R, Szczurek M M, Wegrzynski L, Fok T and Fiedorowicz H 2013 Water-window microscopy using a compact, laser-plasma SXR source based on a double-stream gas-puff target *Applied Physics B: Lasers and Optics* **111** 239–47

[10] Wachulak P W, Torrisi A, Bartnik A, Adjei D, Kostecki J, Wegrzynski L, Jarocki R, Szczurek M and Fiedorowicz H 2015 Desktop water window microscope using a double-stream gas puff target source *Applied Physics B* **118** 573–8

[11] Wachulak P, Torrisi A, Nawaz M F, Bartnik A, Adjei D, Vondrová Š, Turňová J, Jančarek A, Limpouch J, Vrbová M and Fiedorowicz H 2015 A Compact "Water Window" Microscope with 60 nm Spatial Resolution for Applications in Biology and Nanotechnology *Microscopy and Microanalysis* **21** 1214–23

IOP Conf. Series: Journal of Physics: Conf. Series **849** (2017) 012051 doi:10.1088/1742-6596/849/1/012051

Introducing high efficiency image detector to X-ray imaging tomography

Kentaro Uesugi, Masato Hoshino and Akihisa Takeuchi

Japan synchrotron radiation research institute, JASRI/SPring-8, Kouto 1-1-1, Sayo, Hyogo 679-5198 Japan

ueken@spring8.or.jp

Abstract. A high efficiency X-ray image detector was introduced in an x-ray imaging micro-tomography system. The detector is a visible light conversion type which consists of phosphor screen, straight-fiber (1:1) optics and scientific CMOS device. The obtained CT images show high reproducibility of X-ray linear absorption coefficient of test material (Cu/Al concentric pattern). The x-ray intensity could be reduced about 10 % of previous system with lens-coupled image detector system.

1. Introduction

An imaging (full-field) X-ray microscope using Fresnel zone plate (FZP) has been developed in some facilities [1-3]. They achieve the spatial resolution of better than 1 micrometer. Increasing the efficiency of objective zone plate and image detector is essential to get better quality of images in x-ray imaging tomography. Those affect scan time, radiation damage on the sample. Concerning the image detector, the signal-to-noise ratio and frame rate are rapidly modified by cutting edge technology such as scientific CMOS (sCMOS) device. The image detector which commonly used in high energy X-ray region is visible light conversion type. The detector consists of phosphor screen, visible light optics and image device. The most used visible light optics (coupling techniques) are lens- and fiber-couplings (fiber optics plate, FOP). The advantage of FOP system is high efficiency (conversion gain) with high numerical aperture optics. In this study, we have introduced FOP-sCMOS detector in the imaging X-ray microscope optics.

2. System

The experiments were carried out at BL47XU in SPring-8. A schematic illustration of the experimental setup is shown in Figure 1. The developed system consists of an "in-vacuum type" undulator, a Si(111) double crystal monochromator cooled with liquid nitrogen, a double Rh-coated total-reflection mirror system to suppress the higher harmonics, X-ray microscope optics with sector condenser zone plate (sCZP), objective Fresnel zone plate (FZP), Zernike phase plate and an X-ray image detector. The zone material of the sCZP, the FZP, and the Zernike phase plate is tantalum. The parameters are summarized in Table 1. The inner diameter, outer diameter and pattern thickness of the Zernike phase plate are 151 μm, 159 μm and 0.96 μm, respectively. High precision stages are made by Kohzu precision Co., Ltd. All automatic translation stages have accuracy better than 1μm. The sample rotation stage (RD55-01) has been specially developed for high load capacity. The effective pixel size at 8 keV is about 90 nm/pixel. The detail of the X-ray microscope optics was described in elsewhere [4-6].

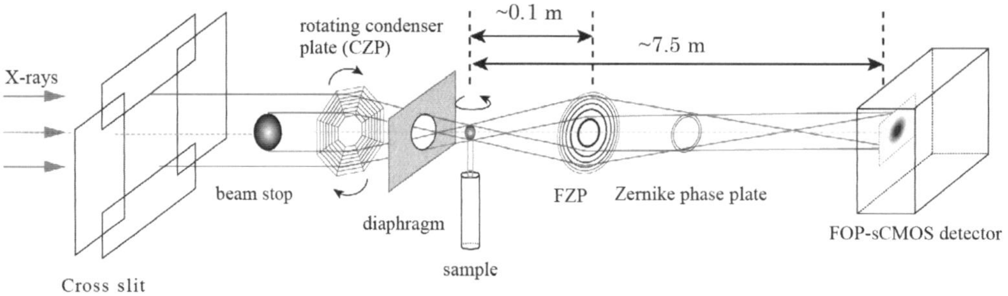

Figure 1. A schematic illustration of the system. Light source, double crystal monochromator and total reflection mirrors are not shown.

TABLE 1. Parameters of CZP and FZP. Both zone plates are made of Tantalum on SiC membrane (2µm-thickness). The ZPs were fabricated by means of the electron-beam lithography technique at NTT-AT, Japan.*This number shows the distance from CZP to sample which has some uncertainty because of the size of beam stop in front of the CZP.

	CZP (sector type)	FZP
Number of zones	2250	1550
Diameter (µm)	1000	310.0
Outermost zone width (nm)	100	50
Focal length at 8keV (mm)	*400	100
Thickness of zones (µm)	1.0	1.0

2.1. X-ray image detector

We have tested some detectors to fit the X-ray full field microscope optics. The specifications of detectors for imaging microscope were measured at experimental hutch 1 of BL20B2 in SPring-8 using 10keV X-ray, which is summarized in Table 2. All of the experiments were done under the "photon limited condition", therefore the S/N of the images was determined by the number of X-ray photon. Type (a), (b) and (c) has same sCMOS chip (FL-400, Hamamatsu Photonics) inside. Phosphor screen and sCMOS chip are connected by 1:1 fiber optics plate in type (a) and (b). Type (c) consists of a phosphor screen (P43 10um-thickness), a tandem lens system of two camera lens (two AI AF Nikkor 50mm f/1.4D) and sCMOS camera (C11440-22C, Hamamatsu Photonics). All of them have the magnification factor of one. Therefore, the effective pixel sizes are the same as pixel size of sCMOS chip. The type (a) has highest conversion gain and the highest modulation transfer function (MTF) (Figure 2). Therefore the type (a) detector (Hamamatsu Photonics, C12849-1015S) was used in this study.

Table 2. Characteristics of each detector.

	(a)	(b)	(c)
Optics type	FOP	FOP	Lens
Phosphor	$Gd_2O_2S:Tb^+$	$CsI:Tl^+$	$Gd_2O_2S:Tb^+$
Thickness (µm)	10	50	10
Exposure (msec)	20	10	60
Pixel size (µm)	6.5 x 6.5	6.5 x 6.5	6.5 x 6.5
Format	2048 x 2048	2048 x 2048	2048 x 2048
Frame rate (Hz)	30	30	100
Conversion gain (e-/photon)	22.5	20.2	3.5

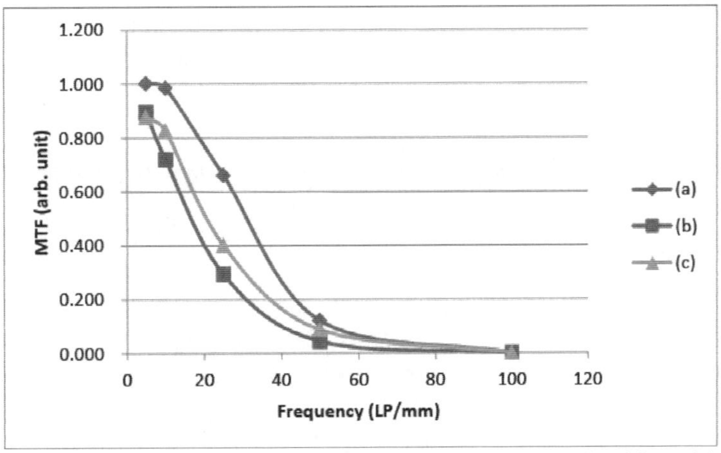

Figure 2. Modulation transfer function (MTF) of each detector summarized in Table 2. The MTF was measured at BL20B2 of SPring-8 using 10keV energy X-ray.

3. Results

3.1. Evaluation of the system

For the evaluation of the spatial resolution and quantitative properties of the X-ray linear absorption coefficient (LAC) measured by tomography, the artificial test patterns of Aluminium (Al) and copper (Cu) multilayer were fabricated at AIST Kansai [7,8]. Cu and Al were deposited onto rotating Al wire substrate by DC magnetron sputtering apparatus.

300 nm pitches of Al and Cu (Each line width is 150 nm.) are clearly resolved. It shows that the spatial resolution is higher than 300 nm. Potentially it would be around 200 nm because the spatial resolution (5% MTF) of the image detector is 2.2 pixels (197 nm in this case) and wobble of the rotation stage (Kohzu precision RD55-01) is less than 100nm.

As shown in Figure 2, FOP system has better MTF than lens system with the same phosphor screen. The line profiles of CT images also show the same result (Figure 3 (c)).

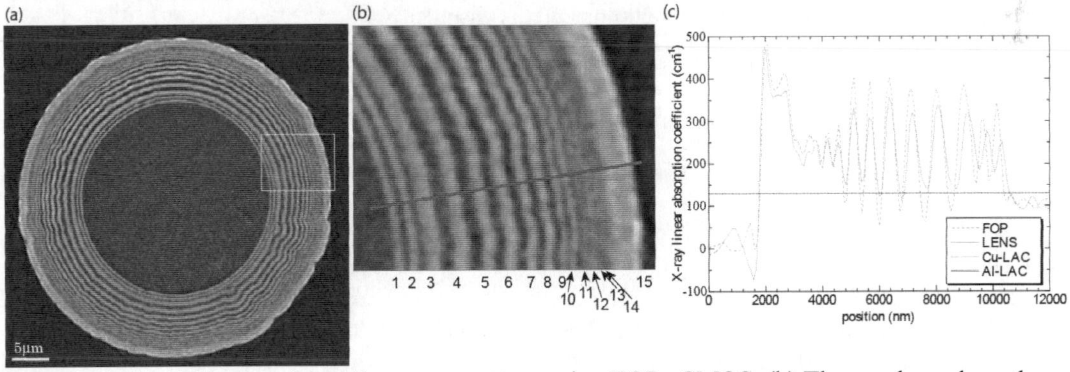

Figure 3. (a, b) CT image of multilayer test pattern using FOP-sCMOS. (b) The numbers show the thickness of Cu layer. 1,2: 250 nm, 3,4:500 nm, 5,6:400 nm, 7,8:300 nm, 9,10:200 nm, 11,12,13,14:150 nm, 15:1000 nm. (c) Line profiles of test pattern using lens-coupled and FOP detectors. Energy:8 keV, Pixel size:89.7 nm/pixel, Field of view:ϕ115 μm, Exposure time:250 msec/projection, Number of projection:900.

3.2. Demonstration

As a demonstration, a proturan head was imaged with the developed system. The proturan was fixed and dehydrate with Bouin Solution and ethanol, respectively. Although the scan time was only 15 minutes, the image could not obtain using Lens-coupled detector system because of the deformation of the sample during the scan. The proturan head was successfully observed by changing the system from lens-coupled detector to FOP detector. The X-ray intensity was reduced to 10% of lens-coupled detector and the scan time was reduced to 5minutes. Therefore the total dose was reduced to approximately 3% of the Lens-coupled detector system's.

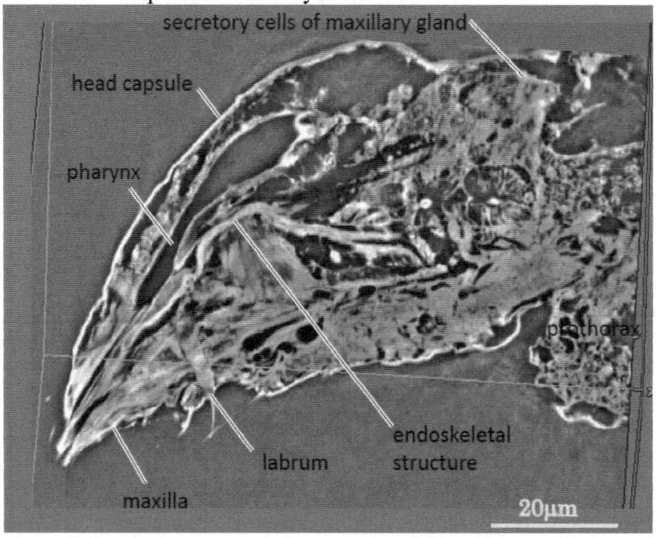

Figure 4. A sagittal view of the proturan head. Energy:8 keV, Pixel size:91.9 nm/pixel, Field of view:ϕ115 μm, Exposure time:250 msec/projection, Number of projection:900. Zernike phase plate was used to obtain phase contrast images.

4. Summary

Incident X-ray intensity has been reduced to 1/10 of previous system using high efficiency X-ray image detector (FOP-sCMOS) in X-ray imaging (full-field) microscope, which is essential for avoiding radiation damage on the samples. Furthermore the scan time is only 5 minutes for 900 projections. The sample deformation and drift of imaging optics could be negligible small during the short scan time. The spatial resolution is 200 nm in this case.

Acknowledgements

The proturan was provided by Dr M. Fukui (Ehime Univ. Japan). The experiments have been done with the approvals of SPring-8 Proposal Review Committee (Proposal No. 2016A1852, 2015B1058, 2013A1448). KU was supported by JSPS KAKENHI Grant Number 15H0569. We thank Tomoki Fukui for his technical supports.

References

[1] B. Lai et al., Rev. Sci. Instrum., **66,** 2287 (1995)
[2] K. Uesugi, A. Takeuchi and Y. Suzuki, Proc. SPIE, **6318**, 63181F (2006).
[3] C. Rau et al., Proc. SPIE, **6318**, 63181G (2006).
[4] A. Takeuchi, K. Uesugi, and Y. Suzuki: J. Phys. Conf. Ser., 186, 012020 (2009).
[5] A. Takeuchi, Y. Suzuki, and K. Uesugi: AIP Conf. Proc., 1365, 301 (2011).
[6] Y. Suzuki, A. Takeuchi, K. Uesugi, and M. Hoshino: AIP Conf. Proc., 1365, 160 (2011).
[7] K. Uesugi et al., AIP Conf. Proc. 705, pp. 1316-1319 (2004).
[8] S. Tamura et al., Journal of Synchrotron Radiation 9, pp. 154-159, (2002).

IOP Conf. Series: Journal of Physics: Conf. Series **849** (2017) 012052 doi:10.1088/1742-6596/849/1/012052

Hard X-ray multilayer zone plate with 25-nm outermost zone width

H Takano[1,5], K Sumida[1], H Hirotomo[1,6], T Koyama[2], S Ichimaru[3], T Ohchi[3], H Takenaka[4] and Y Kagoshima[1,*]

[1] Graduate School of Material Science, University of Hyogo, 3-2-1 Kouto, Kamigori, Ako, Hyogo 678-1297, Japan

[2] JASRI/SPring-8, 1-1-1 Kouto, Sayo, Hyogo 679-5198, Japan

[3] NTT Advanced Technology Corporation, 3-1 Morinosato, Wakamiya, Atsugi, Kanagawa 243-0124, Japan

[4] TOYAMA Co., Ltd., 3816-1, Kishi, Yamakita, Ashigarakami, Kanagawa 258-0112, Japan

[*]E-mail: kagosima@sci.u-hyogo.ac.jp

Abstract. We have improved the performance of a previously reported multilayer zone plate by reducing its outermost zone width, using the same multilayer materials ($MoSi_2$ and Si) and fabrication technique. The focusing performance was evaluated at the BL24XU of SPring-8 using 20-keV X-rays. The line spread function (LSF) in the focal plane was measured using a dark-field knife-edge scan method, and the point spread function was obtained from the LSF through a tomographic reconstruction principle. The spatial resolution was estimated to be 30 nm, which is in relatively good agreement with the calculated diffraction-limited value of 25 nm, while the measured diffraction efficiency of the +1st order was 24%.

1. Introduction

X-ray microscopy is a powerful analytical tool for nondestructive observation of the internal structure of an object with high spatial resolution. Recently, X-ray focusing devices such as Fresnel zone plates (FZPs) and Kirkpatrick–Baez mirrors have achieved focused beam sizes of around 10 nm, becoming key technologies in X-ray microscopy [1-5]. An FZP is a circular transmission diffraction grating consisting of alternate transparent and opaque zones. Its spatial resolution δ is determined by the outermost zone width (Δr_N), namely $\delta = 1.22\Delta r_N$, where N is the total number of zones. Hence, the narrower the width Δr_N, the better the spatial resolution that can be achieved. On the other hand, diffraction efficiency is primarily determined by the thickness T of the zones in the X-ray propagation direction. In the hard X-ray region and for the zone materials generally used, the optimal thickness is greater than a few microns. Therefore, in that region, the zone structure should have an extremely high aspect ratio $T/\Delta r_N$ (much larger than 100) to achieve high efficiency. However, zone structures with

[5] Present affiliation: Institute of Multidisciplinary Research for Advanced Materials, Tohoku University

[6] Present affiliation: SPring-8 Service Co., Ltd.

such high aspect ratios cannot currently be realized by electron beam lithography, which is the most frequently employed technique for FZP fabrication. To overcome this problem, multilayer zone plates (MZPs) were first proposed by Rudolph *et al.* [6] and subsequently investigated by several groups. An MZP is fabricated by stacking alternate layers on a cylindrical wire-like substrate and reducing the cross-section of the layers to an optimal thickness. It should be mentioned that there have been some outstanding works achieving sub-5 nm [7] and sub-30 nm spatial resolution [8].

We also developed an MZP fabrication technique and continued upgrading its optical performance [9–11]. A $MoSi_2$/Si multilayer system was chosen for the MZP because of its superior heat resistance and diffusion at the interface [12]. The multilayer pairs were deposited on a glass fiber using magnetron sputtering, and the diameter of the fiber was measured by a laser scanner before the multilayer deposition. Multilayer thicknesses were designed based on the measured fiber diameter. The sectioning, polishing, and thinning of the MZP was done mechanically.

The most important performance parameter of an MZP as an optical device is its spatial resolution δ and, as mentioned above, it is determined by the outermost zone width Δr_N. In our previous works, we attained increasingly narrower values of Δr_N (40 nm in [9, 10] and 30 nm in [11] and confirmed that a near diffraction-limited spatial resolution was attained. According to results calculated using coupled wave theory [13, 14], the local maximum diffraction efficiency as a function of Δr (the local grating half-pitch) starts to gradually decrease below ~30 nm for a fixed photon energy of 20 keV [15]. Thus, we have almost reached a spatial resolution limit in the case of the present cylindrical flat layer MZP structure, where neither a tilted, nor a wedged, nor a curved-type multilayer structure [16] is adopted. However, values of Δr below 30 nm may still yield a practical diffraction efficiency. In this paper, we report the focusing performance of our upgraded MZP with $\Delta r_N = 25$ nm using 20-keV X-rays.

2. Design parameters

A second important performance parameter for practical uses is the diffraction efficiency. It primarily depends on the zone thickness T because an MZP works as a phase zone plate under the assumption that the zone width Δr is sufficiently large to neglect dynamical diffraction effects. However, as Δr becomes smaller than ~30 nm, less X-rays propagate through the zones due to dynamical diffraction effects, which leads to a decrease in the diffraction efficiency. Figure 1(a) shows the maximum local diffraction efficiency of diffraction gratings consisting of the same multilayer material ($MoSi_2$ and Si) as a function of the pitch for 20-keV X-rays, which was calculated by using coupled wave theory [13, 14]. In this calculation, T was chosen in order to maximize the diffraction efficiency for each value of

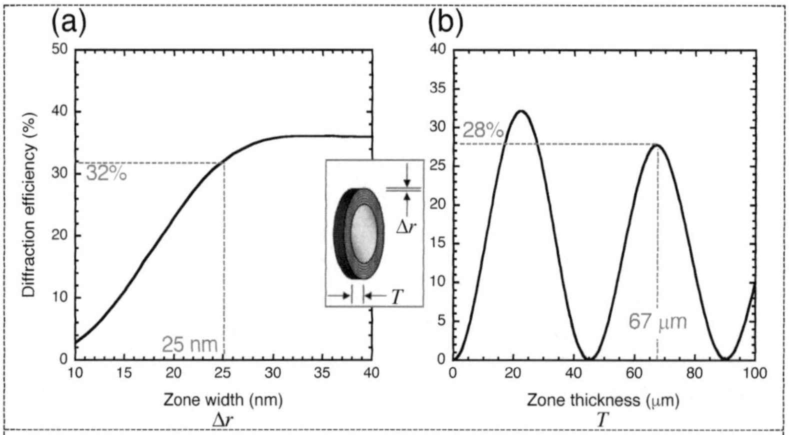

Figure 1. Maximum diffraction efficiency for a photon energy of 20 keV (calculated using coupled wave theory) as a function of (a) the grating half-pitch Δr and (b) the zone thickness T for $\Delta r = 25$ nm.

the grating pitch $2\Delta r$. A zone width Δr of 25 nm yields a diffraction efficiency of 32%, which is not so low as compared to the maximum value of 36.1%. Figure 1(b) shows the diffraction efficiency as a function of T. The thickness that produces the maximum diffraction efficiency is 22 μm, which is too small for the mechanical thinning. For this reason, we chose the second best diffraction efficiency of 28%, which occurs for T = 67 μm. The core of the MZP was a glass fiber, whose roundness was measured with a laser scanner (KEYENCE LS-7000). The fiber was rotated at 5° intervals, and for each angle, we measured the diameter at three points separated 5 mm and calculated the average. Taking again the average of all these average values, we obtained a fiber diameter of 50.99 ± 0.04 μm. The design parameters for this new MZP are shown in Table 1. Figure 2 shows the scanning electron microscopy (SEM) images of the fabricated MZP.

Table 1. Design parameters.

Layer material	$MoSi_2/Si$
Outermost zone width Δr_N	25.2 nm
Total number of zones N	178
Diameter D	60.4 μm
Zone thickness T	67 μm
Focal length f	24.5 mm (at 20 keV)

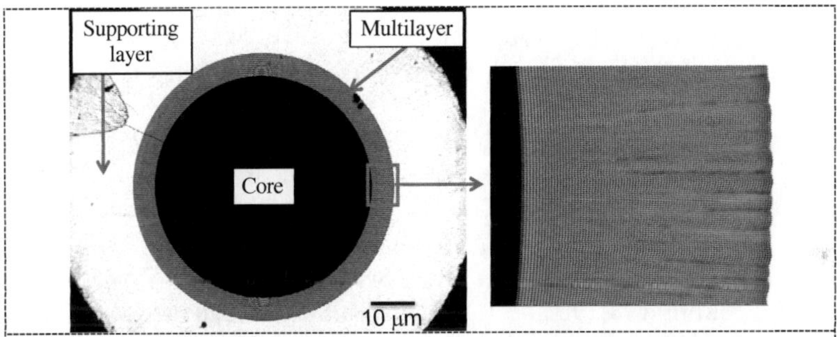

Figure 2. SEM image of the fabricated MZP.

3. Focusing performance

The line spread function (LSF) in the focal plane was measured at the BL24XU of SPring-8 using 20-keV X-rays. The optical system was the same as in [9]. Owing to the narrow annular aperture, subsidiary maxima of the point spread function (PSF) become much more intense than in the case of a circular aperture. As a result, the full-width at half-maximum (FWHM) of the LSF necessarily becomes much wider than FWHM of the PSF because an LSF convolutes the subsidiary maxima of the PSF. The calculated and measured LSFs are shown in Figure 3(a), where the latter was obtained using a dark-field knife-edge scan method [17]. The period of the ripples in the measured LSF is in good correspondence with that in the ideal LSF, although the base intensity is roughly two times higher. These oscillating structures come from the subsidiary maxima of the Airy pattern produced in the focal plane. The high base intensity may be caused by waviness of the outer zone layers. The distance between the first minima is 39 nm, which is in relatively good agreement with the diffraction-limited value of 36 nm. In order to evaluate the spatial resolution, one should not use the LSF but the PSF. In order to obtain the PSF from the measured LSF, we introduced a tomographic reconstruction principle [9, 10]. A single experimental LSF profile (dots in Figure 3(a)) was used as the representative LSF for the tomographic reconstruction. The half-distance between the first minima of the experimental PSF was estimated to be 30 nm, as shown in Figure 3(b). As the corresponding distance in the theoretical

PSF was 25 nm, we can conclude that near diffraction-limited performance was achieved by the MZP. Further, the measured diffraction efficiency of the +1st order was found to be 24%.

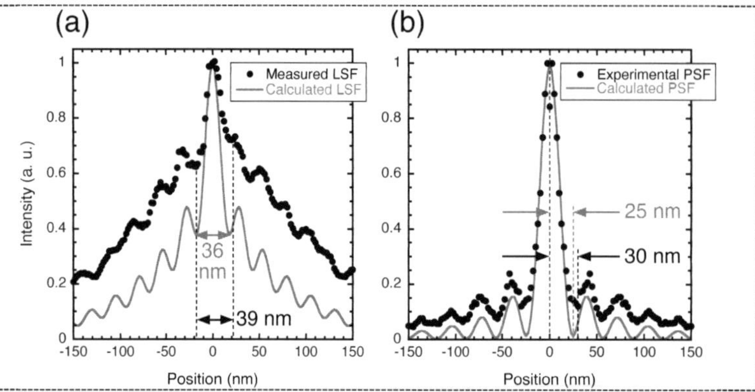

Figure 3. (a) Ideal (red line) and measured (black dots) LSF profiles. (b) Ideal (red line) and experimental (black dots) PSFs obtained through tomographic reconstruction.

Acknowledgments

This work was supported by JSPS KAKENHI Grant Number 24510125. The synchrotron radiation experiments were performed at the BL24XU of SPring-8 with the approval of the Japan Synchrotron Radiation Research Institute (JASRI. Proposal Nos. 2014A3200, 2014A3203, 2014B3200 and 2014B3203). We would like to thank Editage (www.editage.jp) for English language editing.

References

[1] Yan H, Conley R, Bouet N and Chu Y S 2014 *J. Phys. D: Appl. Phys.* **47** 263001
[2] Yamauchi K, Mimura H, Kimura T, Yumoto H, Handa S, Matsuyama S, Arima K, Sano Y, Yamamura K, Inagaki K *et al.* 2011 *J. Phys.: Condens. Matter* **23** 394206
[3] Rehbein S, Heim S, Guttmann P, Werner S and Schneider G 2009 *Phys. Rev. Lett.* **103** 110801
[4] Vila-Comamala J, Jefimovs K, Raabe J, Pilvi T, Fink R. H, Senoner M, Maaßdorf A, Ritala M and David C 2009 *Ultramicroscopy* **109** 1360
[5] Chao W, Kim J, Rekawa S, Fischer P and Anderson E. H. 2009 *Optics Express* **17** 17669
[6] Rudolph D, Niemann B and Schmahl G *Proc. SPIE* 1982 **316** 103–5
[7] Eberl C, Döring F, Liese T, Schlenkrich F, Roos B, Hahn M, Hoinkes T, Rauschenbeutel A, Osterhoff M, Salditt T *et al.* 2014 *Appl. Surf. Sci.* **307** 638–44
[8] Keskinbora K, Robisch A-L, Mayer M, Sanli U T, Grévent C, Wolter C, Weigand M, Szeghalmi A, Knez M, Salditt T *et al.* 2014 *Opt. Express* **22** 18440–53
[9] Koyama T, Takano H, Konishi S, Tsuji T, Takenaka H, Ichimaru S, Ohchi T and Kagoshima Y 2012 *Rev. Sci. Instrum.* **83** 013705
[10] Takano H, Konishi S, Koyama T, Tsusaka Y, Ichimaru S, Ohchi T, Takenaka H and Kagoshima Y 2014 *J. Synchrotron Rad.* **21** 446–8
[11] Hirotomo T, Takano H, Sumida K, Koyama T, Konishi S, Ichimaru S, Ohchi T, Takenaka H, Tsusaka Y and Kagoshima Y 2016 *AIP Conference Proceedings* **1696** 020017
[12] Takenaka H, Kawamura T, Ishii Y and Asagiri S 1995 *J. Appl. Phys.* **78** 5227–30
[13] Schneider G 1997 *Appl. Phys. Lett.* **71** 2242–4
[14] Maser J and Schmahl G 1992 *Opt. Commun.* **89** 355–62
[15] Koyama T, Ichimaru S, Tsuji T, Takano H, Kagoshima Y, Ohchi T and Takenaka H 2008 *Appl. Phys. Express* **1** 117003
[16] Yan H, Conley R, Bouet N and Chu Y S 2014 *J. Phys. D: Appl. Phys.* **47** 263001
[17] Suzuki Y, Takeuchi A, Takano H and Takenaka H 2005 *Jpn. J. Appl. Phys.* **44** 1994–8

Circular Unit Cell Gratings for X-ray Dark-Field Imaging

Matias Kagias[1,2], Amogha Pandeshwar[2], Zhentian Wang[1,2], Pablo Villanueva-Perez[1,2], Konstantins Jefimovs[1], and Marco Stampanoni[1,2]

1)Swiss Light Source, Paul Scherrer Institute, Villigen, Switzerland
2)Institute for Biomedical Engineering, University and ETH Zurich, Zurich, Switzerland

Abstract. Dark-field imaging has been demonstrated to provide complementary information about the unresolved microstructure of the investigated sample. The usual implementation of a grating interferometer, which can provide access to the dark-field signal, consists of linear gratings limiting the sensitivity to only one direction (perpendicular to the grating lines). Recently, a novel grating design, composed of circular unit cells, was proposed allowing 2D-omnidirectional dark-field sensitivity in a single shot. In this work we present a further optimisation of the proposed grating by changing the arrangement of the unit cells from a Cartesian to a hexagonal grid. We experimentally compare the two designs and demonstrate that the latter has an improved performance.

1. Introduction

Grating interferometry (GI) is a phase and scattering sensitive method capable of providing complementary information to standard absorption imaging of a sample [1, 2, 3]. A number of different possible applications have been unlocked by these capabilities, covering both the industrial [4] and medical fields [5, 6]. The main strength of scattering or so called dark-field imaging is that it can reveal information about the unresolved microstructure of the investigated sample [7, 8, 9, 10, 11]. The usual realisation of a grating interferometer is based on linear gratings. Such a design allows differential phase and dark-field sensitivity in only the direction perpendicular to the grating lines. In the case of samples with highly anisotropic scattering properties, a strong correlation of the retrieved dark-field signal and the relative orientation of the grating lines and the sample can be observed [12]. In order to achieve a more complete understanding of the unresolved microstructure multiple measurements by rotating the sample or the interferometer are necessary [13]. Recently, a method based on a single phase grating, capable of being sensitive to scattering in all the directions of the imaging plane has been proposed [14]. This was achieved by utilising a phase grating that is composed by a mosaic repetition of circular gratings. We will call this kind of design as square packing of circular unit cells. The circular design of the unit cells allows 2D-omnidirectional sensitivity, by repeating the unit cells the omnidirectional sensitivity can be obtained over the full field of view (FOV). The omnidirectional images that are obtained have a pixel matrix defined by the mosaic repetition of the unit cells. A square packing of the unit cells is undoubtedly the most straightforward implementation of such a grating design since the layout of the unit cells defines the pixel matrix. However, such an implementation comes with a number of limitations. A square packing of circles on a plane does not have an optimum fill factor since dead space is created in the corners of the cells. Moreover, each unit cell will interfere with its neighbouring, therefore it is favourable that the layout has as many axes of symmetry as possible in order to increase the uniformity of the resulting interference pattern. A hexagonal packing of the unit cells has a higher fill factor and more axes of symmetry than square packing. In this work, we fabricate such a grating and perform experimental comparison to demonstrate the improved performance of the design.

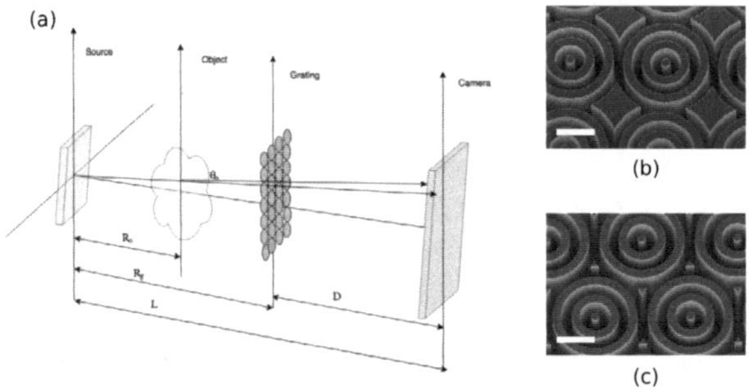

Figure 1. (a) Schematic of the used setup. (b) SEM image of the square packed grating (SPG). (c) SEM image of the hexagonally packed grating. The scale bar both in (b) and (c) is 10 μm

2. Materials and methods

The method in use has been fully documented and developed in [14]. The key points of [14] are repeated here for clarity. A general schematic representation of the setup is shown in figure 1 (a). The proposed grating design exploits its local circular symmetry in order to achieve a sensitivity in all the directions of the imaging plane. In more detail, the interference fringe is recorded with sufficient resolution and then analysed. The phase grating is designed to introduce a phase shift of $\pi/2$ for the X-ray wavelength λ. For a grating period g_1 the intensity interference pattern can be observed at the following distances from the grating

$$D_n = n\frac{g_1^2}{2\lambda}, n = 1, 3, ... \quad (1)$$

In order to observe the interference pattern at these distances sufficient coherence is required. There is a direct connection between beam coherence and fringe visibility of the interference pattern. When a scattering sample is introduced the angular distribution of the beam will be broadened which in turn will cause a reduction in the visibility of the pattern. By checking the visibility reduction at every angle of each circular pattern directional dark-field images can be produced.

The experimental procedure consists of recording two images, one with only the grating in the beam (flat), and one with the grating and the sample (sample). The analysis of the recorded images can be broken down into the following steps. Initially, the generated interference fringe is recorded. A unit cell finding algorithm is then used to detect and crop out each cell. The detection criterion is based on the maximum intensity that is observed in the centre of the cell due to the circular symmetry [15]. The next step consists of transforming each unit cell from cartesian to polar coordinates, symmetry criteria can be used to yield a more accurate interpolation than a direct conversion of coordinates. Finally, the visibility reduction is calculated for each angle.

The square packing (SPG) of the circular unit cells has a fill factor of $\frac{\pi}{4} \approx 0.785$. If the unit cells are arranged in a honeycomb manner (hexagonal packing) (HPG) the packing ratio increases to $\frac{\pi}{2\sqrt{3}} \approx 0.906$ leading to a higher unit cell density in the final image. This will result in a higher photon utilisation for dark field imaging.

To evaluate the performance of the proposed hexagonal packing in comparison to the previously used square packing we fabricated gratings with the two different designs. The two designs have an identical period g_1 and each unit cell contains the same number of periods. Both gratings are fabricated on the same Si chip, allowing easy processing but also a straightforward change of grating during the experiment. Scanning electron microscopy images (SEM) of the fabricated gratings can be seen in figures 1 (b) and (c).

3. Results

Two types of experiments were performed to compare the two grating designs. The first one was a straightforward comparison of the angular visibility of the flat images. The second comparison focused on the angular sensitivity that each grating can achieve for a known scattering sample. The sample was composed of 0.6 μm wide Au lines in Si that were repeated with a period of 1.3 μm. The Au lines are expected to scatter in the direction perpendicular to them. The sample was rotated covering the angles for 0° to 180° with a step size of 1°. By using a model proposed in [12] the direction of the lines and consecutively the scattering direction was estimated. We call this the most prominent scattering direction (MPSD). The standard deviation of the distribution of angles over the FOV was used as an accuracy merit of the estimation.

The experiments were performed at the TOMCAT beamline [16], of the Swiss Light Source, Switzerland. The selected energy was 25 keV to minimise the absorption of the Au lines. The gratings were fabricated in-house with the following parameters: period 5μm, 25μm unit to unit distance and 17μm depth. A 20 μm thick LuAG:Ce scintillator coupled with a 10 fold objective was placed 25 cm from the phase grating. The magnified visible light image was then recorded by a pco.edge 4.2 CMOS camera. The effective pixel size was 0.65 μm which allowed the recording of the 5 μm period fringes.

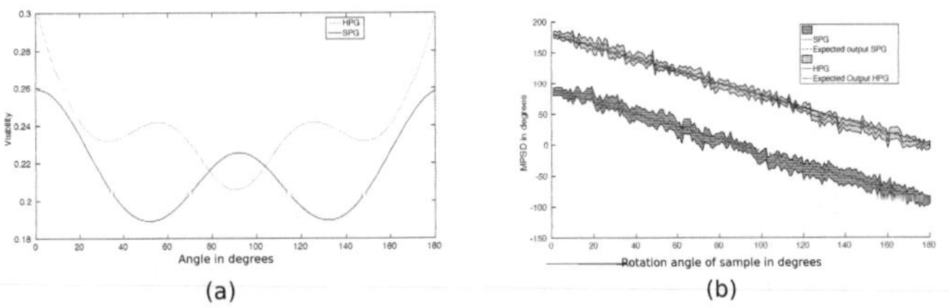

Figure 2. (a) Visibility curves for HPG and SPG. (b) MPSD distributions for SPG and HPG.

The calculated visibilities for each grating design are shown in figure 2 (a). Both visibilities depend on the angle, and two reasons contribute to this modulation. First, the photon source is not symmetric and therefore it is expected to have different visibilities in x and y directions. Secondly and most importantly the configuration of the unit cells play a vital role in the angular visibility. In the case of SPG (red line) we expect maxima at 0°, and 90° and minima at 45° and 135° which are observed in our experimental measurements. Analogously, for the HPG we expect maxima at 0°, 60°, and 120° and minima at 30°, 90°, and 150°. In figure 2 (b) we compare the estimated most prominent scattering direction (MPSD) again for both designs. It should be noted that the curves are shifted by 90° in order to make the comparison easier. The width of the bands follows the visibility curves, on average the HPG delivered 25% higher angular sensitivity compared to the SPG. To close, we show an imaging example of a trabecular bone sample embedded in PMMA. In figure 3 (b) and (c) we can see the estimated scattering angle for the SPG and HPG respectively. The noise is reduced in the case of the HPG. For completeness the transmission image is shown in figure 3 (a).

(a) (b) (c)

Figure 3. (a) Transmission and maximum scattering angles for SPG (b) and HPG (c) of trabecular bone sample.

4. Conclusion

Two gratings designs for 2D-omnidirectional dark-field imaging were compared. The hexagonally packed grating (HPG) had an improved visibility compared to the square packed one. Additionally, the angular sensitivity was improved by 25% by switching to the HPG. This change in the design of the grating can boost the performance of the method and increase its applicability in more demanding applications.

5. Acknowledgments

Part of this work has been supported by ERC Grant No. ERC-2012-StG 310005-PhaseX. We would like to thank Dr. Marios Georgiades for providing the bone sample.

References

[1] T. Weitkamp, A. Diaz, C. David, F. Pfeiffer, M. Stampanoni, P. Cloetens and E. Ziegler, Opt. Express **13**, 6296-6304 (2005).
[2] A. Momose, S. Kawamoto, I. Koyama, Y. Hamaishi, K. Takai, and Y. Suzuki, Jpn. J. Appl. Phys. **42**, L866-L868 (2003).
[3] C. David, B. Nöhammer, H. H. Solak, and E. Ziegler, Appl. Phys. Lett. **81**, 3287-3289 (2002).
[4] V. Revol, I. Jerjen, C. Kottler, P. Schütz, R. Kaufmann, T. Lüthi, U. Sennhauser, U. Straumann, and C. Urban, J. Appl. Phys. **110**, 044912 (2011).
[5] M. Stampanoni, Z. Wang, T. Thüring, C. David, E. Roessl, M. Trippel, R.A. Kubic-Huch, G. Singer, M.K. Kohl, and N.Hauser, Invest. Radiol. **46**, 801-806 (2011).
[6] D. Stutman, T.J. Beck, J.A. Carrino, and C.O. Bingham, Phys. Med. Biol. **56**, 5697-5720 (2011).
[7] Z. Wang, and M. Stampanoni, Phys. Med. Biol. **60**, 4123-4135 (2015)
[8] Z. Wang, N. Hauser, G. Singer, M. Trippel, R.A. Kubik-Huch, C.W. Schneider, and M. Stampanoni, Nat. Commun. **5**, 3797 (2013).
[9] M. Bech, A. Tapfer, A. Velroyen, A. Yaroshenko, B. Pauwels, J. Hostens, P. Bruyndonckx, A. Sasov and, F. Pfeiffer, Sci. Rep. **3**, 3209 (2013).
[10] S. Schleede, F.G. Meinel, M. Bech, J. Herzen, K. Achterhold, G. Potdevin, A. Malecki, S. Adam-Neumair, S.F. Thieme, F. Bamberg, K. Nikolaou, A. Bohla, A. Ö. Yildirim, R. Loewen, M. Gifford, R. Ruth, O. Eickelberg, M. Reiser, and F. Pfeiffer, PNAS **109**(44), 17880-17885 (2012).
[11] K. Hellbach, A. Yaroshenko, F.G. Meinel, A.Ö. Yildirim, T.M. Conlon, M. Bech, M. Mueller, A. Velroyen, M. Notohamiprodjo, F. Bamberg, S. Auweter, M. Reiser, O. Eickelberg, and F. Pfeiffer, Invest. Radiol. **50**(7), 430-435 (2015).
[12] T.H. Jensen, M. Bech, O. Bunk, T. Donath, C. David, R. Feidenhans'l, and F. Pfeiffer, Phys. Rev. B **55**, 3317 (2010).
[13] T.H. Jensen, M. Bech, I. Zanette, T. Weitkamp, C. David, H. Deyhle, S. Rutishauser, E. Reznikova, J. Mohr, R. Feidenhans'l, and F. Pfeiffer, Phys. Rev. B **82**, 214103 (2010).
[14] M. Kagias, Z. Wang, P. Villanueva-Perez, and M. Stampanoni, Phys. Rev. Lett. **116**, 093902 (2016)
[15] D.E Silva, Appl. Opt. **11**, 2613-2624, (1972).
[16] M. Stampanoni, A. Groso, A. Isenegger, G. Mikuljan, Q. Chen, A. Bertrand, S. Henein, R. Betemps, U. Frommherz, P. Bohler, D. Meister, M. Lange, and R. Abela, AIP Conf. Proc., 879, 848 (2007).

IOP Conf. Series: Journal of Physics: Conf. Series **849** (2017) 012054 doi:10.1088/1742-6596/849/1/012054

Current status of X-ray phase imaging at SPring-8: Toward 4D X-ray phase tomography for biological samples

Masato Hoshino, Kentaro Uesugi and Naoto Yagi

Japan Synchrotron Radiation Research Institute, 1-1-1 Kouto, Sayo, Hyogo 679-5198, Japan

hoshino@spring8.or.jp

Abstract. Measurement conditions in X-ray phase tomography using a grating interferometer were evaluated to achieve fast measurements toward 4D X-ray phase tomography for biological samples. X-ray phase tomographic images obtained from different exposure time and the number of projections were examined to determine required statistics for high density resolution while keeping a fast measurement. Application of a high-efficiency fiber-coupled X-ray imaging detector to X-ray phase tomography was also discussed from a comparison between the fiber-coupled and lens-coupled detectors.

1. Introduction

X-ray phase contrast imaging enables us to measure weak absorption materials such as a biological sample with much higher image contrast compared to the absorption-based imaging. In particular, use of an X-ray interferometer is helpful for the quantitative phase measurement. A grating interferometer is one of the X-ray interferometer, which has been used at not only synchrotron radiation facility but also the laboratory-based X-ray source [1 3].

At the bending magnet beamline BL20B2 in SPring-8, X-ray phase tomography using the grating interferometer has been routinely applied to quantitative measurements of biological samples. Recently, fast and high-throughput X-ray phase tomography has been developed to realize quantitative dynamic measurements. 4D X-ray phase tomography based on the simple uniaxial stretch-and-release cyclic scan was demonstrated by optimizing the measurement procedure [4]. To achieve such a fast and high-throughput measurement, some optimizations in the measurement condition and apparatus were carried out. In this paper, their results and discussions to achieve fast and high-throughput X-ray phase tomography using a grating interferometer are presented. In addition, application of a high-efficiency fiber-coupled X-ray imaging detector to the X-ray phase tomographic measurement is also discussed.

2. Measurement setup for X-ray phase tomography

An X-ray phase tomographic system using a grating interferometer was developed at the bending magnet beamline BL20B2 in SPring-8. An upstream experimental hutch located at 43m from the source was used. The X-ray energy was set to 18.8keV (3×10^9 photons/sec/mm^2). The grating interferometer is composed of a phase grating (G1) and an absorption grating (G2). G1 is made of Ta and the pattern thickness is 2.1μm, which produces a phase shift of $\pi/2$ at the X-ray energy of 18.8keV. G2 is made of Au and the pattern thickness is 16.6μm. The grating pitch is 10μm in both gratings. The distance between G1 and G2 was set to 1st order fractional Talbot distance. G2 was set with an inclined condition to increase the effective absorption at the grating pattern. The detail of the inclined condition is described elsewhere [5]. A sample was measured in a specially designed water bath filled with normal saline. In out measurement system, the biological sample is suspended in the water bath and rotated for the tomographic measurement. The thickness of the water bath is 15mm in which more than 70% X-ray photon are absorbed. A phase stepping method was used for the phase retrieval. G2 set on the Piezo stage was translated with a triangle-shaped analog signal from a waveform generator. Detail of the fast phase stepping method used in the measurement is described in our recent work [4].

3. Results and discussion

3.1. Application of high-efficiency X-ray imaging detector to X-ray phase tomography

To achieve an efficient measurement, we have introduced a fiber-coupled X-ray imaging detector (C12849-101U, Hamamatsu Photonics). Here, the performance of the fiber-coupled detector in X-ray phase tomography was evaluated by comparing it with a conventional lens-coupled detector. The fiber-coupled detector is composed of P43 scintillator with 10µm in thickness, a straight fiber and a scientific CMOS sensor. On the other hand, the lens-coupled detector is composed of a beam-monitor (AA40, Hamamatsu Photonics) with P43 scintillator with 10µm in thickness and a scientific CMOS camera (ORCA Flash4.0 ver.2, Hamamatsu Photonics). AA40 has a lens with the focal length of 50mm (Nikon, AI AF Nikkor 50mm f/1.4D). Another lens with the focal length of 50mm (Nikon AI Nikkor 50mm f/1.2S) is installed in front of the camera to realize a 1:1 optical system. This optical configuration in the lens-coupled detector has been typically employed in the X-ray micro-imaging at BL20B2. The effective pixel size and field of view in both detectors are 6.5µm and 13.3mm (H)×13.3mm (V), respectively. Measured conversion gains from a single X-ray photon to the digitalized counts in the fiber-coupled and lens-coupled detectors were 40 counts/photon and 8.8 counts/photon, respectively, while both sensors had a 16bit analog-digital convertor resolution. Although two different sCMOS sensors were used in this comparison, more than 4 times efficiency could be confirmed in the fiber-coupled detector at the same pixel size.

X-ray phase tomographic images of a formalin fixed mouse fetus measured with the fiber-coupled and lens-coupled detectors are shown in Fig. 1(a). The measurement condition is as follows; the number of projection 900, 5-step fringe scan for the phase retrieval and the exposure time 60ms. Total measurement time was less than 8min. The fiber-coupled detector provides a sharper image compared to that obtained from the lens-coupled detector. Magnified images at the rectangles indicated in Fig. 1(a) are shown in Fig. 1(b) and 1(c). Small features are clearly depicted in the fiber-coupled detector. On the other hand, standard deviation values at the background were measured to evaluate the density resolution. In this case, phase shift values obtained from tomography were converted into the mass density. Standard deviation values obtained from the fiber-coupled and lens-coupled detectors are 2.4mg/cm^3 and 1.9mg/cm^3, respectively.

To discuss the result shown above, a point spread function (PSF) of each detector was measured using a pinhole with 5µm in diameter. Each PSF is shown in Fig. 2. The PSF in the lens-coupled detector has higher background around the main peak as shown in the magnified profile while the full width at half maximum (FWHM) is almost equal between fiber-coupled (FWHM: 17µm) and lens-coupled (FWHM: 16µm) detectors. The higher background observed in PSF may be due to the lens system and influence the image quality as some sort of smoothing filter.

From these results and discussion, a fiber-coupled X-ray imaging detector is useful to realize high-efficiency detection, and it provides a sharper image in the current X–ray phase tomographic system. In contrast, the density resolution in the lens-coupled detector may be improved by the smoothing effect compared to the fiber-coupled detector. Although a spatial resolution in the grating interferometer is ultimately limited by the grating period, the optical system in the visible light conversion type X-ray detector also affect the image quality in the phase tomogram.

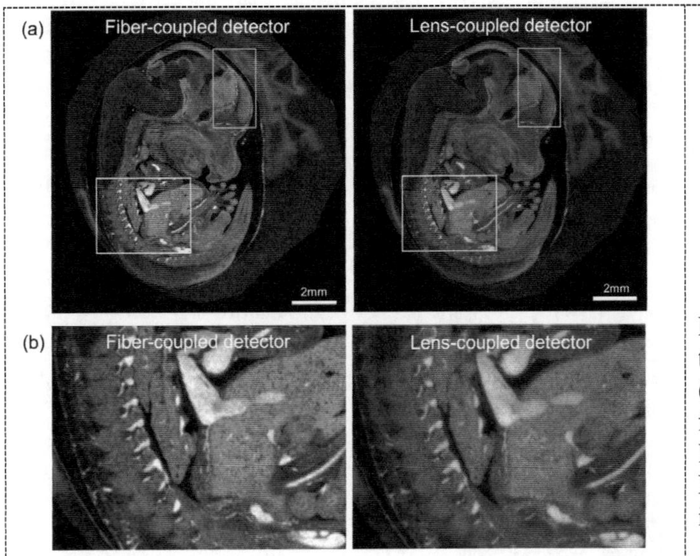

Figure 1. (a) X-ray phase tomographic images of a mouse fetus (gestation day 14) obtained from a fiber-coupled detector (left) and a lens-coupled detector (right). (b), (c) Magnified images at the rectangles indicated in 1(a)

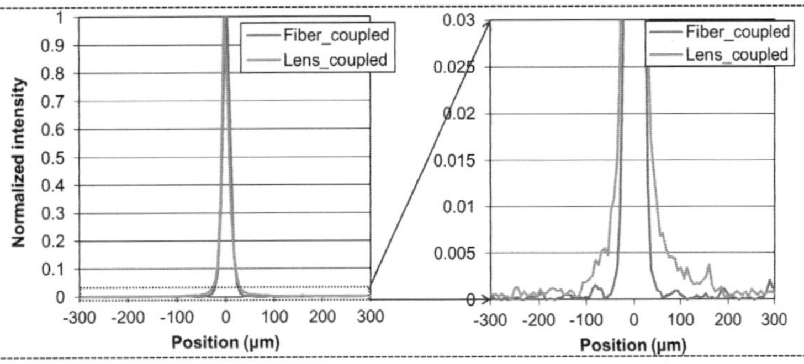

Figure 2. Point spread functions measured with fiber-coupled and lens-coupled detectors.

3.2. Evaluation of measurement condition in fast X-ray phase tomography

X-ray phase tomographic measurements of a standard sample were carried out to evaluate the measurement condition in fast X-ray phase tomography. Here, the fiber-coupled X-ray imaging detector was used. The standard sample was composed of three kinds of water solution. Each water solution was set in the pipette tube made of polypropylene. One of the X-ray phase tomographic images of the standard sample is shown in Fig. 3(a). Since the sample was measured in normal saline, the difference of the density between pure water and the background was approximately 7mg/cm^3. The standard deviation at the background was measured to evaluate the density resolution. The results obtained from five kinds of measurement conditions are summarized in Fig. 3(b). Insets shown in Fig. 3(b) are X-ray phase tomographic images of pure water in the pipette tube. As a result, standard deviation values at the background are drastically improved by increasing the exposure time from 10ms to 60ms. Line profiles measured at inset images in conditions 1 and 3 are shown in Fig. 3(c). In the exposure time of 10ms (estimated contribution of X-ray photons in single pixel of the detector: 237 photons/pixel/10ms), it is difficult to find a clear distinction between pure water and normal saline. On the other hand, the difference of the density is recognized in the exposure time of 60ms (estimated contribution of X-ray photons in single pixel of the detector: 1423 photons/pixel/60ms). In the current system, a binning pixel mode will be useful to shorten the exposure time while keeping high density resolution It will be helpful to ensure X-ray photon statistic. Furthermore, the density resolution may be improved by degrading the spatial resolution.

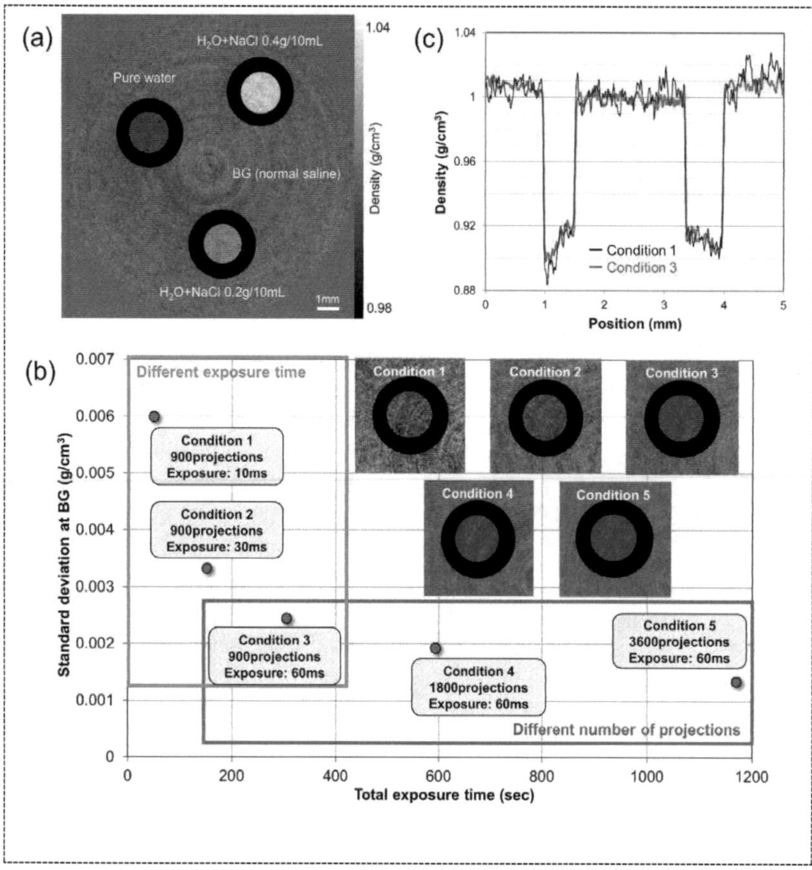

Figure 3. (a) X-ray phase tomographic image of three kinds of water solution in the pipette tubes. The gray-scale represents the mass density converted from phase shift values. (b) Standard deviation values measured at the background of the tomographic images obtained under five different measurement conditions. Insets show sectional images of pure water in the pipette tube under five different conditions. Difference of the density between pure water and background is approximately 7mg/cm^3. (c) Line profiles measured across the sectional images of pure water in conditions 1 and 3.

4. Conclusion

Measurement conditions in fast X-ray phase tomography with a grating interferometer when using a high-efficiency fiber-coupled X-ray imaging detector was evaluated toward 4D X-ray phase tomography. Although it was possible to acquire X-ray phase images in the photon limited condition, a certain level of photon statistic was required to ensure high density resolution. 4D X-ray phase tomography based on the knowledge obtained from this evaluation has already been employed in user applications.

Acknowledgments

These experiments were performed under the approval of the SPring-8 Proposal Review Committee (Proposal Nos. 2014B1329, 2015A1400, 2015B1323, 2016A1285). The experiments were technically supported by Mr. Tomoki Fukui.

References

[1] A. Momose, S. Kawamoto, I. Koyama, Y. Hamaishi, K. Takai and Y. Suzuki, Jpn. J. Appl. Phys. **42** (2003) L866.
[2] T. Weitkamp, A. Diaz, C. David, F. Pfeiffer, M. Stampanoni, P. Cloetens and E. Ziegler, Opt. Express **13** (2005) 6296.
[3] F. Pfeiffer, T. Weitkamp, O. Bunk and C. David, Nature Physics **2** (2006) 258.
[4] M. Hoshino, K. Uesugi and N. Yagi, Rev. Sci. Instrum. **87** (2016) 093705.
[5] M. Hoshino, K. Uesugi, T. Tsukube and N. Yagi, J. Synchrotron Rad. **21** (2014) 1347.

IOP Conf. Series: Journal of Physics: Conf. Series **849** (2017) 012055 doi:10.1088/1742-6596/849/1/012055

Improvement of quantitative performance of imaging x-ray microscope by reduction of edge-enhancement effect

A Takeuchi[1], K Uesugi[1] and Y Suzuki[1,2]

[1]Japan Synchrotron Radiation Research Institute (JASRI) / SPring-8
[2]Graduate School of Frontier Sciences, University of Tokyo

take@spring8.or.jp

Abstract. A Fresnel zone plate (FZP) with a gradually-decreasing zone thickness from the centre to the outer region, called as an Apodization FZP (A-FZP) has been developed for the purpose of the improvements of imaging properties of imaging (full-field) x-ray microscope. Diffraction efficiency distribution of a A-FZP has a sloped-shouldered shape like a Gaussian, resulting a Gaussian-like pupil and Gaussian-like point spread function without any side peaks. Using this A-FZP, imaging properties at the peripheral region of the field of view was remarkably improved by decreasing edge-enhancement contrast and ringing which have been serious problem on the imaging properties with the conventional FZP objective optics.

1. Introduction

High-resolution x-ray computed tomography (CT) by using imaging (full-field) x-ray microscope optics with hollow-cone illumination has been developed (Fig. 1) [1-3], and is now routinely used for various kinds of user experiments at SPring-8. By using a Fresnel zone plate (FZP) as an objective, achieved spatial resolution for three-dimensional CT imaging is approximately 100 nm [4]. For the purpose of quantitative CT measurement, both a high linearity of image contrast and a large field of view are severely required as well as high spatial resolution. In the present situation, however, there remained a serious problem in the imaging properties of the optical system. That is a kind of false image contrast at the peripheral region of the field of view caused by an edge-enhancement and a periodic pattern noise called as a ringing. These problems can be removed either by employing an incoherent illumination or by using only a paraxial region of the optical system. However, in a FZP optics, neither of those is accepted because of necessity of diffraction order sorting. One of the solution of this problem is to employ a Gaussian beam optics by using an apodization filter, which reduces the transmittance of the objective pupil from the central region to the peripheral region gradually. As a result, the pupil function of the objective becomes a sloping-shoulder-shaped function like a Gaussian. Therefore, its point-spread function (PSF) will also have a Gaussian-like shape without any side-peaks. Figure 2 shows results of computer simulation calculating the image profile of a test object under the on/off-axis condition of imaging microscope optics. Upper part of Fig. 2 shows the case of using a conventional objective lens. When an object exactly locates on the optical axis (on-axis condition, upper-left of Fig. 2), calculated image contrast shows of course an ideal image contrast. In the case of off-axis condition (upper-right of Fig. 2), however, image characteristics are seriously deteriorated by fringe patterns due to ringing and edge-enhancement effect. Because such phenomena are caused by the interference between the object and the objective pupil, even an ideal objective lens occurs such edge-enhancement and ringing at periphery of field of view. On the other hand, in the case of using an apodization lens, edge-enhancement and ringing are almost disappeared even in the extreme off-axis condition (lower-right of Fig. 2). It is because the obscure pupil profile of the apodization objective has an effect of reducing the amplitude of the interference fringes causing edge-enhancement and ringing.

We have newly developed an FZP objective with gradually decreasing zone thickness from the center to the outer regions, called as an Apodization FZP (A-FZP) [5]. The diffraction efficiency of A-FZP has a sloping-shoulder-like distribution, which acts as an apodization objective lens.

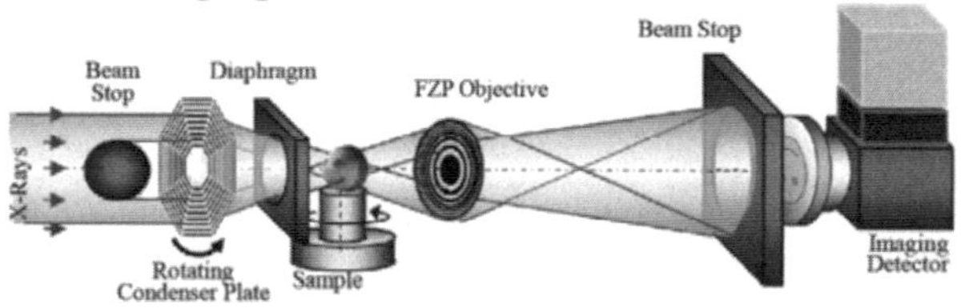

Figure 1. Schematic diagram of imaging x-ray microscope optics with hollow-cone illumination for high-resolution x-ray CT.

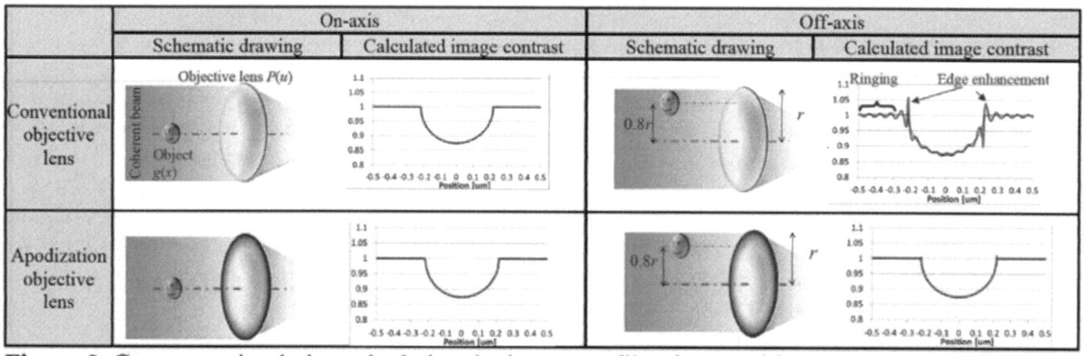

Figure 2. Computer simulation calculating the image profile of a test object under the on/off-axis condition of imaging microscope optics. Calculated and ideal image contrasts are shown with solid lines and with broken lines, respectively. In this simulation, coherent illumination with 8 keV x-rays, a tantalum sphere object with 0.5 μm diameter, and an ideal objective lens with a circular aperture of radius r are supposed. Left two and right two represent when the off-axis distance of the object is 0 (on-axis) and is $0.8r$ (extremely off-axis), respectively.

2. Apodization FZP

Conceptual drawings of A-FZP is shown in Fig. 3. Conventional FZP is also shown for comparison. Conventional FZP has an even zone thickness. Its cross-sectional distribution of the 1st order diffraction efficiency representing a pupil function shows a rectangular shape, and its PSF is well-known as a Bessel function (or a Sinc function) with side peaks. Such side peaks cause the edge-enhancement and ringing at the peripherally of the field of view. On the other hand, A-FZP has a gradually-decreasing zone thickness, resulting a sloping-shouldered pupil function like a Gaussian. Therefore, its PSF has also a Gaussian-like shape without any side peaks.

An A-FZP was fabricated with electron-beam lithography and reactive ion etching techniques at NTT Advanced Technology (NTT-AT, Japan) . Gradually decreasing zone structure (in this case, right-end one in Fig. 3) was realized by using a "micro-loading effect", that is an inherent phenomenon observed in the dry etching process that the etching rate decreases with the reduction of pattern width [6]. This effect causes non-uniformity of etching depth between narrow and wide patterns. Typical parameters of the fabricated A-FZP are as follows; zone material is tantalum deposited on a silicon-carbide membrane with 2 μm thick, thickness of the central zone is 1 μm, outermost zone width is 50 nm, radius of the zone structure is 310 μm. Measured result of the diffraction efficiency distribution of the 1st order at 8 keV x-rays is shown in Fig. 4. Diffraction efficiency decreases from the central region to the peripheral region, which will realize an apodization objective lens. Focusing properties were carefully measured by using a knife-edge scan test in a microbeam optics. Upper part of Fig. 5 shows the focused beam profile of a conventional FZP with 100 nm outermost zone width for comparison. And lower part of Fig. 5 shows that of an A-FZP with the outermost zone width of 50 nm. In spite of the decreasing zone structure of the A-FZP, nearly-diffraction-limited focal spot was obtained. Moreover, focal spot produced by the A-FZP shows almost no side peaks whereas that of the conventional FZP clearly shows the side peaks of Bessel function.

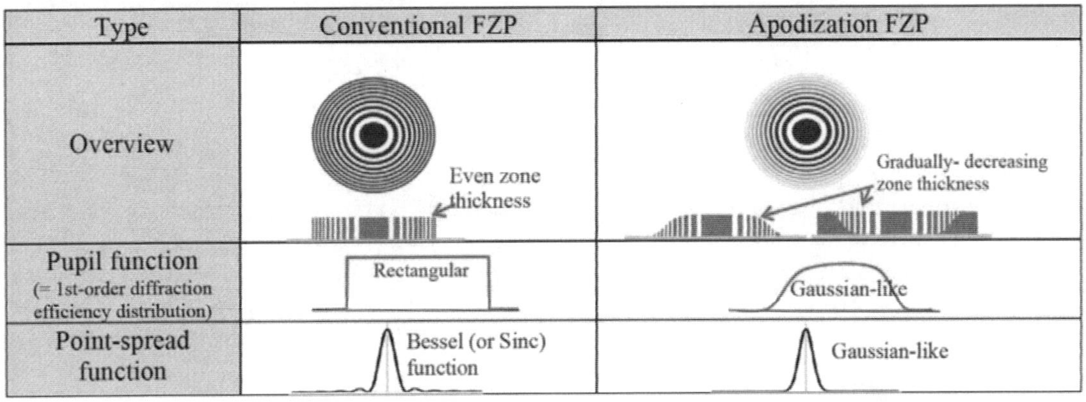

Figure 3. Comparison between a conventional FZP and an apodization FZP.

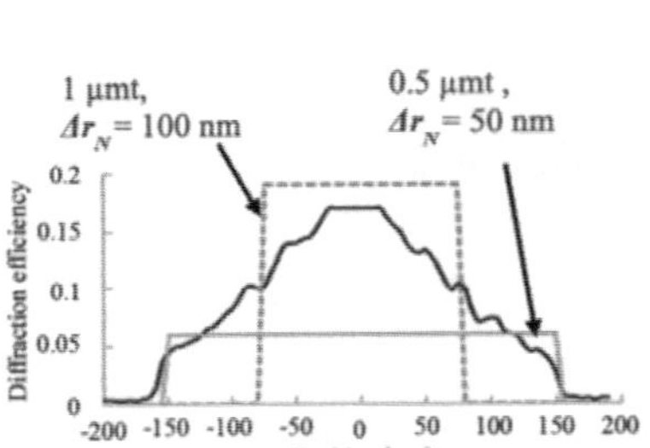

Figure 4. Measured 1st order diffraction efficiency distribution (8 keV). Theoretical values of 1μm and 0.5μm thick FZPs are also shown as broken and grey lines, respectively.

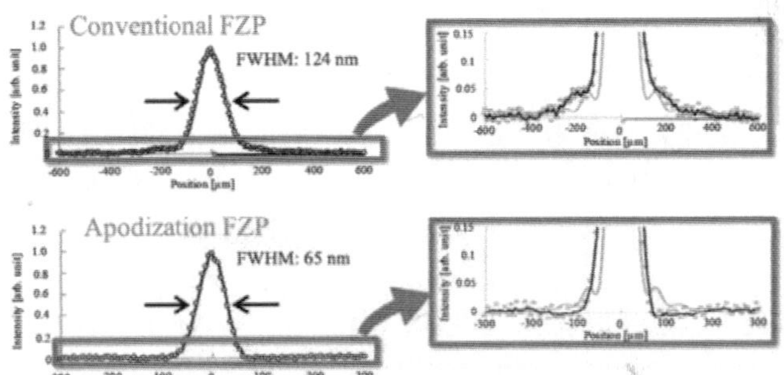

Figure 5. Focused beam profiles measured by knife-edge scan test (8 keV), of conventional FZP with the outermost zone with of 100 nm (above) and of apodization FZP with the outermost zone width of 50 nm (below). Magnified views of side peaks are also shown. Grey lines show the theoretical values of the conventional FZPs. Scan step is 6.25 nm, dwell time is 0.1 sec.

3. Experiments and results of Imaging x-ray microscope / CT

Using the A-FZP as the objective of the imaging x-ray microscope optics, performance test of the high-resolution x-ray CT has been carried out at the undulator beamline BL47XU of SPring-8. The optical setup is shown in Fig. 1, and the details are expressed elsewhere [1-3]. For comparison, CT experiments using the conventional FZP objective were also carried out for the same samples. A concentric Al/Cu multilayer pattern deposited on an Al core wire (Fig. 6(a)) was used as a test sample for the evaluation. Figures 6(b) and 6(c) show the cross-sectional CT images and the line profiles measured by using a conventional FZP objective and by using an A-FZP objective, respectively. In Fig. 6(b), CT image is deteriorated by edge-enhancement effect. The effects are seen at the boundaries of core wire / layer and cover layer / air (shown as arrows). On the other hand, in Fig. 6(c), such effect is rarely seen, and the line profile represents more precise linear absorption coefficient (LAC).

Figure 7 shows the CT images of Kilabo meteorite. While streak patterns and edge-enhancement contrasts are seen in Fig. 7(a) obtained by using a conventional FZP, such false image contrasts are almost disappeared in Fig. 7(b) obtained by using an A-FZP. Comparing their histograms of CT values, Fig. 7(b) shows clearly separated two peaks whereas Fig. 7(a) shows only a single peak with asymmetric lobe. It means quantitative measurement of LAC is improved by employing A-FZP.

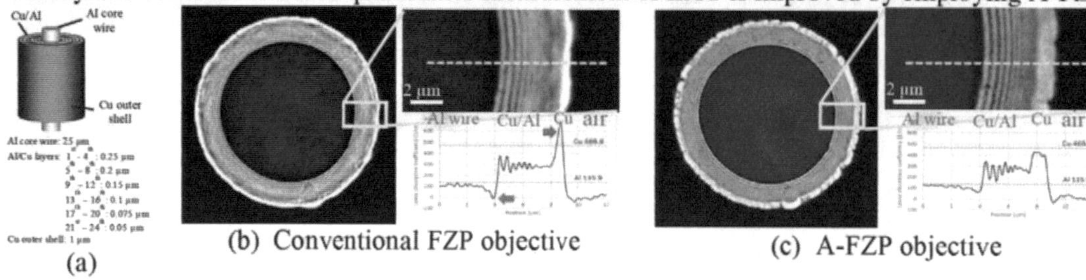

Figure 6. Concentric Al/Cu multilayer pattern sample and its CT images; (a) schematic drawing and parameters of the sample, (b) and (c) whole CT image (left), its partially magnified image (upper-right), and its line profile (lower-right) measured by using conventional FZP, and by using A-FZP as the objective, respectively. X-ray energy is 8keV, voxel size is 83.7 nm, exposure time for single image acquisition is 0.5 sec, and CT scan time is 30 min for 1800 projections.

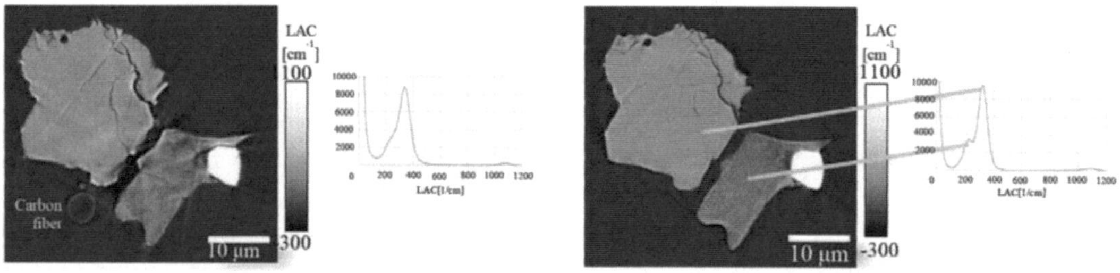

Figure 7. CT images (left) and its CT value histograms (right) of a Kilabo meteorite; (a) measured by using a conventional FZP and (b) by using an A-FZP as the objective. X-ray energy is 8keV, voxel size is 83.7 nm, exposure time for single image acquisition is 0.5 sec, and CT scan time is 30 min for 1800 projections.

4. Conclusions

Apodization FZP, which has gradually decreasing zone thickness from the center to the peripheral region, have been developed to realize a Gaussian beam optics at the x-ray region. Nearly diffraction-limited focal spot size of 65 nm was obtained. Employing the A-FZP as an objective lens of imaging microscope optics realized a remarkable improvement on the imaging properties by reducing edge-enhancement and ringing.

Acknowledgements

The authors thank Tomoki Fukui for his technical supports. The experiments were performed at the BL20XU and BL47XU in SPring-8 under the approval of the Japan Synchrotron Radiation Institute (JASRI) (Proposal Nos. 2014B1153, 2014B1464, and 2016A1359).

References

[1] A. Takeuchi, K. Uesugi, H. Takano, and Y. Suzuki, Rev. Sci. Instrum. 73, 4246 (2002).
[2] K. Uesugi, A. Takeuchi, and Y. Suzuki, SPIE 6318, 63181F (2006).
[3] A. Takeuchi, K. Uesugi, and Y. Suzuki, J. Phys. Conf. Series 186, 012020 (2009).
[4] R Mizutani, R. Saiga, S. Takekoshi, et. al., J. Microsc. 261, 57 (2016).
[5] A. Takeuchi et al. submitted to J. Syncrotron Radiation.
[6] M. Oda, A. Ozawa, and H. Yoshihara, J. Vac. Sci. Technol. B 11(1), 37 (1993).

IOP Conf. Series: Journal of Physics: Conf. Series **849** (2017) 012056 doi:10.1088/1742-6596/849/1/012056

Nanotomography endstation at the P05 beamline: Status and perspectives

I Greving[1], M Ogurreck[1,4], F Marschall[2], A Last[3], F Wilde[1], T Dose[1], H Burmester[1], L Lottermoser[1], M Müller[1], C David[2] and F Beckmann[1]

[1]Helmholtz-Zentrum Geesthacht, Institute of Materials Research, Germany
[2]Paul Scherrer Institut, Laboratory for Micro- and Nanotechnology, PSI, Switzerland
[3]Karlsruhe Institute of Technology, Institute of Microstructure Technology, Germany
[4]Diamond Light Source Ltd, Didcot, UK

imke.greving@hzg.de

Abstract. The Imaging Beamline IBL/P05 at the DESY storage ring PETRA III, operated by the Helmholtz-Zentrum Geesthacht, has two dedicated endstations optimized for micro- and nanotomography experiments [1-3]. Here we present the status of the nanotomography endstation, highlight the latest instrumentation upgrades and present first experimental results. In particular in materials science, where structures with ceramics or metallic materials are of interest, X-ray energies of 15 keV and above are required even for sample sizes of several 10 μm in diameter. The P05 imaging beamline is dedicated to materials science and is designed to allow for imaging applications with X-ray energies of 10 to 50 keV. In addition to the full field X-ray microscopy setup, the layout of the nanotomography endstation allows switching to cone-beam configuration. Kinematics for X-ray optics like compound refractive lenses (CRLs), Fresnel zone plates (FZP) or beam-shaping optics are implemented and the installation of a Kirkpatrick Baez-mirror (KB mirror) system is foreseen at a later stage of the beamline development. Altogether this leads to a high flexibility of the nanotomography setup such that the instrument can be tailored to the specific experimental requirements of a range of sample systems.

1. Introduction

X-ray imaging is a vital tool to study the structure of materials for a wide range of sciences e.g. medicine, biology, archaeology and geology [1,7,8]. Also for engineering materials this technique is important, allowing for a better knowledge of the inner structure, the material porosity or crack propagation, helping to understand failure mechanisms and thus optimizing the production process. Standard micro CT routinely allows resolutions down to about 1 μm.

A range of scientific questions however, require a higher spatial resolution in order to resolve structures in the nanometer range. Scanning electron microscopy allows imaging in this size regime, but is limited to surface analysis. Transmission electron microscopy (TEM) however requires very thin samples of a few hundred nanometres in thickness and only gives very limited information about the bulk structure of a material. Focused Ion Beam (FIB) tomography on the other hand can offer resolutions of better than 5 nm [8] and offers 3D information. Issues like curtaining effects and long-term stability however limit this technique to sample diameters of below 10 μm. In addition no in situ measurements of the sample can be performed and the sample is destroyed during the layer – by – layer decomposition.

X-ray microscopy in contrast is a non-destructive method to acquire 3D information of the bulk sample with resolutions down to below 50 nm in the hard X-ray regime. In addition, the high flux at synchrotron sources enables good time resolution and the high coherence opens up a range of phase contrast methods for x-ray microscopy or holotomography for cone – beam setups. Choosing the right set of optics, also sample environments can be used for in situ investigations at the nanometer scale.

Here we present the status and perspectives of the nanotomography setup at the P05 / Imaging Beamline IBL operated by the Helmholtz Zentrum Geesthacht (HZG) at PETRA III storage ring at DESY in Hamburg (Germany).

2. Experimental Setup

2.1. Layout of the Imaging Beamline IBL / P05

The beamline is dedicated to materials science and allows visualizing structural properties in the range of several micrometers down to the nanometer regime [1-4]. It comprises of two experimental stations: A micro – and a nanotomography station, a sketch of the beamline is given in figure 1 (left). The optics hutch of the IBL beamline is equipped with two monochromators: A Double Chrystal Monochromator (DCM) and a Double Multilayer Monochromator (DMM) both allowing for an energy range of 5 – 50 keV. The DCM has an energy bandwidth of $\Delta E/E \sim 10^{-4}$ while the installed DMM which is currently under commissioning, will allow for higher flux at an energy bandwidth of $\Delta E/E \sim 10^{-2}$ with multilayer coatings produced inhouse by HZG [5].

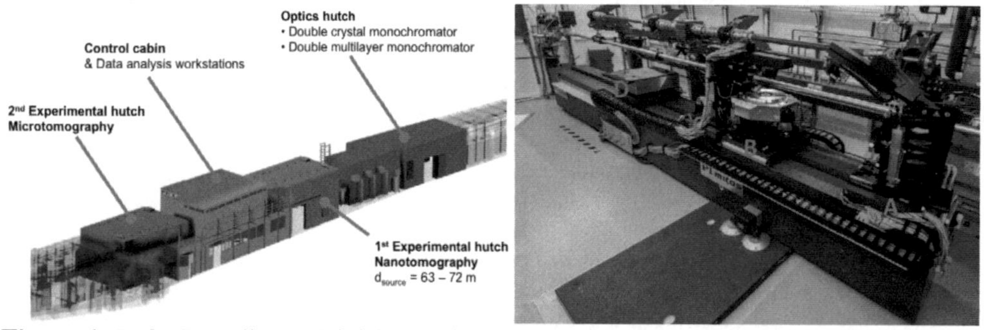

Figure 1. Left: Beamline IBL/P05 overview [2]. Right: Photograph of the experimental hutch with two optics stations and high precision sample rotation stage in between. The four granite slabs can move independently from each other. On the 4th slab a detector system can be mounted.

2.2. Nano Tomography Endstation

The nanotomography endstation shown in figure 1 (right) is a highly flexible setup [1-3] dedicated to full field nano imaging. On top of the 6.8 m long granite substructure four granite slabs are mounted on air bearings and can be moved independently. On the first slab, closest to the source (right hand side in figure 1, A) a six – axis optics kinematics stage is mounted and the extension arm is fitted with high precision apertures. This extension arm can be moved in and out of the beam path e.g. for switching between X – ray microscopy and cone beam geometry. The second slider (B in figure 1, right) holds the sample stage, consisting of a high precision rotation stage and a linear axis to move the sample perpendicular to the incoming beam. The air – bearing rotation stage itself is mounted on three pods allowing for height as well as tip/tilt alignment of the sample stage. Overall a wobble and axial error of less than 30 nm (RMS) is achieved for the sample rotation. The sample itself is mounted on a six – axis kinematics in the center of the rotation stage allowing for precise alignment of the sample in the center of rotation. On the third slider (C in figure 1, right) a second optics station is installed, mirroring the stage on the first slider (A). It is also equipped with a six – axis kinematics and a set of high precision apertures (e.g. used as order selecting apertures for X – ray microscopy using FZP). The extension arm can be varied in length depending on the requirements of the experiment. For precise alignment of the X – ray optics a second camera system (PixeLink) is installed at 300 mm behind the optics stage on the third slider (C in figure 1, right). This alignment camera can be moved remotely in and out of the beam path and is equipped with a 100 μm thick $CdWO_4$ scintillator crystal and yields a 10x light optical magnification, thus achieving an effective pixel size of $\sim 0.6\ \mu$m. The detector system for the tomography setup is mounted on the last slider (D in figure 1, right).

Currently a pco.edge 4.2 is installed with a LuAG scintillator of 30 μm thickness. The sCMOS chip is cooled to 0 °C, consists of 2048 x 2048 pixels with a pixel size of 6.4 μm. Different magnifications of the visible light optics can be chosen M=7 – 12, yielding effective pixel sizes between 0.54 µm and 0.92 µm with an numerical aperture of 0.34.

3. Experimental Results

An X – ray microscope for different energies has been installed at the nanotomography endstation. Two different sets of optics were chosen depending on energy: For the lower energy regime (< 15k eV) a Fresnel zone plate (FZP) was installed as an objective lens and a diffractive beamshaper as condenser optics, both manufactured at the PSI. The design energy for the FZP is 14 keV with an outermost zone width of 80 nm and a working distance of 70 mm. The fields of the beamshaper are 50 μm x 50 μm in size, covering a field of 1.8 mm x 1.8 mm. For the higher energy regime (>15 keV) polymer compound refractive lenses (CRLs) were used as objective lenses and a rolled prism lens as condenser optics [6,9]. The optics were fabricated at the Institute of Microstructure Technology (IMT) at the Karlsruhe Institute of Technology in Karlsruhe (KIT) using X – ray lithography. These CRLs allow for resolving line widths below 100 nm and the VHVHV – design of alternating vertical and horizontal lens elements ensures minimal astigmatism [6]. The working distance of these lenses is around 100 mm.

Figure 2. XRM images of Xradia test pattern. Inner circle 50 – 100 nm lines, 2nd circle 100 – 200 nm lines: Left: using CRL setup at 17.4 keV and a rolled prism lens as a condenser; Right: using FZP setup at 14 keV with a mosaic lens as condenser optics. Middle: SEM image of the test pattern, revealing that the 50 nm structures in the center are collapsed.

Performing a resolution test using an Xradia test pattern shows that with both setups the 100 nm lines of the Siemens star are well resolved (Figure 2). It has to be noted, that the image from the CRL setup was acquired with a pco.4000 CCD camera combined with an LSO scintillator with a thickness of 16 μm. For the FZP setup the pco.edge 4.2 was used (see section 2.2) in combination with the 30 μm LuAg scintillator. The reconstructed data shown in figure 3 have been acquired using the FZP setup at 14 keV. The nano porous gold sample was prepared by Focused Ion Beam (FIB) and the sample cross – section is 8 μm x 8 μm. The SEM image of the sample reveals the nano structure, which is also clearly visible in the reconstructed volume shown in figure 3 on the right hand side. The gold filaments are in the order of 400 nm in width and are clearly resolved in the three dimensional image.

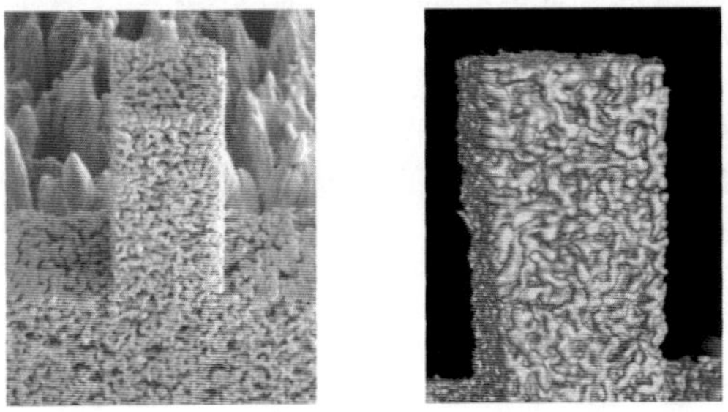

Figure 3. Nano porous gold. Left: SEM image; Right: Rendering of reconstructed volume. The experiment was performed using the FZP setup at an energy of 14 keV (eff. pixel size 17.4 nm). The square column has an edge length of about 8 μm x 8 μm and was prepared by FIB milling. The filament sizes of the NPG sample are in the range of 400 nm in width and are nicely resolved.

IOP Conf. Series: Journal of Physics: Conf. Series **849** (2017) 012056 doi:10.1088/1742-6596/849/1/012056

4. Future Perspectives

In the future better image statistics will be reached by implementing a new visible – light optics design with a higher numerical aperture. This new camera system will house up to three cameras permanently mounted and aligned with respect to the sample stage. A movable mirror will allow for switching in between the different cameras. This way the alignment of the experimental setup and the optics can be done using a fast camera, e.g. a scientific CMOS and the tomography will then be performed using a CCD with high statistics. Another increase in image quality and decrease in scan time will be reached by using the newly installed DMM, once it is fully commissioned.

Furthermore it is planned to commission a cone beam setup and make it available to users. In contrast to X-ray microscopy geometry the cone beam technique offers a higher flexibility, e.g. the FoV depends on the chosen sample to detector distance and not on the chosen objective lens. The technique also allows for the implementation of sample environments, since in X-ray microscopy setups the space around the sample is often very limited. A specifically designed in vacuum six-axis kinematics is installed in the vacuum tank in the optics hutch such that a pre – focusing device can be implemented there. In order to align these pre – focusing optics an additional camera system equipped with a scintillator will also be installed in the vacuum tank and is currently in the design state. This in combination with the DMM will offer a very flexible cone beam setup at the instrument in the future.

Acknowledgements

The authors thank Prof. E. Lilleodden and her group (HZG, Institute of Materials Mechanics) for the nanoporous gold sample imaged at P05 (figure 3). I. G. acknowledges the Virtual Institute of New X-ray Analytic Methods in Material Science (VI – NXMM) for funding. M. O. gratefully acknowledges financial support from the German Research Foundation (DFG) via SFB 986 M3, project Z2. The authors thank Harald Vogt of the Karlsruhe Institute for Microstructurtechnology (IMT) and the Karlsruhe Nano Micro Facility (KNMF) for the design and fabrication of the rolled X-ray prism lenses.

References

[1] Greving I, Wilde F, Ogurreck M, Herzen J, Hammel J U, Hipp A, Friedrich F, Lottermoser L, Dose T, Burmester H, Müller M and Beckmann F. Proc. SPIE 9212, Developments in X-Ray Tomography IX, 92120O-8, 2014

[2] Ogurreck M, Wilde F, Herzen J, Beckmann F, Nazmov V, Mohr J, Haibel A, Müller M, and Schreyer A, Journal of Physics: Conference Series, vol. 425, p. 182002, 2013

[3] Wilde F, Ogurreck M, Greving I, Hammel J U, Beckmann F, Hipp A, Lottermoser L, Khokhriakov I, Lyatev P, Dose T, Burmester H, Müller M, and Schreyer A, AIP Conference Proceedings 1741, 030035, 2016

[4] Ogurreck M, do Rosario J J, Leib E W, Laipple D, Greving I, Marschall F, Last A, Schneider G A, Vossmeyer T, Weller H, J. Synchrotron Rad. 23, 2016

[5] Störmer M, Horstmann C, Häussler O, Spiecker E, Siewert F, Scholze F, Hertlein F, Jäger W, and Bormann R, SPIE, vol. 7077, p. 707705, 2008.

[6] Marschall F, Last A, Simon M, Kluge M, Nazmov V, Vogt H, Ogurreck M, Greving I, and Mohr J, Journal of Physics: Conference Series, vol. 499, p. 012007, 2014

[7] Sakdinawat A and Attwood D, Nature photonics, vol. 4, p 840 – 848, 2010

[8] Möbus G, Inkson B J, Materials Today, vol.10, p 18-25, 2015

[9] Reznikova E, Weitkamp T, Nazmov V, Last A, Simon M, Saile V, Physica Status Solidi, p 2811-2816, vol. 204, 2007

IOP Conf. Series: Journal of Physics: Conf. Series **849** (2017) 012057 doi:10.1088/1742-6596/849/1/012057

CARNAÚBA: The Coherent X-Ray Nanoprobe Beamline for the Brazilian Synchrotron SIRIUS/LNLS

Hélio C. N. Tolentino[1,2], Márcio M. Soares[1], Carlos A. Perez[1], Flávio C. Vicentin[1], Dalton B. Abdala[1], Douglas Galante[1], Verônica de C. Teixeira[1], Douglas H. C. de Araújo[1], Harry Westfahl Jr.[1]

[1] Laboratório Nacional de Luz Síncrotron - LNLS/CNPEM, C.P. 6192, 13083-970 Campinas, Brazil.
[2] Université de Grenoble-Alpes and CNRS, Institut Néel, B.P. 166, 38042 Grenoble, France.
e-mail: helio.tolentino@lnls.br

Abstract. The CARNAÚBA beamline is the tender-to-hard X-ray (2 – 15 keV) scanning nanoprobe planned for the 4th generation storage ring SIRIUS at the LNLS. CARNAÚBA uses an undulator source with vertical linear polarization in a low-beta straight section and grazing incidence-focusing mirrors to create a nanoprobe at 143 m from the source. The beamline optic is based on KB mirrors and provides high brilliance at an achromatic focal spot down to the diffraction limit diameter of ~30 nm with a working distance of ~6 cm. These characteristics are crucial for studying nanometric samples in experiments involving complex stages and environments. The CARNAÚBA beamline aims to perform raster scans using x-ray fluorescence, x-ray absorption spectroscopy, x-ray diffraction and coherent x-ray imaging techniques. Computed tomography will extend these methods to three dimensions.

1. Introduction

The Coherent x-Ray Nanoprobe Beamline (CARNAÚBA) is the tender-to-hard X-ray scanning nanoprobe beamline planned for the SIRIUS storage ring at the LNLS (Laboratório Nacional de Luz Síncrotron), Campinas, Brazil. The CARNAÚBA project aims to deliver to the scientific community a state-of-the-art nanoprobe based on a Kirkpatrick-Baez (KB) achromatic optic thus allowing for spectroscopic, resonant and imaging x-ray techniques. In order to face complex sample environments, the main goal is not to get record small spots but to have a nanoprobe with a stable beam and a comfortable working distance.

CARNAÚBA will deliver the smallest nanoprobe among the first phase SIRIUS beamlines. The beamline fully exploits the high brilliance and coherence characteristics of the X-ray beam delivered by the undulator source both for diffraction-limit focusing and contrast formation. It offers not only X-ray fluorescence (XRF) and X-ray absorption (XANES) spectroscopy, but also the most advanced multimodal coherent imaging techniques. CARNAÚBA is a unique research tool for a wide interdisciplinary user community working in the fields of agriculture, life-, earth-, environmental-, and materials science [1]. The beamline starts to be assembled in 2018. Many ideas and prototypes have been developed. Among them, a permanent magnet undulator with a few periods has been assembled (section 2) and a cryo-cooled four bounces x-ray monochromator has been designed (section 3). The assembling and commissioning phase will take place in 2019, when the first users will be welcome. In this paper we present CARNAÚBA's design characteristics and expected performance.

2. The undulator source

The SIRIUS storage ring [2] has been designed to have a natural horizontal emittance of $\epsilon_x = 245\, pm\, rad$ and a vertical emittance (coupling 1%) of $\epsilon_y = 2.4\, pm\, rad$, without considering further reducing of the emittance by the insertion of undulators. The projected β values in the middle of the low-beta straight sections are $\beta_x = 1.4\, m$ and $\beta_y = 1.7\, m$, in the horizontal and vertical directions, respectively. The similar values of β have been chosen to reduce as much as possible both vertical and horizontal beam-stay-clear values to ~4.5 mm and ~8 mm, respectively, as well as to optimize the photon flux emittance for an undulator of ~2.0 m. Such a singular condition for the β values allows using an undulator source with vertical linear polarization.

Linearly polarized light will be used for scattering, diffraction and absorption spectroscopy at CARNAÚBA. With linear polarization in the vertical, the scattering plane of the Si(111) crystal monochromator can be in the horizontal - to preserve a good vertical stability - without harming the flux close to the Brewster angle in the tender X-ray range.

The adopted solution for the CARNAÚBA insertion device has been a modified Delta undulator. The essential difference in the Delta design compared to the conventional one is that there is no gap variation: the field, or the K value, varies by changing the phase, i.e. by sliding one set of magnets along the beam related to the other [3]. Our modified Delta design has four sets of permanent magnets that can slide independently to provide a variable orientation magnetic field. A short prototype has been assembled and measured. Refined parameters for the final prototype will be available soon.

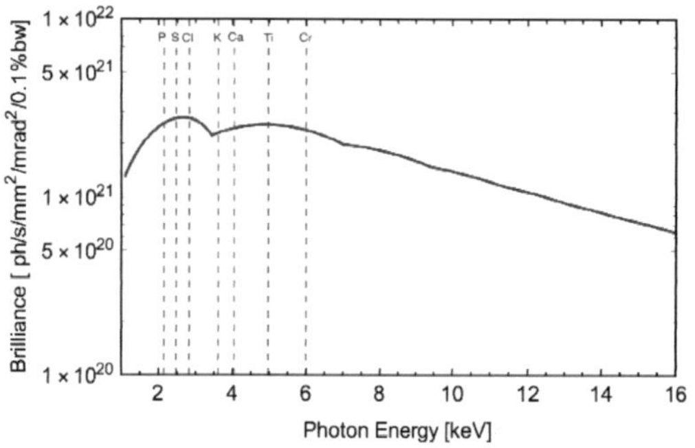

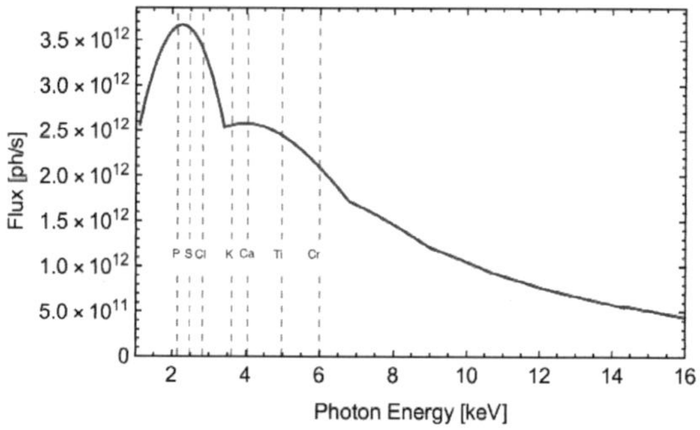

Figure 1. Left: Brilliance of the CARNAÚBA undulator with K_{max}=2.3 and 100 poles of 20 mm. Right: flux at the focal spot, within 0.01% bw and collecting a divergence of about 6x10 μrad^2 in the horizontal and vertical, respectively. The machine current is 350 mA. Vertical lines represent some light element K-edges.

To avoid gaps between the first and third harmonic of the undulator, a K_{max} of 2.05 or higher is required. Taking the beam-stay-clear values into consideration, a two-meter long undulator with 20 mm period and K_{max} of 2.3 is a possible configuration. The brilliance, as well as the total flux at the focus after the optics, of such undulator is presented in Figure 1. The undulator matches well the main applications of the CARNAÚBA beamline delivering energies from 2 keV up to 15 keV with very high brilliance. The vertical dashed lines represent the energy positions of some light elements K-edges (P, S, Cl, K, Ca, Ti, Cr), covered by the 1st and 3rd harmonics of the undulator. The total flux, taking into account the optics described in the next section, i.e., a bandwidth of the four crystal setup (≈0.01%bw) with the machine running at 350 mA and a KB system collecting a divergence of 6x10 μrad^2, overpasses 10^{12} photons/sec at the focal spot. We should point out that the collected divergence selects the beam coherent fraction, meaning that the total flux corresponds to the fully coherent flux.

3. The optics

Source characteristics in 4th generation storage rings are normally limited by diffraction in the vertical (typical FWHM sizes of 10 μm) while still limited by emittance in the horizontal (about 50 μm at the SIRIUS undulator). With the available distance from the source at CARNAÚBA, 143 m, focusing to smaller than 20 nm can be done in one step in the vertical by a mirror M_V with a demagnification of about 500, leaving a working distance of about 30 cm. In the horizontal, this is not possible, and demagnification has to be done in two steps by using a secondary source aperture (SSA), which can then be slit down conveniently. With this choice, the horizontal mirror M_H requires a larger demagnification than the vertical and has to be placed downstream in the sequential mounting in a KB system.

We used simple analytical expressions for geometrical optics beyond the thin lens approximation and a correction to account for the diffraction limit [4] to design and optimize the CARNAÚBA optical layout (Fig. 2). After white beam slits (WBS) defining a 50x50 μrad² divergence, the mirror M_1 (at 27 m) creates a secondary source SSA (at 54 m) in the horizontal in a 1:1 configuration. The bremsstrahlung radiation shield (BRS) (at 30 m) stops the bremsstrahlung cascade 3 m after the M_1 mirror. The mirror M_2 (at 56 m) bounces the beam back, parallel to the original propagation. Both mirrors are used for harmonic rejection and may eventually be used to steer the beam at the SSA and at the experimental station, respectively. The harmonic rejection scheme of the primary optics, with the mirrors working at fixed grazing incidence of 5 mrad, employs three stripes, Rh, bare Si and Ni that can be selected by a vertical translation. A horizontal deflecting 4-bounces crystal monochromator (4CM) (at 136 m) close to the experimental station, select the photon energy with $E/\Delta E \approx 10000$. A prototype has been designed and will be assembled and tested in 2017. Pink beam mode may be used by sliding laterally the 4CM by a few mm. The nanoprobe station (at 143 m) holds the KB mirror system (with M_V at 142.64 m and M_H at 142.88 m), sample holder environment and detectors.

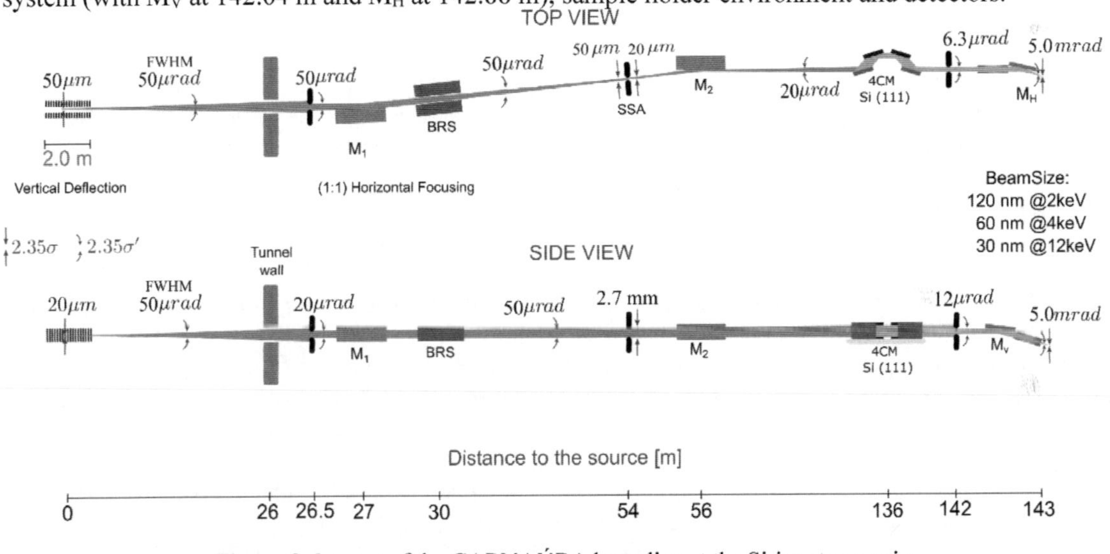

Figure 2. Layout of the CARNAÚBA beamline at the Sirius storage ring.

Considering, as a first approximation, ideal elliptical mirrors in the strong focusing approximation, the focus size is given by a convolution of the diffraction limit and the demagnified source. Trading between geometric demagnification and diffraction limit at the same time as the flux lead us to the optimized layout shown in Figure 2. The vertical and horizontal focus can be made very similar in size and divergence by adjusting the beam fraction collected at the SSA slit. This is shown in Figure 3 where the SSA aperture is set to 20 μm. One may note that for energies above 6 keV they are similar and kept below 50 nm. This is essentially due to the large numerical aperture of the KB mirror system (~5 mrad) working at a maximum incidence angle of 5.6 mrad. The drawback of such a large numerical aperture is a limiting energy of about 15 keV due to the mirror cut-off. In the lowest energy side, below 6 keV, the focal spot increases, clearly limited by diffraction. In this range, the SSA aperture can be increased to gain more flux without enlarging the focus spot. For instance, using a 50 μm aperture, the flux density should increase by a factor 2 or more.

After the analytical ray tracing optimization, a hybrid ray tracing code was used to fully take into account all optical elements, including errors in the mirrors and the wave character of the propagation. Both calculations agree very well to each other, giving similar results in terms of flux of photons, divergences and nanoprobe sizes [4]. From these calculations, the primary optic, M_1 and M_2 mirrors, has to have slope figure errors smaller than 250 nrad. The thermal bump in the first mirror is negligible and does not affect the image at the SSA aperture. As far as the KB mirrors are concerned, the specifications are more stringent, in particular for the higher energies above 12 keV. The mirrors have specifications with slope figure errors of 50 nrad and, in addition, special care has to be taken with the spatial frequency domain that may cause some blurring at focus. The beam size at lower energies is almost diffraction limited and these errors showed to be negligible.

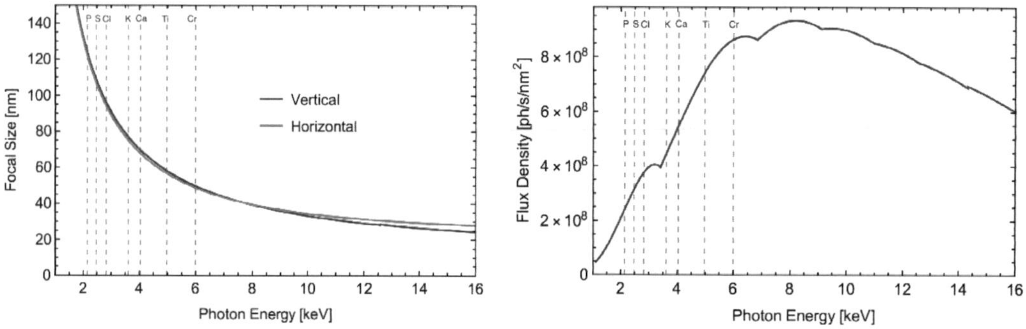

Figure 3. Left: Energy dependence of the vertical and horizontal focus size. Right: Flux density at the focus, within 0.01% bw, collecting a divergence of about 6x10 μrad² in the horizontal and vertical, respectively, and with SSA aperture of 20 μm, with 350 mA in the machine. Vertical lines represent some light element K-edges.

4. Final remarks on the nanoprobe

CARNAÚBA will tackle questions in areas of fundamental importance to science, as in life, geological, environmental and materials science. Its nanoprobe will allow answering many questions regarding the nature of these heterogeneous samples, such as local composition, chemistry, atomic structure, crystal orientation, strain, among other important characteristics, aiming to go down to spatial resolution of few nanometers. When the sample is raster scanned in front of the beam to form an image, the resolution is directly related to the spot size, which depends on energy (Fig. 3-Left). Techniques collecting the transmitted beam through the sample, the fluorescence or the diffraction will be available at the beamline with that spatial resolution. For reciprocal space imaging, resolution depends on the coherent dose on the sample and the flux density is a more relevant parameter (Fig. 3-Right). The attainable resolution will depend on how efficiently the coherent beam is focused into the sample, without too much deforming the wave-front, on stability of the sample to the optics and on the availability of fast area detectors to handle the amount of photons.

Acknowledgments

Technical support from all LNLS groups are acknowledge, in particular, the assistance and calculations done by Sérgio A. L. Luiz and Bernd C. Meyer from the Optics Group.

References

[1] Hélio C.N. Tolentino, et al., (2015). "CARNAÚBA CDR", LNLS Internal Report.
[2] Lin Liu, Natalia Milas, Afonso H. C. Mukai, Ximenes R. Resende and Fernando H. de Sa, "The Sirius project", (2014). J. Synchrotron Rad. **21** 904-911.
[3] Alexander B. Temnykh, "Delta undulator for Cornell energy recovery linac", Phys. Rev. ST Accel. Beams **11**, 120702 (2008)
[4] Harry Westfahl Jr., et a. (2017). J. Syncrotron Rad. to be submitted.

Design and Fabrication of Wolter-type 4-mirror system

N Watanabe[1], S Aoki[1, 2] and N Yamaguchi[2]

[1]Faculty of Pure and Applied Sciences, University of Tsukuba, Tsukuba, Ibaraki, 305-8573, Japan
[2]CROSS (Comprehensive Research Organization for Science and Society), 1601 Kamitakatsu, Tsuchiura, Ibaraki, 300-0811, Japan

watanabe@bk.tsukuba.ac.jp

Abstract. A Wolter-type 4-mirror system was designed and evaluated by ray-trace calculation. The design and fabrication process were developed. The master mandrel of oxygen-free copper was shaped by diamond turning. The Pyrex glass replica was made by a vacuum replication technique. A visible light image of a metal mesh could be obtained using this replica mirror.

1. Introduction

A Wolter type-I mirror is a grazing-incidence optics, which has been used in x-ray astronomy, x-ray microscopy, and neutron imaging. A Wolter mirror for microscopy is a 2-mirror system which is a combination of axisymmetric hyperboloidal and ellipsoidal mirrors with a common focus [1] as shown in Fig. 1(a). On the other hand, a 2-mirror system composed of two hyperboloidal mirrors with a common focus can produce a virtual image as shown in Fig. 1(b). By combining these two types of Wolter mirrors, a 4-mirror imaging optics can be designed. The first Wolter mirror forms a virtual image of an object and the second Wolter mirror forms its real image.

As shown in Fig. 1(a), the sum of aperture angle α of object side and aperture angle β of image side is 2 $\times$ (number of reflection) $\times$ grazing-incidence angle. Therefore, the numerical aperture of a Wolter mirror is limited to 4-times of the grazing-incidence angle. The grazing-incidence angle must be smaller than a critical angle of total reflection, which is less than 10 mrad for hard x-rays at the energy above 5 keV. Then, the numerical aperture is very small. The advantage of this 4-mirror system is that the numerical aperture can be approximately doubled compared to a 2-mirror system [2].

A disadvantage of the 4-mirror system is that the influence of surface roughness and shape error is larger than that of the 2-mirror system. It is necessary to sufficiently reduce surface roughness in the 4-mirror system. However, shape error of the second Wolter mirror is not so serious compared to the first Wolter mirror because the object has been already enlarged.

The design and fabrication process are described in this paper.

2. Design of Wolter-type 4-mirror system

A wolter-type 4-mirror system of the average grazing-incidence angel of 5.5 mrad is designed. The critical angle of x-ray total reflection of Pyrex glass is calculated to be 6.1 mrad at 5 keV from the refractive index [3]. Therefore, the mirror made of glass can be used for x-rays at the energy below about 5 keV. The reflectivities of 2-times and 4-times reflection are calculated to be 79 % and 63 % at 5 keV with no surface roughness.

IOP Conf. Series: Journal of Physics: Conf. Series **849** (2017) 012058 doi:10.1088/1742-6596/849/1/012058

Figure 2 shows design parameters of the 4-mirror system. Region surrounded by a square dashed line in Fig. 2 shows a 2-mirror system that corresponds to Fig. 1(a). First, the parameters of the 2-mirror system are determined. The ellipsoid and hyperboloid shapes can be determined by the following 5 parameters: (1) the radius of the mirror at the intersection plane of the hyperboloid and ellipsoid surfaces (R), (2) the grazing-incidence angle for the ellipsoid surface at the intersection (θ_1), (3) the grazing-incidence angle for the hyperboloid surface at the intersection (θ_2), (4) the magnification ratio of the mirror (M), and (5) the length of the ellipsoidal mirror along the optical axis (L). The length of the hyperboloidal mirror (hyperboloid 0) is determined by ray-tracing calculation from the image point so as to enter all the rays reflected by the ellipsoidal mirror. These parameters are optimized as a two mirror system. After determining the parameters of the 2-mirror system, the shape of hyperboloid 1 can be calculated from the grazing-incidence angles θ_3 in Fig. 2. The shape of hyperboloid 2 can be also calculated from θ_4 [2].

The grazing-incidence angle is changed by the reflecting position. By slightly changing the parameters repeatedly, the average grazing-incidence angle is made close to 5.5 mrad. The total mirror length is 38 mm. The magnification ratio is 9.9. The numerical aperture is 4.0×10^{-2} and the theoretical resolution for Rayleigh criterion is 2.4 nm at 5 keV. On the other hand, the numerical aperture of the 2-mirror system which consists of the ellipsoid and hyperboloid mirrors of the 4-mirror system is 1.8×10^{-2} and the theoretical resolution is calculated to be 5.2 nm.

The geometrical aberration was estimated from ray-trace calculation. The off-axis resolution was estimated from 2.35 × standard deviation from the gravitation center of 1000 rays as shown in Fig. 3.

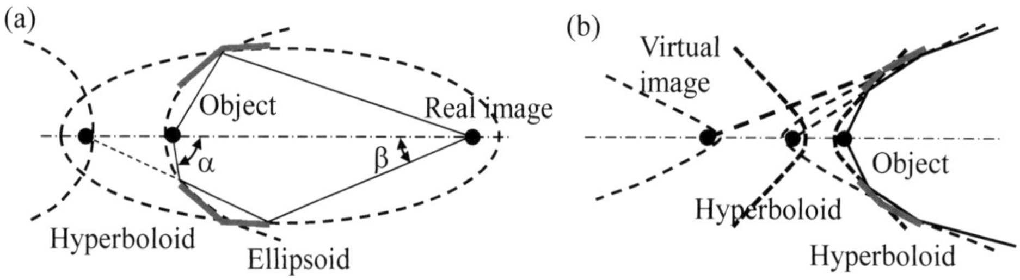

Figure 1. (a) Axisymmetric hyperboloidal and ellipsoidal mirrors with a common focus can produce a real image. (b) Axisymmetric two hyperboloidal mirrors can produce a virtual image.

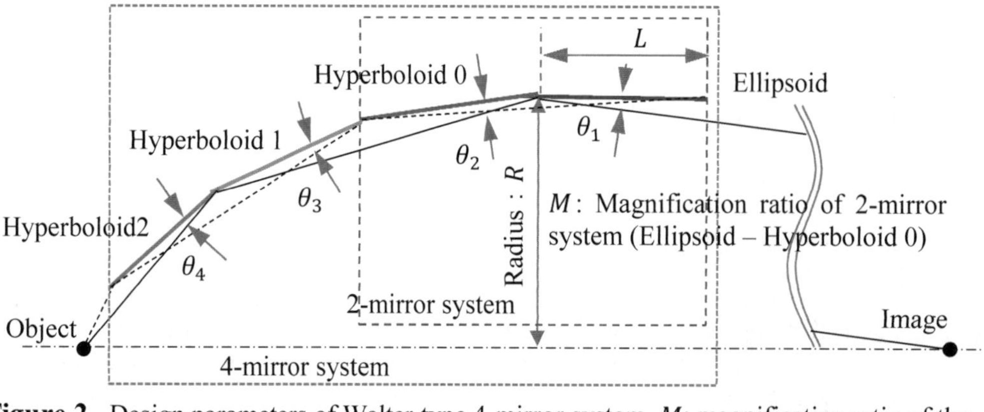

Figure 2. Design parameters of Wolter-type 4-mirror system. M: magnification ratio of the 2-mirror system, R: radius, L: length of the ellipsoidal mirror, θ_1, θ_2, θ_3, θ_4: the grazing-incidence angles.

Assuming that the spatial resolution is 50 nm, the off-axis distance is 20 μm for the 4-mirror system and 40 μm for the 2-mirror system. In the 4-mirror system, the two hyperboloid mirrors of object side enlarges an object image by a factor of about 2. Then, radius of field of view of the 4-mirror system is about half compared with the 2-mirror system. To obtain the 2.4 nm resolution of the Rayleigh criteria, the field of view for the 4-mirror system is calculated to be 6 μm in diameter.

3. Fabrication process of 4-mirror system

Fabrication process of the 4-mirror system was tested by using diamond turning and vacuum replication method [4] as shown in Fig. 4. An oxygen-free copper rod of 10 mm in diameter was used for the master mandrel. The position of each mirror surface of the 4-mirror system was divided into 1000 points. Using a numerically controlled lathe, the mandrel was formed by a single crystal diamond bite. Gold layer of about 100 nm thick was deposited on the mandrel surface as the parting agent. An electric furnace was used in the replication process. The mandrel was put into a Pyrex glass tube and heated up to 655 degrees Celsius in 30 min. After maintaining this condition for 75 min, the furnace was turned off and left about 5 hours to cool down. The coefficient of thermal expansion of the mandrel is larger than that of the glass tube so that the glass is separated from the mandrel after cooling down. Figure 5(a) shows the fabricated Wolter-type 4-mirror system made of Pyrex glass.

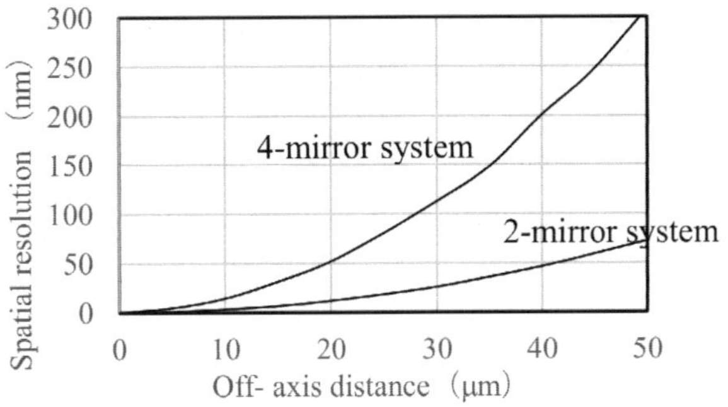

Figure 3. Geometrical aberration calculated from spot diagrams of the 4-mirror and 2-mirror systems [2].

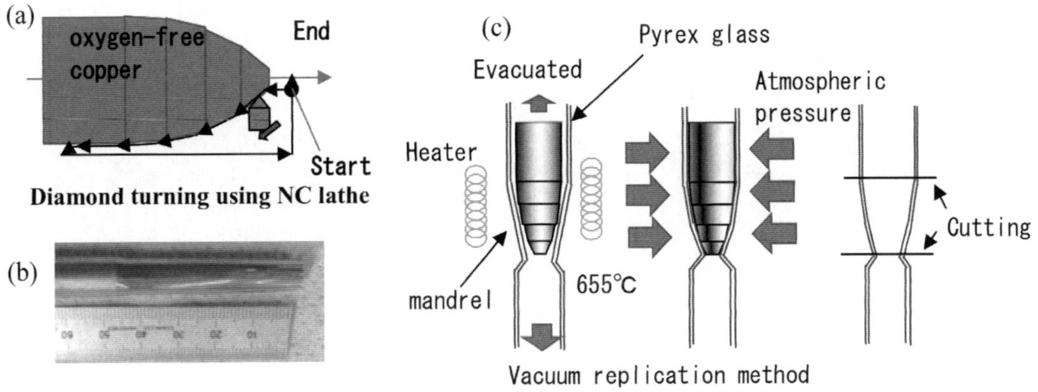

Figure 4. Fabrication process of the 4-mirror system. (a) A mandrel of oxygen-free copper rod was formed by diamond turning. (b) Au layer of about 100 nm was deposited as release agent. (c) The replica mirror of Pyrex glass was made by the vacuum replication method.

It was necessary to polish the surface for x-ray imaging, so that the preliminary test was performed using visible light. Figure 5(b) is a visible light image of a metal mesh. A mesh of 250 μm pitch could be imaged by the replica mirror. As a next step, tungsten carbide was used as material of master mandrel for more precision machining. The replication process is currently under way.

The Wolter-type 4-mirror system has about two times the numerical aperture of the 2-mirror system. This also means that the focal length is about half of the 2-mirror system and the magnification ratio is about 2 times that of the 2-mirror system under the constant object-image distance. Therefore, the 4-mirror system is considered to be effective in high-energy x-ray and neutron imaging, in which the critical angle of total reflection is small.

This work was supported by Grant-in-Aid for Scientific Research (No. JP25108003) from the Ministry of Education, Culture, Sports, Science and Technology and Grant-in-Aid for Scientific Research (No. JP23656056) from the Japan Society for the Promotion of Science.

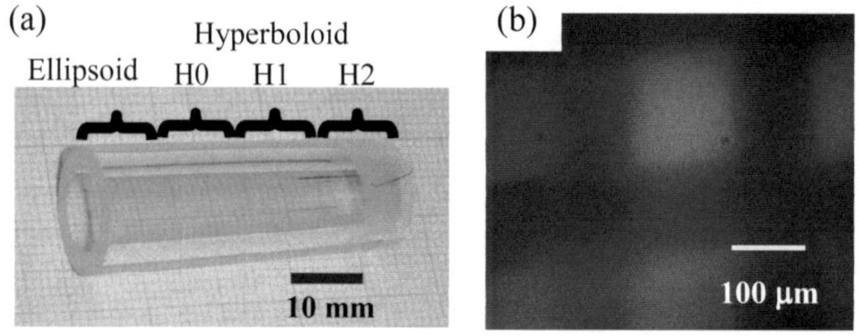

Figure 5. (a) Fabricated 4-mirror system made by Pyrex glass. (b) Visible light image of a metal mesh.

References

[1] Wolter H 1952 *Ann. Phys.* **10** 94
[2] Aoki S, Watanabe N, Asami H and Shimada A 2016 *Opt. Rev.* **23** 161
[3] http://henke.lbl.gov/optical_constants/
[4] Sakayanagi Y and Aoki S 1978 *Appl. Opt.* **17** 601

IOP Conf. Series: Journal of Physics: Conf. Series **849** (2017) 012059 doi:10.1088/1742-6596/849/1/012059

dada – a web-based 2D detector analysis tool

Markus Osterhoff

Institute for X-ray physics, University of Göttingen, Friedrich-Hund-Platz 1, D–37077 Göttingen

E-mail: `mosterh1@gwdg.de`

Abstract. The *data daemon*, *dada*, is a server backend for unified access to 2D pixel detector image data stored with different detectors, file formats and saved with varying naming conventions and folder structures across instruments. Furthermore, *dada* implements basic pre-processing and analysis routines from pixel binning over azimuthal integration to raster scan processing. Common user interactions with *dada* are by a web frontend, but all parameters for an analysis are encoded into a Uniform Resource Identifier (URI) which can also be written by hand or scripts for batch processing.

1. Introduction

Basic analysis of modern (2D) pixel detector images is hindered by seemingly trivial technical problems: varying file names and different formats between beamlines, detectors, and sometimes these change over time. In fact, just reading a detector image as a university's first-time bachelor student can be rather difficult, although it should be a standard task. Then, the correct image orientation (flipping, rotating) has to be found, before basic pre-processing like binning and region of interest (ROI), and empty division can be applied. As a small university group with fluctuating students taking data at different end-stations, the sheer number of file name formats, folder structures, and detector formats can be quite overwhelming when analysis scripts are adapted from older beamtimes.

Comparing *dada* to other common software: FabIO [4] is a Python library for 2D detector images; here we add also file name unification and basic analysis. Dawn [6] has emerged as a very powerful Eclipse based analysis program that features an impressive set of tools for crystallography and fluorescence mapping. Here we focus on holographic imaging and scanning applications. Another famous program is of course ImageJ [5], which has a large user base not only in the optical microscopy community. Oposed to these programmes, *dada* can be run both on a local PC/laptop, on a dedicated analysis machine, and on small clusters. Long-running jobs continue when the interface is closed.

On the other hand, certain pre-processing steps are common to different experiments; more advanced analysis methods that have been investigated during students' thesis works are now understood and need to be frozen in a specific version, with a minimum of interaction (i.e. change of parameters). With our *data daemon* – or *dada* for short – we are solving these two problems in one modularised programme, which is entirely written in C. In the following section 2, we describe more detailed the steps of file name translation, file interpretation, image orientation and basic pre-processing; in section 3 some analysis routines are described and illustrated with

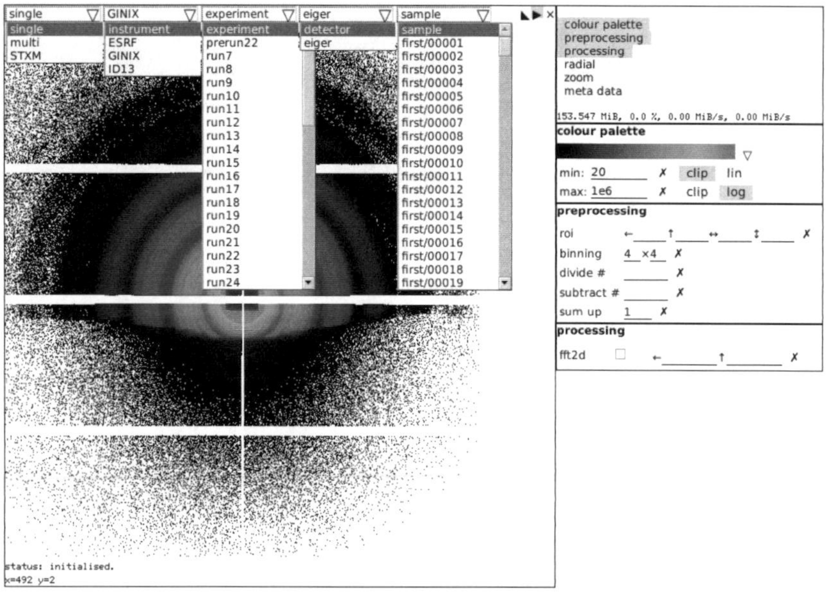

Figure 1. Screenshot of *dada*'s web frontend. In the top row, the selection menus for instrument, experiment, detector and sample are shown; on the right panel, the colour palette can be defined, along with certain pre-processing steps.

exemplary data. Section 4 briefly describes the web-frontend (see Fig. 1) that the user interacts with. We conclude with a summary and outlook in section 5.

2. Image read-in, pre-processing

A single 2D-detector image is defined by the names of instrument (beamline), experiment, sample and detector, and the image number. The "messy" translation into the correct file name is then carried out in the backend and can easily be adapted for new setups. The image data is then read in, dealing with decompression together with interpretation, correcting the image orientation, and applying specific bad pixel masks. These steps are coded in three functions (*filename.c*, *open,c*, *pixelmask.c*); generic versions apply for one detector family, but the subtle differences between experiments are accounted for with specific versions of the routines in instrument/experiment specific subfolders.

Pre-processing starts with 2D region-of-interest (ROI) and binning (pixel decimation) and can include division by and subtraction of (empty) images, and stack operations like adding up sequential images, or finding the maximum value of each image pixel of the stack. These operations are carried out in the *show_single_image* module.

For quick visual inspection, the image data can then be colour-coded into a "common" image file (PNG, or PDF with basic annotations) and delivered to the user, e.g. via web browser. For the colour code, linear and logarithmic scaling and overriding of min/max is possible; also different colour palettes can be chosen. The final image files are cached by the backend for faster access in the future.

A "live version" of *dada* can be used at the beamline, where the latest detector image is shown on the screen, with feedback of basic statistics (total counts, min/max) and the possibility to directly print the image for the analogue lab book.

A 2D array lay-out of raster scanned images is possible using the module *show_composite_image*: it places N_y by N_x of pre-processed single images in a regular grid;

with reasonable ROI and binning, a large 2D raster scan of 2D image data can then be assessed visually. This method can also be used to visualise 1D scans, see an example in the following section.

All detector routines and the *show_** modules are loaded as dynamic libraries. This makes it possible to adjust the programme during execution: detectors can be added at run-time, processing can be adapted and developed without the need to re-start *dada*.

So far, we have implemented data read-in routines for Pilatus detectors (cbf[1] and tiff, also cbf.gzip), Maxipix and other EDF-detectors, certain CMOS detectors (tiff), and recently Eiger and Lambda (HDF5[2]).

Instead of rendering a colour-encoded visual image file, the pre-processed data can be requested as an array of *doubles*[3]; this allows for a unified access of detector images across file name conventions and detector formats from third-party software.

3. Routine Analysis Methods

The most basic analysis for single images is of course min/max and total intensity (sum over all pixel values) after pre-processing. In the "live version" for the beamline, these values can be used as feed-back to the control software SPEC[4] via network, so quantitative scanning on 2D detector ROIs is possible with a typical overhead of for example 15 ms for a Pilatus 300k detector. A quick azimuthal integration for radial beam intensity profiles with user-defined beam centre is possible, and intensity between q-ranges (given as circular discs, in pixel units) is reported. Also the anisotropy of scattering signals is quantified using the principal component method developed in [1].

The aforementioned statistics can then be used in the STXM module (scanning transmission x-ray microscopy) as the observables on a regular 2D grid to analyse raster scans. Thus, a 2D colour-coded image is formed based on e.g. the total intensity of detector images (after ROI, binning, and further pre-processing as described above) for a 2D scan of the sample. Other observables are dark-field (scattering outside a specified region), centre-of-mass in horizontal and vertical direction (to quantify e.g. differential phase contrast); intensities in circular discs; anisotropy and orientation from the principal component analysis to investigate ordered structures in the sample.

4. Web Interface

The *dada* backend uses an *HTTP* server[5] as its primary interface. Thus, all requests (single image as raw data or colour-coded visual image, STXM analysis, meta data) are made via a Uniform Resource Identifier (URI), e.g. of the form

http://dada-demo.roentgen.physik.uni-goettingen.de/demo/stxm/GINIX/run42/p300/cm3/
1?horz=201&vert=201&signal=omeg
&palette=wb&scale=lin&cmin=0.0&cmax=0.7
&radial.pb.x=232&radial.pb.y=314&radial.r1=72&radial.r2=93&radial.r3=200

which requests a STXM analysis of orientations in a 201 × 201 raster scan with user-defined beam centre and colour palette.

A screenshot of the web frontend showing the selection menus for instrument, experiment, detector and sample is shown in Fig. 1.

[1] See, e.g., `https://www.dectris.com/technical_pilatus.html?file=tl_files/root/support/technical_notes/pilatus/Pilatus_CBF_Header_Specification.pdf` or `http://www.esrf.eu/computing/Forum/imgCIF/cbf_definition.html` for definition / examples.

[2] C.f. `https://support.hdfgroup.org/HDF5/`.

[3] So far, this is just a plain memory dump; but more accessible formats are foreseen, e.g. an HDF5 export.

[4] SPEC, by Certified Scientific Software.

[5] The GNU libmicrohttpd library, see `https://www.gnu.org/software/libmicrohttpd/`.

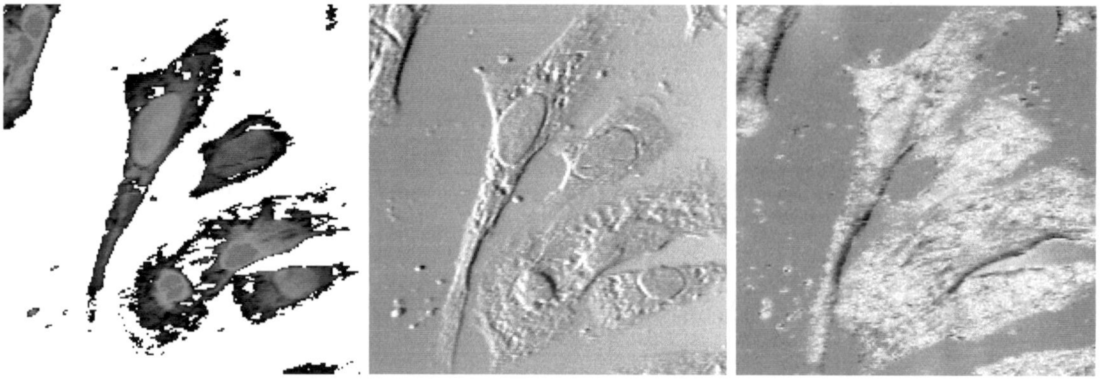

Figure 2. Three different observables calculated for a 2D STXM taken with a Pilatus detector show total scattered intensity (darkfield), centre-of-mass (comparable to differential phase contrast) and the order parameter ω [1, 2] to visualise actin fibre bundles within a cardiomyocyte cell.

In principle, all analyses are requested from URIs in this form; in fact, this allows for a batch analysis by scripts generating such URIs. For most uses, however, a web interface based on jQuery is available that builds up the URI from the user's choices. We stress that ongoing long-running jobs continue after the user has closed the web frontend, and the results can be accessed later on; the web frontend is reachable world-wide (authentication and encryption are possible separately), while the computations are carried out on dedicated machines.

In addition, also an image stack can be viewed in sequence, and individual images of a STXM scan are accessed by clicking the pixel. For radial analysis (azimuthal integration, intensity inside q discs, anisotropy), the beam centre is be defined via mouse clicks, and an overlay of circles in q-space can be enabled. Currently, everything works on a pixel-basis with no sub-pixel interpolation; also the beam centre is defined by the user in full pixels (after binning). Refinements might be added in future.

5. Summary and Outlook

In summary, the *data daemon* (*dada*) allows for a unified access to x-ray pixel detector data, that (i) acts as a central backend hiding technical trivial, but tedious problems from the user, and (ii) implements standardised pre-processing and analysis algorithms. The web frontend helps users in building the request URI that fully determines the job, but for batch analysis this URI may also be generated from users' scripts. Long running jobs continue even if the web frontend is closed, and are shown later on.

Currently, various detectors like Pilatus and Eiger, Maxipix and Lambda, and certain TIFF-writing sCMOS cameras are supported; file name generation is implemented for several ESRF beamlines, the cSAXS at SLS, and the GINIX end-station at PETRA III. The list will certainly grow and be updated.

Also, further analysis routines such as the automized detection of diffracton streaks typical for fibre scattering (streak-finder algorithm, [3]), Fresnel propagation and (holographic) phase retrieval methods, as well as rudimentary tomographic functions (sinogram generation, filtered back projection) will be added as analysis modules.

The *dada* will be published to the user community in 2017 on an open license.

Acknowledgements
We acknowledge funding by Deutsche Forschungsgemeinschaft through SFB 755, INF project, and thank Tim Salditt for his continuous support and feedback.

References

[1] M. Bernhardt, M. Priebe, M. Osterhoff, C. Wollnik, A. Diaz, T. Salditt, and F. Rehfeldt, "X-Ray Micro- and Nanodiffraction Imaging on Human Mesenchymal Stem Cells and Differentiated Cells", Biophys. J. 110, 680–690 (2016).

[2] M. Bernhardt, J.-D. Nicolas, M. Eckermann, B. Eltzner, F. Rehfeldt, T. Salditt: "Anisotropic X-Ray Scattering and Orientation Fields in Cardiac Tissue Cells" (in review)

[3] M. Priebe, M. Bernhardt, C. Blum, M. Tarantola, E. Bodenschatz, and T. Salditt, "Scanning X-Ray Nanodiffraction on Dictyostelium discoideum", Biophys. J. 107, 2662–2673 (2014).

[4] E.B. Knudsen, H.O. Sørensen, J.P. Wright, G. Goret, J. Kieffer: "FabIO: easy access to two-dimensional X-ray detector images in Python", Journal of Applied Crystallography 46 (2013).

[5] Rasband, W.S., ImageJ, U. S. National Institutes of Health, Bethesda, Maryland, USA, http://imagej.nih.gov/ij/, 1997-2016.

[6] M. Basham, J. Filik, M.T. Wharmby, P.C.Y. Chang, B. El Kassaby, M. Gerring, J. Aishima, K. Levik, B.C.A. Pulford, I. Sikharulidze, et al., J. Synchrotron Rad. 22 (2015).

IOP Conf. Series: Journal of Physics: Conf. Series **849** (2017) 012060 doi:10.1088/1742-6596/849/1/012060

MMX-I: A data-processing software for multi-modal X-ray imaging and tomography

A Bergamaschi[1], K Medjoubi[1], C Messaoudi[2, 3], S Marco[2, 3] and A Somogyi[1]

[1]Synchrotron SOLEIL, BP48, Saint-Aubin 91192 Gif sur Yvette, France
[2]UMR 9187, Université Paris-Saclay, CNRS, Université Paris-Saclay, Orsay, F-91405, France
[3]U1196, Institut Curie, INSERM, PSL reseach University, Orsay, F-91405, France

Email : antoine.bergamaschi@synchrotron-soleil.fr

Abstract. Scanning hard X-ray imaging allows simultaneous acquisition of multimodal information, including X-ray fluorescence, absorption, phase and dark-field contrasts, providing structural and chemical details of the samples. Combining these scanning techniques with the infrastructure developed for fast data acquisition at Synchrotron Soleil permits to perform multimodal imaging and tomography during routine user experiments at the Nanoscopium beamline. A main challenge of such imaging techniques is the online processing and analysis of the generated very large volume (several hundreds of Giga Bytes) multimodal data-sets. This is especially important for the wide user community foreseen at the user oriented Nanoscopium beamline (e.g. from the fields of Biology, Life Sciences, Geology, Geobiology), having no experience in such data-handling. MMX-I is a new multi-platform open-source freeware for the processing and reconstruction of scanning multi-technique X-ray imaging and tomographic datasets. The MMX-I project aims to offer, both expert users and beginners, the possibility of processing and analysing raw data, either on-site or off-site. Therefore we have developed a multi-platform (Mac, Windows and Linux 64bit) data processing tool, which is easy to install, comprehensive, intuitive, extendable and user-friendly. MMX-I is now routinely used by the Nanoscopium user community and has demonstrated its performance in treating big data.

1. Introduction

By using scanning hard X-ray microscopy the sample is raster scanned in the focused X-ray beam while collecting the chosen information by dedicated detectors. Each scanned position represents a pixel in the final, reduced and reconstructed image. The width and height of the scanned area defines the field of view of the reduced image.

X-ray scanning microscopy enables to register simultaneously the transmission and the X-ray fluorescence (XRF) signals. The transmitted beam can be recorded by a 2D pixel detector, which produces an image at each position of the raster scan. The absorption, darkfield (DF) and differential phase contrasts (DPC) are retrieved by analyzing the images of the transmitted beam and reducing the information to a scalar value in each pixel of the reconstructed image [1]. Energy resolving detectors register the XRF spectra, which contain the number of detected X-ray photons in the function of the energy. The elemental maps are calculated by numerical integration of the peak areas identified from the XRF spectrum (ROI's). The ROI's contain the specific X-ray fluorescence lines of the selected elements [2].

Combining scanning X-ray microscopy with the fast acquisition infrastructure "FLYSCAN" scheme developed at SOLEIL has tremendously grown the volume of the produced data due to the significantly reduced dwell time per pixel. At Nanoscopium this typically varies in the ~2-500 ms/pixel range depending on the measured samples and the chosen pixel size. The reduction process of such large datasets having up to several hundred GB's volume is challenging. Moreover, sophisticated acquisition techniques, such as scanning tomography, bi-directional raster scanning etc can generate extra complexity in the reduction process, which also has to be addressed.

The main objective for developing MMX-I was to offer to the general user community a software tool, which is adapted to processing such big, multimodal scanning imaging data-sets. This tool intends to help users to reduce the collected data either on site in the time frame of the experiments or off site at their home institutes. The user-community of scanning microscopy is coming from very different scientific fields and thus frequently is expert neither in physics nor in multimodal data processing. As such the software tool must be self-explanatory and easy-to-use. In addition, we wanted to offer the community a free software solution which can also be easily upgraded by developers. For these reasons we have developed the MMX-I freeware [3] for Multi-Modal X-ray Imaging and tomographic reconstruction, which implements a modular friendly user interface.

2. MMX-I key features

MMX-I uses raw data in HDF5 file format for input [4]. Raw data are then reduced into 2D projections of the different modalities (e.g. absorption, dark field, DPC, X-ray fluorescence). These raw projections can then be further processed either by specific reconstruction processes (e.g. phase reconstruction for DPC) or via tomographic reconstruction algorithms. Three parts, HDF5 loading, data reduction, and reconstruction, constitute the base workflow of MMX-I. In the next sections we will highlight some of the key features of MMX-I starting with the HDF5 raw data streaming, then demonstrating the efficiency of the code architecture and finally presenting some of the already implemented reconstruction algorithms.

2.1. HDF5 streaming interface

Multimodal scanning X-ray microscopy associated with fly-scan acquisition [5] leads to the generation of several hundred Gigabytes files in less than 1h. The first step of the MMX-I workflow is to extract the different contrast modalities encoded in these very large files. The time required to perform this step mainly depends on the reading efficiency of the used computer. One could argue that high throughput hardware, such as PCIe SSD's, could be readily used to address this problem. However, as the software is designed to be used by large number of users at a large variety of computers at their home institutes, we avoided to base our read-in process to expensive hardware solutions. Instead we developed the Hdf5Opener API which is a streaming interface for HDF5 files (Figure 1).

Hdf5Opener addresses the problem of reading efficiently very large HDF5 files and also scales its performance depending on the system environment. The first problem is handled by the customized read request sent by MMX-I to the Hdf5Opener enabling to read only the needed information and thus avoiding large reading overheads. Scaling the software performance is possible by storing additional data in the Hdf5Opener storage array. This storage array serves as a buffer for the streaming service.

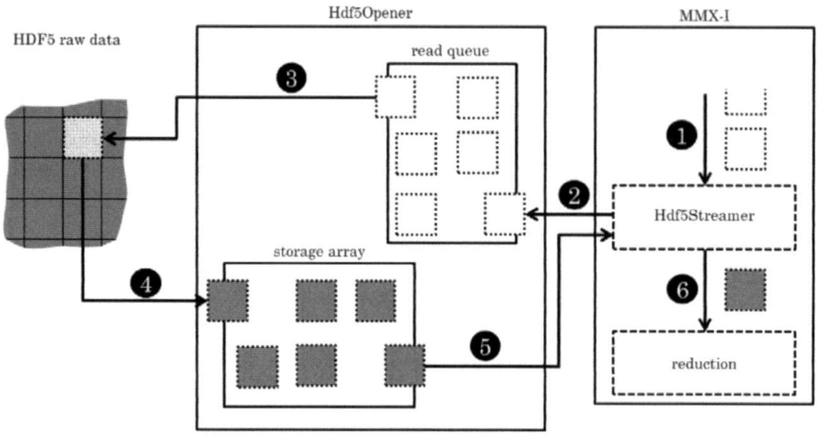

Figure 1. Scheme of the HDF5 raw data streaming process. Empty squares show the read requests, while grey squares represent the retrieved data. **1**: Read requests are generated within MMX-I, **2**: the Hdf5Streamer sends the read requests from MMX-I to the Hdf5Opener, **3**: the Hdf5Opener reads the HDF5 file based on the information contained in the read requests, **4**: raw data are stored in a storage array within the Hdf5Opener, **5**: the Hdf5Streamer retrieves the raw data into the storage array, **6**: MMX-I uses the data to continue the computation process.

2.2. Software architecture

MMX-I intends to be a community driven project, it aims to be used both by non–experienced and expert users and developers. MMX-I is written in Java, and its functional architecture highly profits from the full potential of this Object-oriented language. To ease the work of developers, easy-to-use dedicated Java interface and object-inheritance have been implemented. This allows adding new functionalities or customizing MMX-I to different user experiments in an easy manner. As an example, if a developer wishes to add a reduction function in the MMX-I API, he/she will only need to extend the *BasicComputationFunction* object and then overwrite the dedicated function. Functional annotations, which facilitate the creation of new Objects by generating source code in the compilation pre-processing step, have also been implemented. More information and description are available on the MMX-I Bitbucket home page (https://bitbucket.org/antoinebergamaschi/mmx-i/wiki/Home).

2.3. Key reconstruction functions

Some of the most common reconstruction algorithms have already been implemented into MMX-I and are available for the user's community. Reconstruction methods are either phase contrast or tomographic reconstruction. For phase reconstruction from the measured DPC signal, Fourier [6] and real space [7] integration are available. For tomographic reconstruction, filtered back projection and algebraic reconstruction are provided in MMX-I [8].

3. Results

Tests with beta users in "user beamtime conditions" have been performed at the Nanoscopium [9] beamline (results are not reported here). Moreover, we used a Dell XPS13 Ultrabook for testing the performance of MMX-I in "user laboratory mode". The raw data were stored on a 2 TB Western Digital external hard drive allowing a maximum read rate of 120MB/s through USB3 link.. For each test-sample we compared the acquisition time and the data reduction time to demonstrate the efficiency of the software on small computer. The reduced and reconstructed images are shown in Figure 2 and in Figure 3.

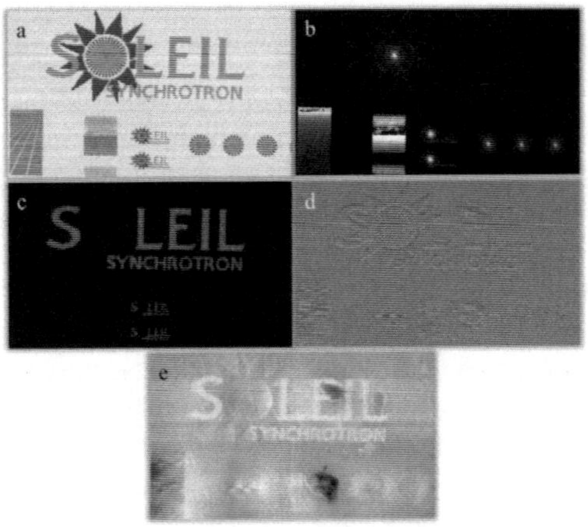

Figure 2. Multimodal 2D scan of a test sample. a) Absorption; b) Darkfield; c) elemental map of Nickel; d) DPC, X direction; e) composite image of a,b,c,d.

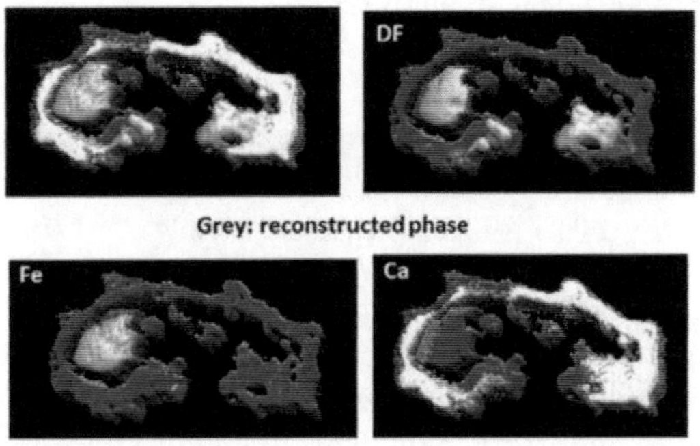

Figure 3. 3D volume rendering of the multimodal tomography of a Foraminifer. In the composite images in yellow is the calcium content of the Ca-carbonate shell, the darkfield is in blue, the iron is in red and the reconstructed phase is in grey.

Figure 2 presents the reconstructed images of a test sample fabricated for multimodal acquisition. The total measurement time was 13 minutes. The reconstructed images contain 1000x500 pixels with 0.5 x 0.5 µm2 pixel-size where for each pixel a full XRF spectrum (X-ray fluorescence modality), a 2D image (transmitted modalities), encoder sample stage position value and intensity monitor value are recorded. A HDF5 file of 80 GB size was generated during this experiment. MMX-I performed the reduction process of this data-set in ~10 min, less than the acquisition duration.

Figure 3 is the 3D volume rendering of the reconstructed tomography modalities of a foraminifer. Foraminifera are used as proxy for paleo-environmental changes. The acquisition of this dataset was fulfilled in 3h. It contains 600 x 200 x 25 pixels (e.g. 25 projections of size 600 x 200) with a pixel size of 1 x1 μm^2. The size of the resulting data-file is 490 GB. The data reduction resulting in the sinograms of the different modalities took 1.25 h by MMX-I on a standard laptop. 80% of this time was due to data reading.

4. Conclusion and outlooks

The first version of MMX-I is now available for the user community. MMX-I aims to process very large HDF5 files even on Ultrabook's. The software has already been tested at Nanoscopium by beta users and has proven to be a reliable and fast multimodal data-processing tool.

Acknowledgments

The authors are grateful for the continuous help, advice and expertise of the SOLEIL Support Groups. Special thanks for Nicolas Leclercq for his work with the development of the FLYSCAN. We acknowledge Isvan Mohacsi and Delphine Desmares for providing the test structures and foraminifera samples.

References

[1] Bunk O, Bech M, Jensen TH, Feidenhans'l R, Binderup T, Menzel A, et al. Multimodal x-ray scatter imaging. New J Phys. 2009 Dec 1;11(12):123016.

[2] Grieken RV, Markowicz A. Handbook of X-Ray Spectrometry, Second Edition,. CRC Press; 2001. 1016 p.

[3] Bergamaschi A, Medjoubi K, Messaoudi C, Marco S, Somogyi A. MMX-I: data-processing software for multimodal X-ray imaging and tomography. Journal of Synchrotron Radiation.

IOP Conf. Series: Journal of Physics: Conf. Series **849** (2017) 012060 doi:10.1088/1742-6596/849/1/012060

2016 May 1;23(3):783–94.
[4] The HDF Group [Internet]. 2015 [cited 2015 Nov 10]. Available from: https://www.hdfgroup.org/
[5] Medjoubi K, Leclercq N, Langlois F, Buteau A, Lé S, Poirier S, et al. Development of fast, simultaneous and multi-technique scanning hard X-ray microscopy at Synchrotron Soleil. Journal of Synchrotron Radiation. 2013 Jan 23;20(2):293–9.
[6] Kottler C, David C, Pfeiffer F, Bunk O. A two-directional approach for gratingbased differential phase contrastimaging using hard x-rays. Opt Express. 2007 Feb 5;15(3):1175–81.
[7] Southwell WH. Wave-front estimation from wave-front slope measurements. J Opt Soc Am. 1980 août;70(8):998–1006.
[8] Kak AC, Slaney M. Principles of computerized tomographic imaging. Philadelphia: Society for Industrial and Applied Mathematics; 2001.
[9] Somogyi A, Medjoubi K, Baranton G, Le Roux V, Ribbens M, Polack F, et al. Optical design and multi-length-scale scanning spectro-microscopy possibilities at the Nanoscopium beamline of Synchrotron Soleil. Journal of Synchrotron Radiation. 2015 Jul 1;22(4):1118–29.